CHEMICAL MODIFICATON OF PROTEINS

chemical modification of proteins

Gary E. Means
University of California, Davis

Robert E. Feeney
University of California, Davis

Holden-Day, Inc.
San Francisco, Düsseldorf, Johannesburg, London,
Panama, Singapore, Sydney, Toronto

The drawing of a hemoglobin molecule on the cover is from "The Hemoglobin Molecule," by M. F. Perutz. Copyright © November 1964 by Scientific American, Inc. All rights reserved.

Copyright © 1971 by Holden-Day, Inc.
500 Sansome Street, San Francisco, California 94111

All rights reserved.

No part of this book may be reproduced in any form, by mimeograph or any other means, without permission in writing from the publisher.

Library of Congress Catalog Card Number: 74-140785

Printed in the United States of America

234567890 MP 79876543

preface

Protein chemists have long been interested in altering the chemical, physical, and biological properties of proteins by chemically changing their structure. One of the first things discovered about proteins was how easily they were changed upon treatment with chemical reagents. Their lability to chemical reagents and reaction conditions has been a serious problem for many purposes. The application of modern knowledge of proteins, new chemical reagents, and more sophisticated analytical techniques, however, has made chemical modification of protein molecules one of the most useful approaches to the study of many of their properties.

One of the most frequent questions asked about biologically active proteins is, "What is unique about the structure that accounts for the particular activity?" Thus, interest in most enzymes is primarily in the "active center," the amino acid side-chain groups of the protein molecule that participate in the activity. The objective of this book is to describe chemical methods used to determine the roles of the individual amino acid side chains in the chemical, physical, and biological properties of proteins. Emphasis is given to the use of these procedures for the identification of side-chain groups participating in the "active center."

There are two main parts of this book plus an appendix. The first part, some 65 pages, is devoted to general discussion of the subject and related topics of importance. The second part, including 145 pages divided into five chapters, describes the reagents used for the nondestructive chemical modification of proteins. Subsections are arranged according to the reagents, with similar or related reagents being grouped to make chapters. Each reagent is described in detail as to its properties and use and the properties of derivatives likely to be formed upon its reaction with a typical protein. Expected changes in properties, side reactions, and analytical considerations needed to characterize the modified proteins are included. Examples are given in most sections illustrating the use of each reagent. The appendix contains experimental procedures taken from the original literature and is intended to enable the reader to perform some of the more important modification reactions.

The book has been developed from the material used for a two-credit one-quarter graduate course intended for students who have had a prior knowledge of general protein chemistry. The course was designed to provide a student with a fundamental knowledge of the chemical modification of proteins with the hope that this knowledge would enable him to select, and even develop, those chemical reactions most suited to his research objectives.

This book would not have been possible without help and advice from many sources. Special appreciation is due to R. D. Cole for a critical appraisal of the initial draft. We would also like to thank the many other friends and colleagues who helped in preparing this material. These include J. R. Whitaker, F. Wold, M. Friedman, F. H. Carpenter, N. E. Gentner, F. C. Greene, S. Govons, Ahmed Ahmed, J. Vandenheede, Ruth Uy, C. Ho, D. Osuga, W. Benisek, G. Blankenhorn, and Y. Lin for reviewing parts of the manuscript and offering helpful criticisms. We also express our appreciation to Judy Miller, Kathy Else, Judy Tweedie, Carolyn Brumley, Sue Brown, and the publisher's staff, especially Sally Anderson, for their valuable assistance in literature searchings and in editing and proofreading the manuscript.

The main researches that laid the foundation for the preparation of this book were supported by a research grant, HD-00122, from the National Institutes of Health, United States Public Health Service.

Gary E. Means
Robert E. Feeney

contents

| Part I | The Chemistry and Chemical Reactions of Proteins | 1 |

CHAPTER 1	PROTEINS AND THEIR PROPERTIES	3
1-1	Historical Developments	3
1-2	Chemical Modifications for Analytical and Industrial Purposes	5
1-3	Current Status	6
1-4	Protein Structure	7
1-5	Protein Stability	9
1-6	Relative Reactivities of Amino Acid Side Chains	11
1-7	Active Sites	20

CHAPTER 2	MODIFICATION OF GROUPS ESSENTIAL FOR ACTIVITY	24
2-1	Substrate Protection of Active-Center Groups	24
2-2	Active-Site Selective Reagents	25
2-3	Determination of the Number of Essential Residues	30
2-4	Reversible Modification	33

CHAPTER 3	MODIFICATIONS TO CHANGE PROPERTIES	35
3-1	Chemical Modifications to Change Physical Structure	35
3-1.1	Changing the Ionic State and Conformation	35
3-1.2	Reduction and Reoxidation of Disulfide Bonds	37
3-1.3	Chemical Cleavage of Peptide Bonds	38
3-1.4	Cross-Linking Reagents	39
3-1.5	Attachment to Solid Supports	40
3-2	Introduction of Special-Purpose Groups	43
3-2.1	Changing the Susceptibility to Proteolysis	45
3-2.2	The Augmentation of Immunogenicity	46
3-2.3	Environmentally Sensitive Labels	47
3-2.4	Isomorphic Replacement	51

CHAPTER 4	SPECIAL PROBLEMS IN ANALYSIS OF CHEMICALLY MODIFIED PROTEINS	55
4-1	Extent of Modification	55
4-2	Side Reactions	58
4-3	Changes in Conformation	60
4-4	Other Considerations	61
4-5	The Use of Variations in Protein Structure to Study Protein Function	63

Part II	The Chemistry and Chemical Reactions of the Reagents	66
CHAPTER 5	ACYLATING AND SIMILAR REAGENTS	68
5-1	Acylating Agents	68
5-1.1	Acetic Anhydride	69
5-1.2	N-Acetylimidazole	72
5-1.3	Succinic Anhydride	74
5-1.4	Maleic Anhydride	76
5-1.5	N-Carboxyanhydrides	78
5-1.6	Ethyl thiotrifluoroacetate	79
5-1.7	Diketene	80
5-1.8	Ethoxyformic Anhydride (Diethylpyrocarbonate)	81
5-1.9	Miscellaneous Acylating Agents	83
5-1.10	Quantitation	83
5-2	Cyanate	84
5-3	Amidination	89
5-3.1	Imidoesters	89
5-3.2	Guanidination	93
5-3.3	Related Reagents	95
5-4	Sulfonyl Halides and Sulfonates	96
5-4.1	Sulfonyl Halides	97
5-4.2	Sulfonate Esters	98

CHAPTER 6	ALKYLATING AND SIMILAR REAGENTS	105
6-1	Haloacetates	105
6-2	Maleimides	110
6-3	Acrylonitrile	114
6-4	Ethylenimine	117
6-5	Aryl Halides	118
6-6	2-Hydroxy-5-nitrobenzyl Bromide	123
6-7	Formaldehyde and Some Other Carbonyl Compounds	125
6-7.1	Formaldehyde	125
6-7.2	Other Aldehydes and Ketones	128
6-8	Reductive Alkylation	130
6-9	Pyridoxal Phosphate	132

CHAPTER 7		ESTER- AND AMIDE-FORMING REAGENTS	139
	7-1	Esterification	139
		7-1.1 Methanol/HCl	139
		7-1.2 Diazoacetates	141
	7-2	Carbodiimide-Promoted Amide Formation	144
CHAPTER 8		REDUCING AND OXIDIZING REAGENTS	149
	8-1	Reducing Agents	149
	8-2	Sulfite	152
	8-3	5,5′-Dithiobis(2-nitrobenzoic acid) (Ellman's Reagent)	155
	8-4	o-Iodosobenzoate	157
	8-5	Tetrathionate	159
	8-6	Performic Acid	160
	8-7	Hydrogen Peroxide	162
	8-8	Photooxidation	165
	8-9	N-Bromosuccinimide	169
CHAPTER 9		ELECTROPHILIC REAGENTS	175
	9-1	Iodine	175
	9-2	Tetranitromethane	183
	9-3	Diazonium Ions	186
CHAPTER 10		MISCELLANEOUS REAGENTS	194
	10-1	Dicarbonyls	194
		10-1.1 1,2-Cyclohexanedione	194
		10-1.2 Glyoxal	195
		10-1.3 Phenylglyoxal	195
		10-1.4 Diacetyl Trimer	197
		10-1.5 Malonaldehyde	197
	10-2	Mercurials	198
	10-3	Cyanogen Bromide	204
	10-4	Sulfenyl Halides	206
	10-5	Nitrous Acid	207

Part III		Appendix	212

SELECTED TECHNIQUES FOR THE MODIFICATION OF PROTEIN
SIDE CHAINS   214

	A-1	Amino Groups	214
	A-2	Sulfhydryl Groups	217
	A-3	Disulfide Groups	221
	A-4	Carboxyl Groups	222
	A-5	Guanidino Groups	223
	A-6	Phenolic Groups	225

A-7	Imidazole Groups	226
A-8	Indole Groups	227
A-9	Thioether Groups	228

Author Index 231

Subject Index 241

the chemistry and chemical reactions of proteins

1
proteins and their properties

Proteins are extremely complex molecules. This is illustrated by the model of a relatively simple protein, chicken egg white lysozyme, shown in Figure 1-1. Its complexity is due to its relatively great size and to the unique three-dimensional arrangement of its many functional groups. The fascinating physical, chemical, and biological properties of these indispensible components of living matter are subjects of great interest. Relating their various properties to specific structural features is the basic goal of protein chemistry.

1-1 HISTORICAL DEVELOPMENTS

Early History. The early history of protein chemistry was primarily concerned with the chemistry of the individual amino acids and with the synthesis and properties of small peptides. Fischer's work on amino acids and peptides, including some of the first chemical syntheses of peptides, is one of the outstanding classical stories of organic chemistry. Sumner's demonstration in the 1920's that enzymes are proteins stimulated interest toward identifying particular amino acids responsible for their catalytic activity. Many of the chemical methods for proteins developed during this period, however, were for quantitatively determining the amounts of individual amino acids in proteins, rather than for modifying them.

A great acceleration in the development of chemical methods for the study of proteins occurred during and immediately following World War II. Important developments of this period were the reviews of protein modification written in 1947 (Olcott and Fraenkel-Conrat, 1947; Herriott, 1947). Some of the procedures described in these publications are still very much in use. An important landmark a few years later was the demonstration by Balls and Jansen (1952) that the specific

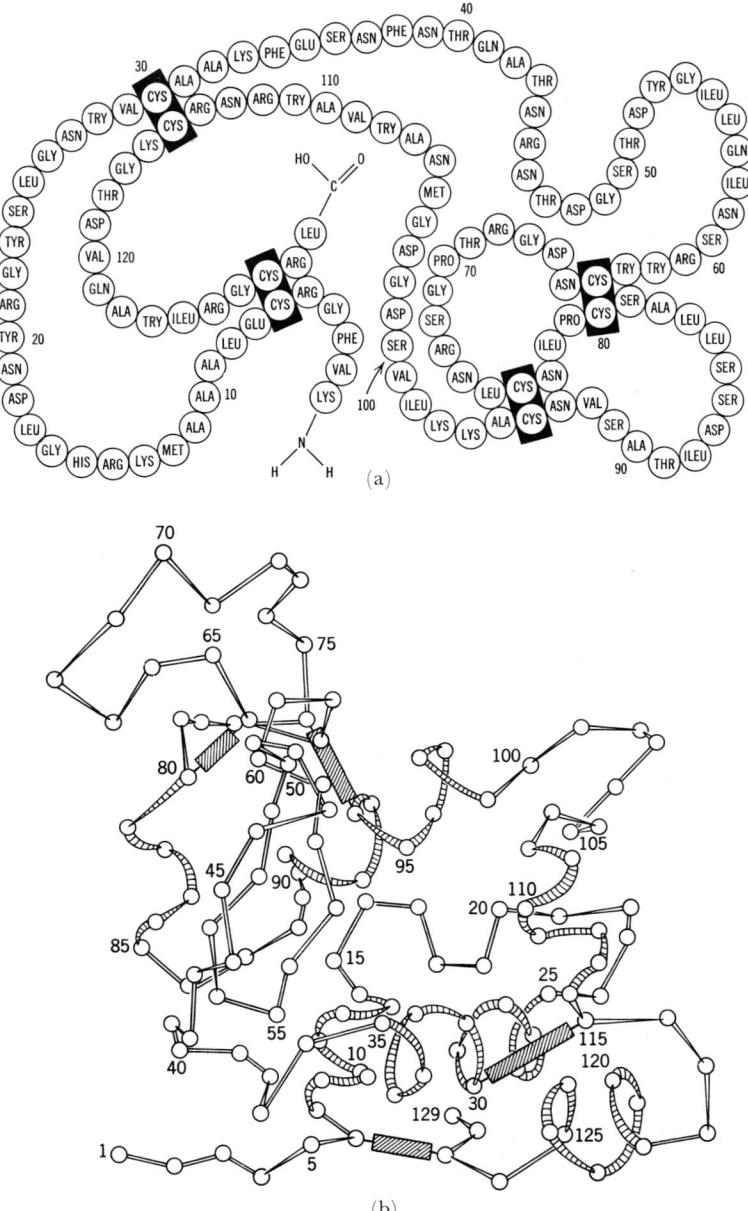

FIGURE 1-1 (a) The structure of egg white lysozyme indicating the positions of the four disulfide bonds. ASN and GLN denote asparagine and glutamine, respectively. (From Canfield and Liu, 1965.) (b) Schematic drawing of the main chain conformation of lysozyme. (By W. L. Bragg, from Blake et al., 1965.)

inactivation of certain proteases by diisopropylfluorophosphate involved reaction with specific serine residues presumably in their active sites.

Scientific developments during the following years led to increased interest in several physical-chemical techniques for studying proteins which, in turn, stimulated interest in protein modification. The use of X-ray diffraction for the study of proteins, for example, promoted interest in methods for isomorphic introduction of heavy atoms into proteins (see Section 3-2). For other studies, techniques for enhancing the fluorescence of proteins by introducing various fluorescent groups were developed. Improvements in many physical-chemical techniques have both contributed to, and made use of, protein-modification techniques.

The development of more accurate and more sensitive analytical techniques has similarly been a major contributor to, and beneficiary of, advances in protein modification. Development of the automated amino acid analyzer has been one of the most important improvements, having contributed immeasurably to all of protein chemistry.

Recently, great interest has been directed to the search for "affinity-labeling" or "active-site-directed" reagents. These reagents are designed to react preferentially with only those parts of protein molecules that are physically in the vicinity of particular biochemically active sites. Their use gives a highly selective way of chemically modifying a protein (see Section 2-2).

1-2 CHEMICAL MODIFICATIONS FOR ANALYTICAL AND INDUSTRIAL PURPOSES

For many years a motivating factor in the study of the chemical modification of proteins was the desire to determine quantitatively the amounts of proteins or their component amino acids. Many methods have been developed for such purposes. Because the intent was not to preserve the integrity of the protein, many of these methods are very harsh and in this way differ from most procedures described in this book. A few methods originally designed strictly for quantitative purposes have, however, also proved useful for selective chemical modification of proteins. The use of nitrous acid for the determination of amino groups in proteins, for example, is also of value for their selective modification.

Only a few methods are suitable for both modifying and determining amino acid side chains in proteins. Two such methods now in common usage are the reaction of sulfhydryl groups with p-mercuribenzoate and the reaction of amino groups with trinitrobenzenesulfonic acid. Both reagents can be used under relatively mild conditions which do not damage most proteins, and it is easy to measure the numbers of groups modified.

Commercial applications of the chemical modification of proteins have a long history related to the pharmaceutical, dyeing, and clothing industries. An early application in the pharmaceutical industry was the use of formaldehyde to modify bacterial toxins and viruses; similar procedures are still important commercially. The purpose of this treatment is to kill, inactivate, or so change the virus or toxin as to render it incapable of eliciting its toxic or pathological response, while retaining its ability to elicit an immunogenic response when injected into an animal. The bacterial toxins, when so modified, are known as *toxoids*.

One of the oldest processes involving protein modification is the treatment of animal hides or hair for human use, as in the tanning of leather. Increased knowledge has led to recent improvements in these ancient procedures. For example, glutaraldehyde, used for cross-linking of proteins, is now also used for tanning leather. It apparently functions similarly by cross-linking collagen in the leather. Similarly, several different modifications are now used to give wool fibers superior performance for clothing. Chlorination or treatment with polyepoxides is being used commercially. The latter primarily react with amino groups. Use has also been made of reagents splitting disulfide bonds for the purpose of obtaining "permanent press" in finished clothing items.

1-3 CURRENT STATUS

The current interest in chemical modifications of proteins is indicated by the many reviews and books on its various aspects. Baker (1967), for example, has published a book dedicated to the organic chemistry of the active site (of enzymes), and Hirs (1967) has edited a detailed compilation of laboratory procedures for proteins. Other reviews, including general discussions of chemical reagents and their reactions, have been published by Cohen (1968), Glazer (1970), Vallee and Riordan (1969), Stark (1970), Shaw (1970), and Spande et al. (1970).

There is now a long series of chemical reagents and reactions used for: (a) synthesizing peptides and proteins (Stewart and Young, 1969), (b) sequential and stepwise degradation of proteins to determine their structures (Stark, 1970), and (c) preparation of derivatives of amino acids to increase their volatilities and detectabilities by vapor phase chromatography (Gehrke et al., 1968). The first type under (a) includes methods which must maintain the integrity of the polypeptide chain and the structures of the side groups of the amino acids, because the objective is to end with a normal polypeptide or protein. Special reagents are used to block or stabilize the partially synthesized protein. In general, however, the methods are milder than (b) and (c) types. This is not the case for the methods used for preparing derivatives for vapor phase chromatography. For this purpose, the only requirement is maintenance of the structure of the individual amino acids. Consequently, the methods used for these derivatives usually involve much harsher conditions than are used for intact proteins.

Many of the methods in use today are sufficiently simple and convenient to be widely employed in general studies by the person who is not a specialist in chemical modifications. The utility of chemical modifications is greatly extended by its use in conjunction with sophisticated physical methods. For those proteins whose total conformation is known from X-ray crystallographic studies, the X-ray studies and chemical modifications mutually supplement one another. Similarly, results from many other procedures, such as absorption spectra or pH titrations, may sometimes be better interpreted if chemical modification data are available.

1-4 PROTEIN STRUCTURE

The three-dimensional structure of a protein is governed by its primary structure (sequence of amino acids) and its environment. Changes in either of these can have important effects on its properties. Modifying any of its amino acid residues necessarily changes the primary structure of a protein. The conditions to which it is subjected in bringing about the desired chemical change are often quite harsh. Consequently, one of the most important objectives of a protein chemist is to avoid or minimize undesired structural changes. In some cases limited changes in structure through chemical modification can be a desired objective. To better understand the potential of protein modification, as well as its limitations, in regard to both interests, it is necessary to know something about the forces involved in determining protein structure.

Covalent Bonds. Two kinds of covalent bonds are commonly found in proteins: peptide bonds between the amino acid residues and disulfide bonds. Both are subject to specific chemical modifications.

Peptide bonds are subject to cleavage by many proteolytic enzymes. By using an enzyme with a particular specificity, it is often possible to effect the selective cleavage of particular types of peptide bonds. The use of proteolytic enzymes to cleave peptide bonds has been an invaluable tool in the determination of primary sequences of proteins. Alternately, it is now possible to bring about the cleavage of certain peptide bonds through action of specific chemical agents. Thus, peptide bonds involving carboxyl groups of methionine are subject to specific cleavage by cyanogen bromide, and those of tryptophan, and to a lesser extent of tyrosine, are subject to cleavage by N-bromosuccinimide. Chemical agents for peptide cleavage have been described in several other recent publications (Witkop, 1968; Spande et al., 1970).

Disulfide bonds occur in greatest number in extracellular proteins and appear to confer upon them a considerable amount of stability. Chemical modification of disulfide bonds is for the most part a matter of their cleavage, either by oxidation or reduction. In most proteins, they are the only groups susceptible to mild reduction. Their reversible reduction-reoxidation properties have been widely studied in attempts to discover the important factors which determine the three-dimensional conformations of proteins. Reagents and methods for the reduction of disulfide bonds are described in Section 8-1. The use of performic acid for the oxidative cleavage of disulfide bonds is described in Section 8-6.

Noncovalent Bonds. Noncovalent interactions between various parts of a peptide chain and between peptide chains and solvent, or substances dissolved therein, determine a protein's three-dimensional structure (e.g., see Ramachandran and Sasisekharan, 1968). Environment provides the framework within which such structure can be maintained, and changes in environment can result in changes in conformation, which may or may not be of major consequence to a protein's various properties. Such changes in conformation are, in some cases, reversible upon return to the original environment. Conditions favoring structural or conformational changes are frequently desirable to achieve more complete reaction during chemical modification. The use of agents such as urea to disrupt the normal noncovalent bonds and to increase the reactivity of proteins is discussed later in this chapter.

The noncovalent forces usually cited as important contributors to protein structure are the result of (a) electrostatic interactions, (b) hydrophobic interactions, and (c) hydrogen bonding. Electrostatic and hydrophobic interactions and the forces arising therefrom result from the amino acid side chains. Hydrogen bonds are an important structure-determining factor of the peptide chain and also arise from side-chain interactions. Modification of a protein must necessarily effect one or more of these structure-determining forces. If, as is often the case, changes in structure are undesirable, then the usefulness of a modification will depend upon its having negligible effects on these forces.

Changes in electrostatic relationships are a frequent result of protein modification. By using a number of different reagents it is sometimes possible to determine somewhat the extent of such change. Succinylation of positively charged amino groups, for example, produces in their places negatively charged groups, giving a net change of two charge units per amino group $(+1 \rightarrow -1)$, a change which in many cases has important structural consequences. Acetylation of an amino group results in a net change of one unit $(+1 \rightarrow 0)$, and in most cases results in correspondingly smaller structural changes. Modification by any of several methods which retain the positive charge will usually have little effect on conformation. Chemical modification of proteins involving changes in ionic or electrostatic relationships, in order to effect their dissociation into subunits, or to effect certain other conformational changes, is discussed in Subsection 3-1.1.

Hydrophobic forces are extremely important for maintaining the structure of proteins. Hydrophobic amino acid residues are in general, however, not amenable to chemical modification. Leucine, isoleucine, valine, and phenylalanine residues, for example, cannot be modified by any of the presently known protein reagents. Two hydrophobic residues, tryptophan and tyrosine, are susceptible to several chemical alterations, but even these residues, because they are hydrophobic and frequently located in the interior of proteins, are relatively protected from reagents dissolved in the solvent. The absence of procedures to modify many hydrophobic residues constitutes an important limitation to the study of proteins by chemical modification. Studies of the extent or degree of reactivity of tryptophan and tyrosine residues with various solvent-borne reagents have been used to measure the distribution of these residues between the exterior and the hydrophobic interior of many proteins. This application is discussed later in this chapter.

Although hydrogen bonds are individually relatively weak, because there are so many such bonds, they contribute extensively to the total force determining protein conformation. The importance of hydrogen bonds for the maintenance of peptide-backbone structures, including so-called α-helix and β-structures, is well known. No chemical methods specifically affect these bonds, except perhaps those methods involving peptide-bond cleavage. Although deuterium- and tritium-exchange studies in the broad sense involve peptide-bond modification, they are not discussed (see Hvidt and Nielsen, 1966).

The presence of hydrogen bonds between protein carboxyl groups and tyrosine residues has been used to explain the anomalous titration behavior of these groups. In some cases, the identity and sequential position of such groups have been determined using chemical modification. Three of the eleven carboxyl groups in pancreatic ribonuclease involved with the abnormal titration behavior of three tyrosyl groups, for example, were identified following esterification with anhydrous methanol/HCl (Riehm et al., 1965). Identification of these residues was made on the basis of their resistance to esterification.

1-5 PROTEIN STABILITY

Proteins vary greatly in stability. In general, smaller proteins are more resistant than large complex proteins to harsh physical or chemical treatment or, at least, are more able to resume their native structures afterwards. Extracellular proteins often contain relatively high percentages of disulfide cross-links and are relatively more stable than intracellular proteins. These smaller and more stable proteins have been the most intensively studied, apparently because of their stability.

Mobility of Structure. Proteins are not immobile rigid rods or spheres; rather, they are mobile species capable of assuming a number of forms in solution. Many investigators believe the torsion of the various chemical bonds and the mobility of structure to be important for the functioning of many biologically active proteins. Such flexibility is implicit in the so-called "induced fit" of enzymes with substrates as postulated by Koshland (Koshland, 1959), or for the allosteric effects postulated by Monod and coworkers (Monod et al., 1965).

Although plasticity of protein structure appears important for many functions, some rigidity is required for most theories of enzyme catalysis. A recently proposed mechanism for chymotrypsin catalysis, for example, postulated a facilitated proton transfer along rigidly and accurately held hydrogen bonds in the enzyme-substrate complex (Wang, 1968). A rigid structure is essential to such a model.

Flexibility of structure is a limited property of proteins. The ability to expand, contract, fold, or unfold is not without bounds. Certain conformations, it is obvious, are not allowed within the framework of their native state. Within these limits, factors affecting the relative reactivity of groups in a protein or of a reagent can be varied to obtain a desired state or degree of modification. Beyond these bounds, irreversible damage to the protein is likely to take place.

The Effect of External Environment upon the Structure of Proteins. Chemical modification of proteins often entails the use of distinctly harsh conditions. Extremes of pH, for example, may be desirable to maintain a reagent or amino acid side chain in a reactive state, but can have undesirable effects on protein structure. In some cases harsh conditions result from the use of agents (e.g., urea, guanidine hydrochloride) not involved in the chemical reaction but necessary to unfold the protein, or to otherwise increase the accessibility and reactivity of its various side chains. Ideally, removal of the perturbing agent (and the modifying reagent) results in reversion of the protein to its original conformation.

Organic solvents may be necessary in reaction solutions to solubilize a reagent, or they may serve as a reactant in high concentration in order to favorably displace the equilibrium. The esterification of proteins in anhydrous methanol/HCl is an example of the latter, where extensive reaction requires a high concentration of methanol.

The rather harsh conditions tolerable for many of the more stable proteins are not suitable for a majority of enzymes. Moderation of some of the harsher techniques appears to be the best hope in many cases. Fortunately, a more important problem with many labile proteins frequently concerns controlling the extent of reaction and its effects. Thus, it is only through the use of limiting amounts of a reagent that loss of native character is avoidable with many enzymes.

Denaturation. Distinctions between conformational changes and denaturation are for the most part unclear. In many cases, the distinction is only a matter of degree. Conformational changes are sometimes considered to be reversible changes in structure, whereas denaturation more often implies irreversibility. Irreversible changes in structure, or irreversible denaturation, is perhaps the most difficult problem associated with protein modification, and in every case deserves serious consideration.

Literally, denaturation is any change from the natural, or so-called native state (Tanford, 1968). It is also frequently defined as a change from an ordered to a disordered state. Neither definition is entirely adequate from the standpoint of protein modification. Modification necessarily produces a change in the native state of a protein. If we take the second definition, it is necessary to define an additional state including ordered but conformationally distinct forms produced by chemical modification.

Because a chemically modified protein differs chemically, then so must the relative stability of its various conformations also differ from those of the native protein. Differences in relative stability under different conditions is an additional complicating factor.

For most purposes, denaturation can usually be recognized by changes in rather obvious physical properties. Decreased solubility is a classic indicator of denaturation. The loss of biological activity is also frequently taken as a sign of denaturation. Such indications are relatively crude signs of conformational changes, and relatively more refined techniques are preferable following chemical modification. No single criterion is adequate to detect all structural changes, and the possibility exists that even a slight change may have important biological consequences. To distinguish between denaturation and those changes due to modification of functionally important residues is of utmost importance. A number of rather sensitive techniques are now available for detecting altered conformations. Some of the methods useful for detecting denaturation are mentioned in Chapter 4.

1-6 RELATIVE REACTIVITIES OF AMINO ACID SIDE CHAINS

The reactivity of one substance with another depends upon the respective properties of each and upon the environment in which they are placed. The reactivity of a group in a protein with a particular reagent depends, among other things, upon the effect of the environment on the group, and upon the ability of the reagent to enter that environment. Because proteins are quite large, many of their constituent amino acid residues are partially shielded from solvent, and hence are relatively unreactive with reagents dissolved therein. Conversely, some residues, perhaps as a result of their presence in a catalytic center, possess unusually high reactivities. Determination of relative reactivities toward a number of reagents can thus help to define the position of various residues in the three-dimensional structure of a protein. Such information is extremely valuable to the X-ray crystallographer for the interpretation of his results, and, until crystallographic information is available, affords the clearest source of information on three-dimensional structure. On the basis of such information, for example, a model of pancreatic ribonuclease A was proposed which proved remarkably similar to that later proposed on the basis of crystallographic data (Hammes and Scheraga, 1966). The determination of relative reactivities is currently one of the most active applications of protein modification.

Side-Chain Specificity. The different chemical properties of the various amino acid side chains provide a basis for their differential modification. Most protein reagents react with more than one side chain (Table 1-1). Lack of specificity limits

TABLE 1-1 Side-chain reactivities[a]

Reagent[b]	—NH$_2$	—SH	—OH (phenol)	imidazole —NH	—NH—C(=$^+$NH$_2$)NH$_2$	—COOH	indole	—S—S—	—SCH$_3$
Acetic anhydride (5-1, A-1)	+++	++++[c]	++++[d]	+++[c]	—	—	—	—	—
N-Acetylimidazole (5-1, A-6)	±±	++++[c]	++++[d]	++++	—	—	—	—	—
Acrylonitrile (6-3)	+++	+++	—	—	—	—	—	—	—
Aldehyde/NaBH$_4$ (6-8, A-1)	+++	—	—	+	—	—	—	—	—
N-Bromosuccinimide (8-9, A-8)	—	+++	++	+	—	—	+++	—	—
N-Carboxyanhydrides (5-1)	+++	—	—	—	—	—	—	—	—
Cyanate (5-2, A-1)	+++	++++[c]	++[c]	+[c]	—	+[c]	—	—	—
Cyanogen bromide (10-3)	—	+	—	—	—	—	—	—	+++
1,2-Cyclohexanedione (10-1, A-5)	±	—	—	—	+++	—	—	—	—
Diacetyl trimer (10-1, A-5)	+	—	—	—	++	—	—	—	—
Diazoacetates (7-1)	—	++	+	—	—	+++	—	—	—
Diazonium salts (9-3, A-7)	+++	+++	+++	+++	—	++	+	—	—
Diketene (5-1)	+++[d]	—	+	++	—	—	—	—	—
Dinitrofluorobenzene (6-5)	+++	+++	++	—	—	—	—	+	—
5,5'-Dithiobis(2-nitrobenzoic acid) (8-3, A-2)	—	++++[d]	—	—	—	—	—	—	—
Ethoxyformic anhydride (5-1, A-7)	+++	—	—	+++[c]	—	—	—	—	—
Ethylenimine (6-4)	—	+++	—	—	—	—	—	—	—
N-Ethylmaleimide (6-2, A-2)	±±	+++	—	—	—	—	—	+	—
Ethyl thiotrifluoroacetate (5-1)	+++[c]	—	—	—	—	—	—	—	—
Formaldehyde (6-7, A-1)	+++	+	+	+++	+	—	++	—	—
Glyoxal (10-2)	++	—	—	—	+++	—	—	—	—
Haloacetates (6-1, A-2)	+	+++	—	+	—	—	—	—	+
Hydrogen peroxide (8-7, A-9)	—	+++	—	—	—	—	+++	+	+++
2-Hydroxy-5-nitrobenzyl bromide (6-6, A-8)	—	++	—	—	—	—	+++	—	—
Iodine (9-1, A-6)	—	+++	+++	++	—	—	++	—	—
o-Iodosobenzoate (8-4)	—	+++	—	—	—	—	—	—	—

Reagent										
Maleic anhydride (5-1, A-1)	+++[d]	+++[c]	—	—	—	—	—	—	—	—
p-Mercuribenzoate (10-2, A-2)	—	+++	—	—	—	—	—	—	—	—
Methanol/HCl (7-1, A-4)	—	—	—	+++	—	—	—	—	—	—
2-Methoxy-5-nitrotropone (6-7)	+++[d]	—	—	—	—	—	—	—	—	—
Methyl acetimidate (5-3, A-1)	+++	—	—	—	—	—	—	—	—	—
O-Methylisourea (5-3, A-1)	+++	—	—	—	—	—	—	—	—	—
Nitrous acid (10-5)	—	+++	±	—	—	+	—	—	+++	—
Performic acid (8-6)	—	+++	—	—	—	+++	—	—	+++	—
Phenylglyoxal (10-1, A-5)	++	—	—	+++	—	—	—	—	—	—
Photooxidation (8-8, A-7)	—	+++	±±	++	—	±	—	—	±	—
Sodium borohydride (8-1)	—	—	—	—	—	+++	—	—	+++	—
Succinic anhydride (5-1, A-1)	+++	+++[c]	++[c]	—	—	—	—	—	—	—
Sulfenyl halides (10-4)	—	++++	—	—	—	+++	—	—	—	—
Sulfite (8-2, A-3)	—	±±±[d]	—	—	—	—	—	+++[d]	—	—
Sulfonyl halides (5-4)	+++	+++	+++	—	—	—	—	—	—	—
Tetranitromethane (9-2, A-6)	—	+++	—	+++	—	+	—	—	—	+
Tetrathionate (8-5)	—	+++	—	—	—	—	—	—	+++	—
Thiols (8-1, A-3)	—	—	—	—	—	—	—	—	—	—
Trinitrobenzenesulfonic acid (6-5, A-1)	+++	++[c]	—	—	—	—	—	—	—	—
Water-soluble carbodiimide and nucleophile (7-2, A-4)	±	—	±	+++	—	—	—	—	—	—

[a] —, +, ++, and +++ indicate relative reactivities. ±, ±±, and ±±± likewise indicate relative reactivities which may or may not be attained depending upon the conditions employed.
[b] Numbers in parentheses are sections where reagent is discussed in most detail.
[c] Spontaneously reversible under the reaction conditions or upon dilution, regenerating original group.
[d] Easily reversible, regenerating original group.

the usefulness of many reagents. Many of the more reactive side chains can exist in both a protonated and unprotonated form depending upon the conditions. Since the two forms have vastly different chemical properties, pH has an important influence upon the chemical modification of such groups. Whether or not a group is protonated under a given set of conditions usually determines its reactivity with a particular reagent. The tyrosine anion, for example, reacts rapidly with iodine, while the unionized residue reacts very slowly or not at all. Similarly, it is the more nucleophilic,

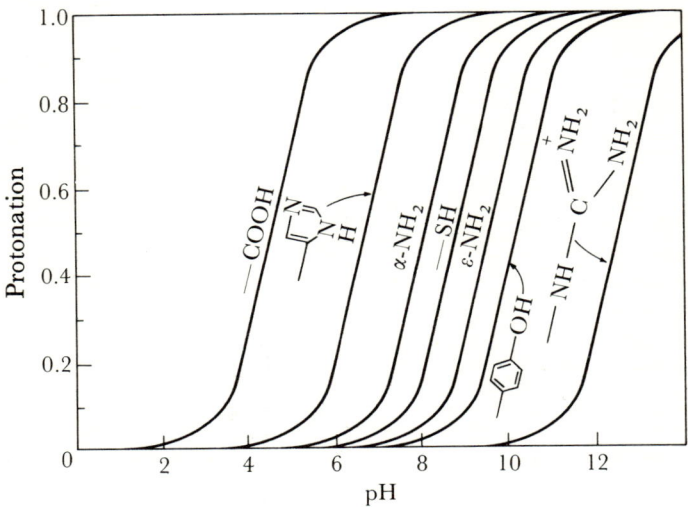

FIGURE 1-2   Theoretical ionic states of amino acid side chains as a function of the pH.

least protonated form of most groups that is most reactive. Protonation of nucleophilic amino or imidazole groups, for example, renders them unreactive with most reagents. There are exceptions, however, as in the esterification of carboxyl groups with methanol/HCl or with diazoacetates (Section 7-1), where the least nucleophilic protonated carboxyl groups are the most reactive. The reaction of iodoacetic acid and other haloacetates with proteins can affect sulfhydryl, amino, imidazole, and methionyl groups to various extents, depending upon the conditions. The reactivity of sulfhydryl groups with haloacetates is so great that under most conditions their modification rates exceed those of all other groups; reactivities of the other groups vary considerably, depending especially on the pH. By taking advantage of the differences in rates of different groups under various conditions, it is often possible to bring about selective modification more or less as desired. Although exerting a great effect upon the reactivity of amino and imidazole groups, pH has little effect on the reactivity of the thioether group of methionine. At low pH's, where both of the former groups are protonated and therefore unreactive, it is possible to selectively modify methionine. At neutrality where amino groups, but not imidazole groups, are protonated, imidazole becomes the most reactive group. At even higher pH's amino groups are the most reactive groups. Variation in reactivity as a function of pH is a convenient way to control the course of many modification reactions. It is unfortunate, however, that the pK's of most of the side-chain groups fall in a rather narrow range so as to minimize the usefulness of this approach (Figure 1-2).

TABLE 1-2   pK's of titratable groups in ribonuclease (from Tanford and Hauenstein, 1956)

Group	Residue	pK_a (expected)	Number of groups (From titration curve)	pK_a (found) (From titration curve)
α-CO$_2$H	Carboxy-terminal	2.1–2.4		
β-CO$_2$H	Aspartate	3.7–4.0	11	4.7
γ-CO$_2$H	Glutamate	4.2–4.5		
-imidazolium	Histidine	6.7–7.1	4	6.5
α-NH$_3^+$	Amino-terminal	7.6–8.0	1	7.8
—SH	Cysteine	8.8–9.1	0	—
ε-NH$_3^+$	Lysine	9.3–9.5	10	10.2
-phenolic	Tyrosine	9.7–10.1	3 + 3[a]	9.95, >12
—NH—C(=NH$_2^+$)NH$_2$	Arginine	>12	4	>12

[a] Six phenolic side chains titrate in two groups, three side chains in each group having the indicated pK values.

The approximate range of pK values expected for low-molecular-weight substituents and the values actually reported for ribonuclease are shown in Table 1-2. The differences are presumably due to perturbing influences of the protein environment. Similar values have been observed with other proteins.

In Part II, most discussions of a reagent begin with an equation illustrating the reaction and showing the reactants and side-chain groups in what is thought to be their reactive forms. It is important to keep in mind, however, that these are not always the major forms present under the conditions indicated for the reaction.

Relative Reactivities of Similar Groups. The reactivities of most small organic compounds with other small molecules usually obey rather simple rules and can be expressed in terms of certain semiquantitative relationships like those developed by

Hammett, Taft, and collaborators (Hammett, 1940; Taft, 1956; Jaffé, 1953). The reactivities of structurally related amino acids with acrylonitrile and several other α,β-unsaturated compounds, for example, can be expressed in terms of a linear free-energy equation, taking both polar and steric factors into account (Friedman and Wall, 1964, 1966). If steric effects are accounted for, logarithms of the second-order rate constants have linear dependencies upon pK's of the amino groups, making

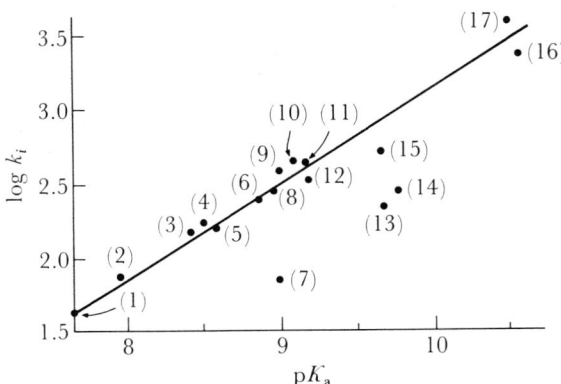

FIGURE 1-3   Correlation of pK_a and nucleophilicity of the amino groups in model compounds. Points correspond to the following amines: (1) glycylglycyl-L-tyrosine, (2) glycine amide, (3) glycyl-L-tyrosine, (4) m-chlorophenylglycine, (5) m-fluorophenylglycine, (6) p-fluorophenylglycine, (7) lysine (α), (8) phenylglycine, (9) p-methylphenylglycine, (10) p-methoxyphenylglycine, (11) tyrosine, (12) phenylalanine, (13) ε-N-acetyl-L-lysine, (14) alanine, (15) glycine, (16) lysine (ε), (17) α-N-acetyl-L-lysine amide. (Data from Freedman and Radda, 1968.)

it possible to predict reaction rates of related compounds. Similar quantitative relationships have been found between the pK's of amines and their reaction rates with cyanate and trinitrobenzenesulfonic acid (Figure 1-3). The more complex situation in proteins, however, has made such analyses difficult to obtain.

Quantitative information about the relative reactivities of specific residues in proteins can be obtained by isolating peptides from partially modified proteins and by comparing the relative yield of each (i.e., if the modified and unmodified fragments are separated, the relative amount of each is the basis upon which the relative reactivity is made). By using a double-labeling procedure, however, it has been possible to eliminate the need for quantitative recovery of each peptide. For example, by modifying the tyrosyl residues of a purified anti-p-azobenzoate antibody with a low level of radioactive iodine (^{125}I), and then combining this with a similar sample modified more extensively with a different radioactive label (^{131}I), hydrolyzing enzymatically, and separating the peptides, it has been possible to use the ratio of the two labels in subsequently isolated peptides to determine relative susceptibilities to iodination (Roholt and Pressman, 1967). Quantitative recovery of each peptide is unnecessary, since the ratio of the two labels in each gives its reactivity.

Studies of the relative reactivities of side chains in elastase has led to interesting attempts to express these reactivities in semiquantitative form (Hartley, 1969). Rate constants have been calculated relative to that of a standard—usually a corresponding simple peptide or amino acid—and differences in rates have been attributed to steric and polar factors. The procedure involves the use of a differential-labeling procedure somewhat similar to that described above. Initial modifications were done to only a very low level with a radioactively labeled reagent and were followed by treatment with a large excess of unlabeled reagent. After degradation to peptides and their isolation, the specific activity of each peptide could be determined and used to calculate relative reactivities. The fractional content of label in each peptide by this procedure is proportional to the relative reactivity of the residue(s) contained therein, so long as a relatively small fraction of each was modified during the initial labeling.

Relative reactivities of groups can be affected either positively or negatively by neighboring groups. Enhanced reactivities are in many cases (although not always) associated with the unique chemistry resulting from the presence of a group in a catalytic site. Such enhanced reactivity with one reagent is not always reflected by an enhanced reactivity with a different reagent. The amino group of the enzymatically crucial lysine-41 in pancreatic ribonuclease has an enhanced reactivity with dinitrofluorobenzene, but reacts only with great difficulty with positively charged reagents like O-methylisourea and methyl acetimidate. Its proximity to the positively charged active center presumably accounts for both unusual reactivities.

Buried and Exposed Groups. The relative positions of various groups are of obvious importance to the physical, chemical, and biological properties of proteins. Residues in the interior, for example, interact to a lesser extent with components of the solvent than do those on the exterior. While the best way to determine the relative location of each residue is by X-ray crystallographic analysis, another good way is in terms of the relative reactivity of such groups to various chemical reagents. Surface groups are expected to have reactivities like the same group in a simple peptide, while buried groups have lower reactivities. Deviations in this simple pattern come about from a variety of additional factors, but, in general, correlate well with the predictions from X-ray analyses.

A variety of methods, including X-ray crystallography, have shown hydrophobic residues tending to occupy the interior of proteins. In agreement with this are results from several physical-chemical procedures which show the hydrophobic chromophores of tyrosine and tryptophan frequently in hydrophobic environments and relatively inaccessible to solvent. By these procedures, tyrosine and tryptophan residues in a large number of proteins have been classified either as "buried" or "exposed" on the basis of their apparent accessibility to the solvent.

Although chemical methods for determining the accessibility of these residues show some variations, they correlate closely with the physical-chemical results (Table 1-3). The reagents most frequently employed to determine the relative numbers of buried and exposed groups, and the groups toward which they are directed,

TABLE 1-3 State of tyrosine residues

Protein	Exposed by physical methods	Reactive in native state		Total tyrosine
		N-acetyl-imidazole	Tetranitro-methane	
Bovine pancreatic carboxypeptidase A	7 groups	6 groups	7 groups	15 groups
Bovine insulin	4	4	2	4
Chicken ovalbumin	2	1.5	5.6	10
Bovine pancreatic ribonuclease	3	3	3	6
Porcine pepsinogen	15	8–10	10.4	17
Sperm whale myo-globin	2	0.8	3	3

are: (a) N-acetylimidazole, tyrosine; (b) cyanuric fluoride, tyrosine; (c) N-bromosuccinimide, tryptophan (or, in its absence, tyrosine); (d) diazo-1-H-tetrazole, histidine; and (e) tetranitromethane, tyrosine.

The number and the nature of buried groups vary greatly with different proteins. Larger and relatively rigid proteins usually have a greater proportion of such groups, while small and loosely folded proteins usually have few. Buried groups can often be exposed by denaturing agents like urea and guanidine hydrochloride, or by extremes of pH. The reactivities of buried sulfhydryl groups in three widely studied proteins and the effects of denaturation upon their reactivities with four different reagents are shown in Table 1-4. The sulfhydryl groups of ovalbumin are unreactive with N-ethylmaleimide and 5,5′-dithiobis(2-nitrobenzoic acid), while three of the four react with p-hydroxymercuribenzoate and iodine. Similarly, those of β-lactoglobulin react readily with the mercurial and iodine, but unless denatured react only weakly with the other two reagents. Since each reagent was employed at a different pH, some of the differences in reactivity may partly reflect the effect of pH on the protein structure. It is obvious that comparisons of group reactivity at different pH's are valid only if the protein has the same conformation at each pH.

As shown in Table 1-4, the numbers of reactive sulfhydryl groups found in proteins frequently differ with different reagents. The hydrophobic character of sulfhydryl groups and their presence, in many cases, in hydrophobic environments (Robinson, 1966), in conjunction with the varied ability of reagents to penetrate such regions, partly account for the differences in reactivity. Thus, it is frequently found that apolar organomercurials like p-hydroxymercuribenzoate are much more effective than hydrophilic reagents like N-ethylmaleimide and iodoacetamide. In a study of yeast alcohol dehydrogenase, sulfhydryl reactivity was shown to parallel increasing chain length of a series of N-alkylmaleimides. N-Ethylmaleimide, for

TABLE 1-4   Sulfhydryl-group reactivities[a]

Reagent	Ovalbumin		Serum albumin		β-Lactoglobulin	
	Native	Denatured	Native	Denatured	Native	Denatured
p-Mercuribenzoate[b]	2.8	4.0	0.45	—	1.9	—
Iodine[c]	2.9	4.0	0.45	0.6	2.1	—
N-Ethylmaleimide[d]	0.4–0.6	3.8	0.45	0.45	0	1.9
5,5′-Dithiobis(2-nitrobenzoic acid)[e]	0	3.8	0.32	0.29	1.0	1.8

[a] Sulfhydryls per mole protein, from Fernandez-Diez et al., 1964.
[b] At pH 4.6.   [c] At pH 6.5.
[d] At pH 7.0.   [e] At pH 8.0.

example, was less reactive than its more apolar homolog N-butylmaleimide (Heitz et al., 1968). Similar results were obtained with porcine heart fumarase, using a similar series of homologous alkyl mercury nitrates (Robinson et al., 1967).

Although the concept of buried and exposed groups was originally employed in regard to hydrophobic residues, it has since been extended to other residues. Reduced reactivity on the part of any group is frequently attributed to its being buried. Changes in the number of reactive groups upon formation of various bimolecular complexes have been used to ascertain the size of combining areas and the magnitude of accompanying conformational changes. Formation of the 1:1 complex of bovine trypsin with an inhibitory protein from bovine pancreas, for example, brings about a reduction in susceptibility of tryptophan residues in the enzyme to N-bromosuccinimide. One or two normally reactive tryptophans are protected in the complex (Spande and Witkop, 1965; Steiner, 1966). It is not known, however, whether protection results from their actual presence in the contact area or from conformational changes accompanying complexation. In an analogous way, the two of twenty-four cysteinyl residues of horse liver alcohol dehydrogenase which react with iodoacetic acid and effect the loss of its enzymatic activity can be protected by prior complexation with DPNH (Li and Vallee, 1965). Variations of such procedures are widely used to identify residues in protein binding sites and are discussed in a later chapter.

1-7 ACTIVE SITES

The investigation of relationships between biological properties and specific residues is currently one of the most active applications of the chemical modification of proteins. A weakness of such studies often results from their failure to distinguish between absolutely vital residues such as those in an active site and those adjacent to an active site, or otherwise indirectly associated with the biological property of interest. Unfortunately, a clear distinction often cannot be made. Indeed, even if the precise nature of the chemical modification could be known, present concepts of active sites include no clear distinction of their boundaries. The tendency to consider active sites as rather limited areas on an enzyme's surface is based on a considerable body of work and appears well founded. It does not necessarily follow, however, that large portions are unimportant to function.

In the past, the principal approach to the study of catalytically important residues involved modification of nearly all residues of a particular type, followed by determination of whether or not such modification caused a loss or change in the characteristic activity. More recently, the trend has been toward less extensive modifications and correlations of this with observed changes in biological properties.

In addition to catalytic sites or so-called "active sites," many biologically active proteins are known to possess other kinds of functional sites. Even the traditional active site is now often discussed in terms of its component substrate-binding and bond-breaking parts. Any discussion of essential groups would be incomplete without an understanding or definition of kinds of functional sites involved. The following operational classification of Vallee and Riordan (1969) is in keeping with subsequent discussion:

Active site: those atoms or side-chain groups of proteins directly involved in the catalytic step, i.e., the process of bond making or breaking

Active center: those groups involved in substrate binding and catalysis. This definition includes the active site

Substrate binding: those groups involved directly in the noncovalent attachment of the substrate to the active center

Effector site: those groups topologically remote from the active center which interact with ligands and thereby influence binding and activity through effects on structure and conformation.

This classification system is intended to apply to proteins having catalytic activity, but it is readily extended to most other proteins. Many proteins, for example, which bind various substances but lack a known catalytic function can be considered in analogous terms. Binding sites of many of these are completely analogous to substrate binding sites; immunoglobulins and various oxygen- and metal-binding proteins are notable examples. Other proteins of interest, but for which the above classification appears at this time to be inadequate, include the various peptide hormones, blood-clotting proteins, proteolytic-enzyme inhibitors, virus-coat proteins, and other structural proteins.

Essential Groups. A concept similar to that of active sites is the concept of "essential groups." Essential groups are those involved in, or in some way required for, a particular property. Active sites as just described are composed of one or more such groups. From the operational point of view, essential groups are those whose modification will bring about a loss of the characteristic property. Its loss is presumed to result from a change in such a group's chemical character. That modification of a functionally vital group might not always eliminate the property is not usually considered. Much more important from a practical point of view are what might be called "false essential groups"—those whose modification results, by virtue of secondary effects, in elimination of the particular property. Identifying such groups as other than essential can be extremely difficult and, in some cases, border on ambiguity; but such identification is necessary in order to understand the relationship between function and chemical structure.

When modification of a group in a protein effects loss of biological activity, it is usually considered essential to that activity. There are many examples, however, where such naive deductions have later proved inaccurate. Acetylation of chicken egg white lysozyme, for example, results in the loss of its catalytic activity as determined under normal assay conditions. This observation led to the erroneous conclusion that amino groups were essential for the activity of lysozyme. When acetylated lysozyme is assayed at lower pH's than normally used, however, most of the initial lysozyme activity can be demonstrated (Yamasaki et al., 1968) (see Figure 1-4). The pH activity profile of the modified enzyme is displaced to lower pH in a manner similar to the shift of its isoelectric point. This illustrates an uncommon but important problem in interpreting the results of chemical modification, namely, that a modified protein may have retained its biochemical activity but in an altered form. By strict definition of the word, such a protein, under the conditions of the normal assay, is inactive. It is still active, however, from the standpoint of its catalytic mechanism, and detection of the catalytic activity is important to the understanding of its chemical mechanism.

A still different phenomenon, observed with certain enzymes having catalytic activity against more than one substrate, is that chemical modification can sometimes cause a greater loss of activity against one substrate than against the others. In some such cases, the loss of activity against one substrate may even be accompanied by an increased activity against another. Acetylation of carboxypeptidase A, for example, results in nearly complete loss of its peptidase activity but increases its esterase activity manyfold (Simpson et al., 1963). This unusual result has been shown to be due to changes in the K_m values for the two substrates, rather than to any change in the "bond-breaking" residues (Bender et al., 1965).

Two explanations frequently invoked to explain the absence of activity following modification of other than active-site groups are based on steric blocking by the added

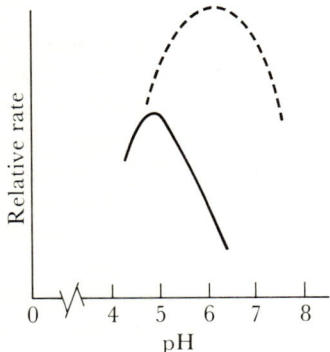

FIGURE 1-4   pH dependence of cell wall lysis by acetylated lysozyme (solid curve) compared to lysozyme (dashed curve). (Adapted from Yamasaki et al., 1968.)

substituents and conformational change. Thus it is thought that merely the presence of a group can sometimes act to block particular functions by its bulk. This plausible hypothesis is in keeping with experimental evidence. Reagents which introduce large substituents have been shown to inactivate certain proteins, while modification of the same groups with less bulky reagents did not. Conformational changes, as discussed earlier (Section 1-5), are a frequent result of protein modification. Both physical and chemical methods are available to detect such changes. Some of these are mentioned in Chapter 4. The investigator, having failed to detect any change in conformation by one or more techniques, frequently must make the decision whether or not to employ yet another.

REFERENCES

Baker, B. R. (1967): *Design of Active-Site-Directed Irreversible Enzyme Inhibitors*, Wiley-Interscience, New York.
Balls, A. K., and E. F. Jansen (1952): *Adv. Enzymol.*, **13**, 321.
Bender, M. L., J. R. Whitaker, and F. Menger (1965): *Proc. Nat. Acad. Sci.*, **53**, 711.

Blake, C. C. F., D. F. Koenig, G. A. Mair, A. C. T. North, D. C. Phillips, and V. R. Sarma (1965): *Nature*, **206**, 757.
Canfield, R. E., and A. K. Liu (1965): *J. Biol. Chem.*, **240**, 1997.
Cohen, L. A. (1968): *Ann. Rev. Biochem.* **37**, 695.
Fernandez-Diez, M. J., D. T. Osuga, and R. E. Feeney (1964): *Arch. Biochem. Biophys.*, **107**, 449.
Freedman, R. B., and G. K. Radda (1968): *Biochem. J.*, **108**, 383.
Friedman, M., and J. S. Wall (1964): *J. Am. Chem. Soc.*, **86**, 3735.
Friedman, M., and J. S. Wall (1966): *J. Org. Chem.*, **31**, 2888.
Gehrke, C. W., D. Rosch, R. W. Zumwalt, D. L. Stalling, and L. L. Wall (1968): *Quantitative Gas-Liquid Chromatography of Amino Acids in Proteins and Biological Substances*, Analytical Biochemistry Laboratories, Inc., Columbia, Mo.
Glazer, A. N. (1970): *Ann. Rev. Biochem.*, **39**, 101.
Hammes, G. G., and H. A. Scheraga (1966): *Biochemistry*, **5**, 3690.
Hammett, L. P. (1940): *Physical Organic Chemistry*, McGraw-Hill, New York.
Hartley, B. S. (1969): personal communication.
Heitz, J. R., C. D. Anderson, and B. M. Anderson (1968): *Arch. Biochem. Biophys.*, **127**, 627.
Herriott, R. M. (1947): *Adv. Protein Chem.*, **3**, 169.
Hirs, C. H. W. (1967): *Methods Enzymol.*, **11**, 27.
Hvidt, A., and S. O. Nielsen (1966): *Adv. Protein Chem.*, **21**, 287.
Jaffé, H. H. (1953): *Chem. Rev.*, **53**, 191.
Koshland, D. E. (1959): *Enzymes*, **1**, 305.
Li, T. K., and B. L. Vallee (1965): *Biochemistry*, **4**, 1195.
Monod, J., J. Wyman, and J. P. Changeux (1965): *J. Molec. Biol.*, **12**, 88.
Olcott, H. S., and H. Fraenkel-Conrat (1947): *Chem. Rev.*, **41**, 151.
Ramachandran, G. N., and V. Sasisekharan (1968): *Adv. Protein Chem.*, **23**, 283.
Riehm, J. P., C. A. Broomfield, and H. A. Scheraga (1965): *Biochemistry*, **4**, 760.
Robinson, G. W., R. A. Bradshaw, L. Kanarek, and R. L. Hill (1967): *J. Biol. Chem.*, **242**, 2709.
Robinson, J. D. (1966): *Nature*, **212**, 199.
Roholt, O. A., and D. Pressman (1967): *Biochim. Biophys. Acta*, **147**, 1.
Shaw, E. (1970): *Physiol. Rev.*, **50**, 244.
Simpson, R. T., J. F. Riordan, and B. L. Vallee (1963): *Biochemistry*, **2**, 616.
Spande, T. F., and B. Witkop (1965): *Biochem. Biophys. Res. Commun.*, **21**, 131.
Spande, T. F., B. Witkop, Y. Degani, and A. Patchornik (1970): *Adv. Protein Chem.*, **24**, 97.
Stark, G. R. (1970): *Adv. Protein Chem.*, **24**, 261.
Steiner, R. F. (1966): *Biochemistry*, **5**, 1964.
Stewart, J. M., and J. D. Young (1969): *Solid Phase Peptide Synthesis*, W. H. Freeman, San Francisco.
Taft, R. (1956): in *Steric Effects in Organic Chemistry*, edited by M. S. Newman, John Wiley & Sons, New York, chap. 13.
Tanford, C. (1968): *Adv. Protein Chem.*, **23**, 121.
Tanford, C., and J. D. Hauenstein (1956): *J. Am. Chem. Soc.*, **78**, 5287.
Vallee, B. L., and J. F. Riordan (1969): *Ann. Rev. Biochem.*, **38**, 733.
Wang, J. H. (1968): *Science*, **161**, 328.
Witkop, B. (1968): *Science*, **162**, 318.
Yamasaki, N., K. Hayashi, and M. Funatsu (1968): *Agr. Biol. Chem. (Tokyo)*, **32**, 64.

2
modification of groups essential for activity

A large portion of our information concerning the chemical basis of enzyme function has been obtained through the application of protein modification. In spite of rather obvious limitations, this approach remains the easiest and most direct means to identify "essential groups." Changes in enzymatic activity are attributed to the modification of active-site residues, but in fact, this is only one of several possibilities. Even if a reagent's specificity is completely known, changes in activity may still result indirectly due to an accompanying structural change, as discussed in the previous chapter, or by modification of residues in allosteric or effector sites. Harsh reaction conditions or extensive modification are particularly likely to bring about unwanted structural changes. If chemical modification, on the other hand, has no effect on the properties of interest, it is clearly reasonable to assume that the modified residues are not essential to the activity under investigation.

When changes in activity accompany the modification of a single residue, conformational changes are less likely to be responsible. Because of the high reactivity of many active-site residues, it has been possible to prepare a relatively large number of monosubstituted enzyme derivatives. The high reactivity of certain of these groups appears related to their catalytic function.

2-1 SUBSTRATE PROTECTION OF ACTIVE-CENTER GROUPS

Protection by substrates or related compounds has been used to selectively prevent the modification of active-center residues in many enzymes. Differential labeling (i.e., modification first in the presence and then in the absence of a substrate) may be used to identify these residues. If enzymatic activity is retained following modification in the presence of substrate but is lost in its absence, it is usually assumed that a group in the active site has been protected by substrate. If, after modification in the presence of substrate followed by its removal, modification is again carried out, this time using a radioactive reagent, it should be possible to determine, from the amount of label incorporated, the number of groups whose modification is responsible for the inactivation. Using the radioactive label it is possible to isolate and to identify active-center fragments resulting from chemical or enzymatic degradation. Designation of such fragments as "active-center" peptides can sometimes be misleading, however, since substrate protection may be conferred upon residues outside the active center but protected by a structural change accompanying combination of the enzyme with substrate.

A similar procedure can be used to specifically label the active site of an individual enzyme in a mixture of many enzymes. It was shown by Cohen and Warringa (1953), for example, that cholinesterase could be specifically labeled with ^{32}P-diisopropylfluorophosphate while in an unpurified mixture of enzymes. Butyrylcholine was used to protect the active site of the enzyme during treatment with nonradioactive diisopropylfluorophosphate, which then reacted with other esterases not protected by butyrylcholine. After removal of butyrylcholine, the unreacted cholinesterase was labeled with ^{32}P-diisopropylfluorophosphate.

2-2 ACTIVE-SITE SELECTIVE REAGENTS

If specific reagents for each of the groups commonly found in proteins were available, the task of identifying important groups would still not be easy. A different approach to the determination of essential groups has led to the development of so-called "site-specific reagents." These reagents are designed to combine noncovalently with particular sites on proteins and, in a second step, to react with residues in or near that site. If the covalent bond formed in the second step is stable to conditions for chemical or enzymatic degradation of the protein, it is possible to identify and characterize fragments originating from the labeled site. In accordance with current usage, three divisions of site-specific reagents will be employed: *substrate labels*, *pseudosubstrates*, and *affinity labels*. This division is based on certain distinguishing characteristics and is useful for descriptive purposes, but it should be clear that there is overlapping of the basic principles. Two reviews have recently dealt with the design and use of such reagents (Baker, 1967; Singer, 1967).

Substrate Labeling. An enzyme's substrate is perhaps the perfect example of a site-specific protein reagent. Unfortunately, bonds formed between enzymes and their substrates are usually labile and transitory. The specificity of their combination, however, has prompted attempts to stabilize these linkages. With acetoacetate decarboxylase, for example, reduction with sodium borohydride in the presence of its substrate acetoacetate has been shown to bring about specific labeling of one lysyl residue with a loss of catalytic activity. Similar results have been obtained with other enzymes and their specific substrates. A single lysine residue is labeled by sodium borohydride reduction of the corresponding aldolase–dihydroxyacetone phosphate intermediate (Figure 2-1). In both of the above cases, the labeled residues have been isolated as peptides after cleavage of the protein, and the amino acids adjacent to the labeled residues have been determined. The amount of substrate introduced in such an experiment can be used as an estimate of the number of active sites. Substrate labeling leaves little question concerning the involvement of the labeled residues but unfortunately is applicable only to a rather select group of enzymes. Its application to aldolases has recently been described (Lai et al., 1967).

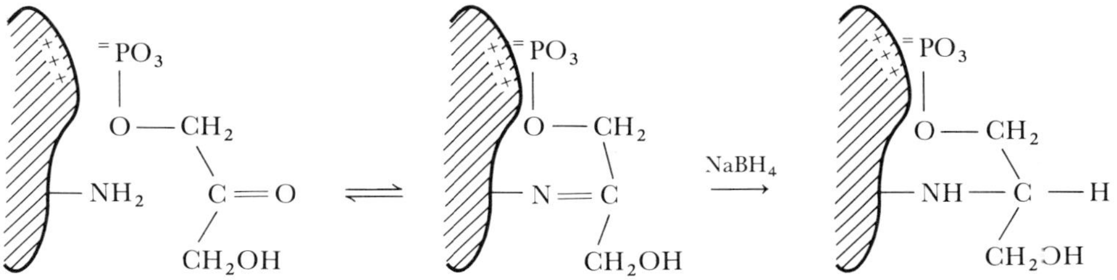

FIGURE 2-1 Active-site labeling of rabbit muscle aldolase.

Many enzymes which bind pyridoxal-5-phosphate, such as glycogen phosphorylase, can be similarly reduced by sodium borohydride. The absorbance and fluorescence properties of the resultant pyridoxyllysine residues make their detection and quantitation relatively simple. The method can be employed even with proteins like bovine serum albumin, aldolase, and pancreatic ribonuclease, which, although not considered pyridoxal proteins, nevertheless form specific complexes with it, and upon reduction give relatively homogeneous derivatives. The labeling specificity appears in most of these cases to be directed by the 5-phosphate group to anion binding sites. This is discussed in greater detail in Section 6-9.

Pseudosubstrates. Acetylcholinesterase, trypsin, chymotrypsin, and many other proteases and esterases are inactivated by diisopropylfluorophosphate (DFP) through its reaction with a specific serine hydroxyl group in each. Similar amino acid sequences have been found around the reactive seryl residue in most of these enzymes.

Serine itself, other serine residues in these proteins, and serine residues in other proteins do not react under the same conditions. The chemical basis of these reactions is not completely understood. The special reactivity is a function of the native structure of the serine enzymes and is abolished upon denaturation. Certain other organophosphates react similarly, with characteristically different rates, with the different enzymes (Mounter et al., 1963). These reagents have been called pseudosubstrates on the basis of certain characteristics they are presumed to share with substrates of these enzymes. In the case of DFP, the electrophilic phosphorus atom reacts with the nucleophilic hydroxyl group of the active-site serine residue, eliminating a fluoride ion and forming a stable diisopropylphosphoryl enzyme, similar in some respects to the acyl enzyme intermediates formed in the normal hydrolytic reactions of these enzymes.

$$\begin{array}{c} CH_3 \\ \diagdown \\ CH-O-\overset{\overset{\displaystyle O}{\|}}{\underset{\underset{\displaystyle O}{|}}{P}}-F \\ CH_3 \diagup \\ | \\ CH \\ \diagup \diagdown \\ CH_3 CH_3 \end{array}$$

pseudosubstrate
diisopropylfluorophosphate
(DFP)

Certain aryl- and alkylsulfonyl halides (for example, phenylmethanesulfonyl fluoride) react with the active-site serine residues of these same enzymes. Like the reaction with DFP, these appear to result from some resemblance of these compounds to substrates of these enzymes. The greater the apparent resemblance to substrates the greater the effectiveness. For example, the reaction of α-chymotrypsin with phenylmethanesulfonyl fluoride is at least 10,000 times faster than its reaction with methanesulfonyl fluoride (Fahrney and Gold, 1963), due apparently to the former compound's greater resemblance to the hydrophobic substrates of α-chymotrypsin.

Affinity Labeling. Utilizing toluenesulfonylphenylalaninechloromethyl ketone (TPCK), Schoellmann and Shaw (1963) were able to effect a highly specific alkylation of an active-center histidine in α-chymotrypsin. The reagent was designed to possess substratelike features in order to localize it at the active center of α-chymotrypsin, where the reactive chloromethyl moiety could irreversibly combine with a reactive group. This reagent has no effect upon trypsin, which hydrolyzes peptide bonds of basic amino acids (lysine and arginine) rather than peptide bonds of hydrophobic amino acids (tyrosine and phenylalanine) which are hydrolyzed by α-chymotrypsin. However, a similar reagent, *p*-toluenesulfonyllysinechloromethyl ketone (TLCK), which resembles a trypsin substrate, specifically and stoichiometrically inactivates trypsin (Shaw et al., 1965). The two reagents are enzyme specific and react only with the undenatured enzymes.

Affinity labeling takes advantage of the normal enzyme-substrate interactions to insure that a large local concentration of reagent exists at the active site. It depends upon the ability of a reactive group in the affinity label to form a stable linkage to the enzyme. Highly reactive α-halocarbonyl groups which, like the more familiar bromo- and iodoacetate, can react with several different nucleophilic groups are found in both TPCK and TLCK. The specificity for a particular histidine suggests the precise stereochemical specificity of the reaction. Such reagents may or may not label residues which participate in the catalytic process but presumably always label

TABLE 2-1 Some active-site-specific reagents for chymotrypsin

Reagent	Labeled residue	References
Tosylphenylalanine-chloromethyl ketone (TPCK)	Histidine	Schoellmann and Shaw, 1963
Diphenylcarbamyl chloride	Serine	Erlanger et al., 1966
α-Bromoacetophenone	Methionine	Schramm and Lawson, 1963
p-Nitrophenyldiazoacetate plus irradiation	Serine, histidine, tyrosine	Shafer et al., 1966

ones near the active center. The affinity-labeling technique has been applied to α-chymotrypsin using many different reagents and, depending upon the reagent, has resulted in the labeling of several different groups (Table 2-1). Although widely separated in the primary sequence, each of the labeled groups is presumably topologically in the vicinity of the active center. Both TPCK and TLCK also react specifically with several proteolytic enzymes from plants, such as papain and ficin (Whitaker and Perez-Villasenor, 1968); in these cases, inactivation occurs by the alkylation of active-site —SH groups. The affinity-labeling technique as applied especially to α-chymotrypsin and trypsin has recently been described (Shaw, 1967).

An interesting variation of the affinity-labeling approach has been applied to trypsin (Inagami, 1965). It was shown that trypsin could be inactivated by iodoacetamide in the presence of methylguanidine, whereas iodoacetamide alone was ineffective. Substitution of butylguanidine or propylguanidine for methylguanidine gave no inactivation. Loss of activity presumably results from alkylation of the same

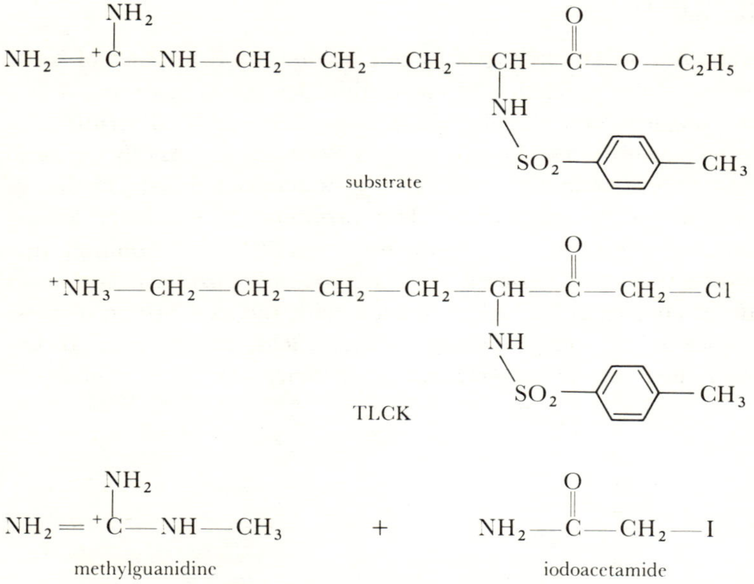

FIGURE 2-2 Comparisons of formulas of inhibitors for trypsin.

active-site histidine as was alkylated by TLCK. The two reagents iodoacetamide and methylguanidine together appear to achieve an overall structure in the active center resembling the more typical affinity-labeling reagent TLCK (Figure 2-2).

The affinity of an antibody for its antigen is similar to that of an enzyme for its substrate. It is not surprising, therefore, that the affinity-labeling technique was applied very early and with considerable success to the determination of residues in antibody combining sites (Wofsy et al., 1962; Metzger et al., 1963). When hapten-specific immunoglobulins were treated with diazonium salts closely resembling their particular hapten, a reaction occurred which thereafter prevented the binding of hapten. The presence of hapten or close analogs prevented the reaction. The formation of 1 mole of azotyrosine accounted for the observed effects; very little reaction took place with other than the combining-site tyrosines. No reaction was obtained with diazonium salts unrelated to the particular hapten. Assuming only one reactive tyrosine in each combining site, its relative reactivity was on the order of 10^4 times that of a typical tyrosine. Spectral changes accompanying the reaction lend favorably to its quantitation. Affinity labeling, especially as applied to the study of antibody combining sites, has recently been reviewed (Singer, 1967).

2-3 DETERMINATION OF THE NUMBER OF ESSENTIAL RESIDUES

Quantitative determination of essential groups can be done in a number of ways. In those cases where substrate protection of an essential residue is observed, differential labeling in the presence and absence of substrate, followed by determination of the substrate-protected residues, can be effective (see Section 2-1). Another general approach depends upon differences in reactivity of active-center side chains as compared to side chains located elsewhere. The variations in reactivity among individual side chains of the same type frequently permit their resolution into classes on the basis of reaction rates (Figure 2-3). It is frequently possible in such cases to correlate the rate of change in biological activity with the rate of one of these classes. For example, when trinitrobenzenesulfonic acid reacts with turkey ovomucoid, a rapid loss of trypsin-inhibitory activity of the ovomucoid

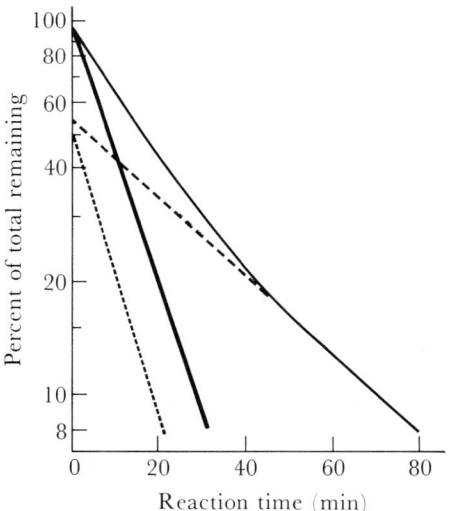

FIGURE 2-3   Semilogarithmic plot for loss of amino groups ["fast" —NH$_2$ groups (dotted line), "slow" —NH$_2$ groups (dashed line), total —NH$_2$ groups (thin line)] and loss of trypsin-inhibitory activity (thick line) in turkey ovomucoid during modification with trinitrobenzenesulfonic acid. (Adapted from Haynes et al., 1967.)

occurs. A similar rapid rate is observed for the reaction of one amino group; this fast-reacting amino group is thus thought to be essential for inhibitory activity (Haynes et al., 1967). Conversely, the modification of a single slow-reacting amino group in pancreatic ribonuclease with O-methylisourea correlates with inactivation of that enzyme (Klee and Richards, 1957).

Upon photooxidation of phosphoglucomutase in the presence of methylene blue, the decrease in activity is so rapid as to suggest more than one essential reactive group per active site, loss of either being sufficient to effect its inactivation. Two groups are thought to be involved, one fast-reacting methionine and one fast-reacting histidine (Ray and Koshland, 1962).

The approach just described is widely applicable, although in some cases the results are subject to ambiguity. When the rate of inactivation correlates with the rate of modification of a single group, clearly either faster or slower than that of all others, the relationship is reasonably evident. Unfortunately, such is frequently not the case. When changes in activity accompany the modification of two or more groups, the relative importance of each is not always clear. If, as in the case of phosphoglucomutase, the inactivation rate exceeds the rate of modification of any individual groups, then more than one may be essential. If inactivation accompanies the modification of many residues, structural changes become more likely. It is clear that residues which are modified more rapidly than the enzyme activity is lost cannot be essential to that activity.

Special Active-Center Modifications. The unique chemistry associated with the active centers of many enzymes has permitted the synthesis of specifically modified derivatives with usually nonspecific reagents. Less frequent is the reverse situation wherein the presence of a group in a catalytic center results in its lowered reactivity with particular reagents. Both represent particular cases of the situation described in the previous section, where the observed reactivity of an active-center group is particularly atypical.

The reactivities of sulfhydryl groups have been shown to be so affected in many enzymes. What appear to be active-center thiol groups in acetoacetate decarboxylase (O'Leary and Westheimer, 1968) and glyceraldehyde-P-dehydrogenase (Mathew et al., 1967), for example, react specifically with several weak acylating agents under conditions where typical sulfhydryl groups do not appreciably react. The resulting acyl enzymes are catalytically inactive. Unique derivatives of this type are the relatively uncommon result of modification with a nonspecific reagent. A few such unique derivatives have been purified and studied in detail.

Specific reaction can result from unusual affinity of a reagent for a particular site. Thus, the anionic alkylating agent iodoacetate appears strongly attracted to the positively charged active centers of both pancreatic ribonuclease A (Stein and Barnard, 1959) and *A. oryzae* ribonuclease T1 (Takahashi et al., 1967). In each case, complete loss of catalytic activity accompanies the reaction. Related alkylating agents lacking a negative charge are without effect. In each case, the residue modified

TABLE 2-2 Reversible protein reagents

Protein group	Reagent	Conditions for reversal
Amino (lysine and amino terminal)	Ethyl thiotrifluoroacetate	1 M piperidine, 0°, 2 hr, or carbonate buffer (pH 10.7), 25°, 30 hr (Sections 5-1, A-1)
	Maleic anhydride	pH 3.5, 60°, 6 hr (Section 6-7)
	Citraconic anhydride	pH 3.5, room temperature, overnight (Sections 5-1, A-1)
	Methyl acetimidate/HCl	Concentrated ammonia–acetic acid, 15:1, or pH 9 in 0.6 to 1.2 M hydrazine (Sections 5-3, A-1)
	Salicylaldehyde	Dialysis or dilution (Sections 6-7, A-1)
	2-Methoxy-5-nitrotropone	1 to 2 M hydrazine (pH 8.5–9), 38°, several hours (Section 6-7)
Phenolic (tyrosine)	Succinic anhydride	Spontaneously deacylates above pH 5 (Sections 5-1, A-6)
	Acetic anhydride	Slowly hydrolyzes above pH 10, or can be removed more rapidly with hydroxylamine (Sections 5-1, A-6)
	Cyanate	Decomposes in alkaline solutions (Section 5-2)
Imidazole (histidine)	Ethoxyformic anhydride	Half-life at 25°C is 55 hr at pH 7, 18 min at pH 10, and is rapidly decomposed at pH 7 by hydroxylamine (Section 5-1)
Thioether (methionine)	Photooxidation Hydrogen peroxide	Reduced by β-mercaptoethanol, thioglycolate, and other thiols (Sections 8-7, 8-8, A-9)
Sulfhydryl (cysteine)	Mercurials	Competitively displaced by high concentrations of thiols (Sections 10-2, A-2)
	Disulfides [especially 5,5'-dithiobis(2-nitrobenzoic acid)]	Competitively displaced by excess of thiols (Sections 8-3, A-2)
	Sulfite (+[O])	Reagent can be displaced by a large excess of thiols (Sections 8-2, A-3)
	Dinitrofluorobenzene	Dinitrophenyl group can be displaced by treating with excess thiol at pH 8.0 (Section 6-5)
Disulfide bridges (cystine)	Thiols Borohydride	Reoxidize in air (Sections 8-1, A-3)
Guanidino (arginine)	Phenylglyoxal	Decomposes slowly at neutral and alkaline pH's (Sections 10-1, A-5)

(i.e., histidine and glutamate, respectively) is one not normally modified by this reagent under the conditions employed. A cationic reagent, p-(trimethylammonium)-benzenediazonium, similarly labels the active center of bovine erythrocyte acetylcholinesterase (Wofsy and Michaeli, 1967). The procedure and reagent employed in the latter case closely resemble those developed for the affinity labeling of antibodies.

2-4 REVERSIBLE MODIFICATION

Good evidence that changes in properties of a protein are primary effects of chemical modification can be obtained by removing the modifying groups and demonstrating a return of the original properties. For this reason, it is desirable to have chemical reagents that form derivatives which can later be easily removed. Such reagents can also be useful to prevent temporarily the reaction of groups with other reagents. Substrate protection and protection by substratelike substances, as discussed in Section 2-1, is a related special kind of reversible modification. While there are a large number of reagents which are potentially reversible, such approaches have not been widely exploited. Some of the most readily reversible reagents are listed in Table 2-2.

α-Chymotrypsin has been reversibly modified with tosyl chloride. With restricted amounts of reagent, tosylation occurs stoichiometrically at the catalytically important hydroxyl group of one serine residue. Treatment at pH 2 converts the tosyl derivative into a cyclic oxazoline which upon subsequent adjustment to pH 7 is converted to active chymotrypsin (Gold and Fahrney, 1964) (see Subsection 5-4.2).

The principle of reversible modification has been incorporated into an interesting technique for characterizing peptides known as diagonal electrophoresis (Perham and Jones, 1967). When a mixture of peptides obtained from a protein treated with a reversible acylating agent (i.e., tetrafluorosuccinic anhydride, maleic anhydride, etc.) is electrophoresed in one dimension and then the labile acyl groups are removed before electrophoresing under the same conditions in the second dimension, all peptides follow a diagonal line except those which had their charge changed by removal of the labile acyl group.

Nitration of tyrosine rings and subsequent reduction of the 3-nitro group to an amino group is a series of reactions which, although not reversible in the usual sense, initially makes the tyrosine more acidic and then less acidic. The procedure is described in Section 9-2.

REFERENCES

Baker, B. R. (1967): *Design of Active-Site-Directed Irreversible Enzyme Inhibitors*, Wiley-Interscience, New York.
Cohen, J. A., and M. G. P. J. Warringa (1953): *Biochim. Biophys. Acta*, **11**, 52.

Erlanger, B. F., A. G. Cooper, and W. Cohen (1966): *Biochemistry*, **5,** 190.
Fahrney, D. E., and A. M. Gold (1963): *J. Am. Chem. Soc.*, **85,** 997.
Gold, A. M., and D. Fahrney (1964): *Biochemistry*, **3,** 783.
Haynes, R., D. T. Osuga, and R. E. Feeney (1967): *Biochemistry*, **6,** 541.
Inagami, T. (1965): *J. Biol. Chem.*, **240,** PC3453.
Klee, W. A., and F. M. Richards (1957): *J. Biol. Chem.*, **229,** 489.
Lai, C. Y., P. Hoffee, and B. L. Horecker (1967): *Methods Enzymol.*, **11,** 667.
Mathew, E., B. P. Meriwether, and J. H. Park (1967): *J. Biol. Chem.*, **242,** 5024.
Metzger, H., L. Wofsy, and S. J. Singer (1963): *Biochemistry*, **2,** 979.
Mounter, L. A., B. A. Shipley, and M. E. Mounter (1963): *J. Biol. Chem.*, **238,** 1979.
O'Leary, M. H., and F. H. Westheimer (1968): *Biochemistry*, **7,** 913.
Perham, R. N., and G. M. T. Jones (1967): *Eur. J. Biochem.*, **2,** 84.
Ray, W. J., and D. E. Koshland (1962): *J. Biol. Chem.*, **237,** 2493.
Schoellmann, G., and E. Shaw (1963): *Biochemistry*, **2,** 252.
Schramm, H. J., and W. B. Lawson (1963): *Hoppe-Seyl. Z. Physiol. Chem.*, **332,** 97.
Shafer, J., P. Baronowsky, R. Laursen, F. Finn, and F. H. Westheimer (1966): *J. Biol. Chem.*, **241,** 421.
Shaw, E. (1967): *Methods Enzymol.*, **11,** 677.
Shaw, E., M. Mares-Guia, and W. Cohen (1965): *Biochemistry*, **4,** 2219.
Singer, S. J. (1967): *Adv. Protein Chem.*, **22,** 1.
Stein, W. D., and E. A. Barnard (1959): *J. Molec. Biol.*, **1,** 350.
Takahashi, K., W. H. Stein, and S. Moore (1967): *J. Biol. Chem.*, **242,** 4682.
Whitaker, J. R., and J. Perez-Villasenor (1968): *Arch. Biochem. Biophys.*, **124,** 70.
Wofsy, L., H. Metzger, and S. J. Singer (1962): *Biochemistry*, **1,** 1031.
Wofsy, L., and D. Michaeli (1967): *Proc. Nat. Acad. Sci.*, **58,** 2296.

3

modifications to change properties

3-1 CHEMICAL MODIFICATIONS TO CHANGE PHYSICAL STRUCTURE

An important objective in modifying proteins is to change their physical structures in ways that will be useful in working with them or interpreting their properties. These modifications therefore are not usually directed toward modifying active sites themselves, although the end result may greatly affect an active site by changing its physical relationship to the rest of the protein.

3-1.1 Changing the Ionic State and Conformation

Proteins are charged molecules having characteristic numbers of both anionic and cationic residues. Modification of the individual charged groups may affect their net ionic charge in a way which is disruptive to their characteristic properties. Decreased solubility and changes in conformation or in the state of aggregation frequently result from such modifications. Slight conformational changes are frequently difficult to detect by physical-chemical means but often produce easily discernible changes in biological properties. Changes in the state of aggregation, on the other hand, are easily detected by a variety of physical techniques. While most protein modification is done hoping to minimize such changes (which are generally referred to as "denaturation"), in some cases changes in conformation or in the state of aggregation can be the desired goal.

While a number of reagents can be useful for this purpose, the most commonly used, and currently one of the best, is succinic anhydride. Succinylation results in the placement of anionic —COO^- groups in place of cationic —NH_3^+ groups, producing a net change of two charge units for each modified group. The resulting electrostatic interactions, somewhat like those obtained in the unmodified protein at very high pH, can result in the disaggregation of many subunit or polymeric proteins. Succinylation of the bovine pancreatic protein complex of procarboxypeptidase A, for example, brings about its dissociation to a succinylated zymogen monomer of carboxypeptidase plus two other proteins. Using this procedure, it was possible to obtain a purified form of the zymogen so as to facilitate the study of its conversion into active enzyme (Freisheim et al., 1967). In this case, succinylation simulated the effect of high pH on the molecular structure of the complex (i.e., polymer → monomer) under conditions wherein its other biological properties were presumably retained. Similar use of succinic anhydride reportedly results in the

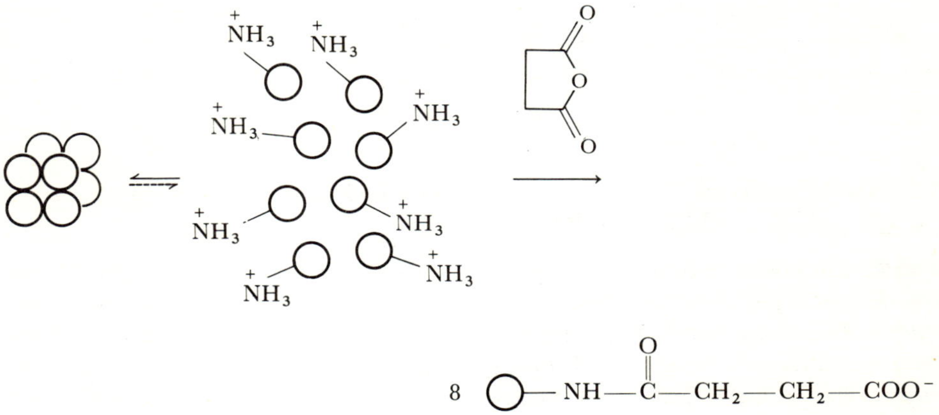

FIGURE 3-1   Schematic representation of the effect of succinic anhydride upon the equilibrium between intact hemerythrin and its subunits.

disaggregation of hemerythrin (Klotz and Keresztes-Nagy, 1963), rabbit muscle aldolase (Hass, 1964), porcine heart aspartate transaminase (Polyanovsky, 1965), and tobacco mosaic virus coat protein (Frist et al., 1965) (Figure 3-1).

Sulfhydryl reagents have been shown in a number of cases to cause the same kind of effect, although the basis of this action is less clear. Thus, the octamer of hemerythrin can be dissociated to monomers by reaction with succinic anhydride or with certain —SH group reagents such as N-ethylmaleimide (Keresztes-Nagy and Klotz, 1965). Dissociation of aspartate transcarbamylase by —SH reagents produces two kinds of subunits which can be individually purified by conventional methods for protein isolation and purification (Gerhart and Schachman, 1965). One of the purified components possesses catalytic activity but lacks the regulatory properties of the intact enzyme. The other component has no catalytic activity but binds the important regulatory compound cytidine triphosphate. By following the loss of regulatory activity upon reaction of the native enzyme with p-mercuribenzoate ion,

Gerhart and Schachman (1968) concluded that dissociation occurs in a concerted manner.

Progressive succinylation brings about an unfolding and expansion of bovine serum albumin, human γ-globulin, and chicken egg white lysozyme, as judged by changes in their Stokes radii and their increased susceptibility to reduction by thiols (Habeeb, 1967). One of two components produced by succinylation of bovine serum albumin shows an increased tendency to aggregate, presumably as the result of its change in conformation. Changes in the conformation of pepsinogen brought about by succinylation, as indicated by its changed optical rotatory properties, decrease its ability to be activated under the normal conditions for activation (Gounaris and Perlmann, 1967). Succinylation of proteins is normally accompanied by a decrease in the sedimentation coefficient as a result of the accompanying general molecular expansion. Such a decrease, therefore, does not necessarily indicate a dissociation into subunits.

3-1.2 REDUCTION AND REOXIDATION OF DISULFIDE BONDS

Only the disulfide bonds of proteins are susceptible to mild reduction. Extensive reduction normally requires complete disruption of tertiary structure by urea or other agents. Treatment of such reduced proteins with iodoacetic acid or acrylonitrile can be used to block the sulfhydryl groups so formed and thereby to prevent reformation of disulfide bonds upon removal of the reducing agent.

Many reduced and fully disordered proteins, if not so alkylated, regain their original disulfide bonds and biological activity upon return to nondenaturing conditions following slow air oxidation (Figure 3-2). The high recovery of activity with many reduced enzymes, upon such treatment, has been cited as evidence that the information for correct folding is contained in their primary structures. Highly successful reoxidations have, however, been thus far limited to relatively small proteins such as lysozyme, ribonuclease A, and ovomucoid. With some other fully reduced and disrupted proteins, insolubility and strong tendencies to aggregate are major obstacles responsible for low yields. This difficulty has been overcome in several cases, however, by reaction of the protein with the N-carboxyanhydride of D,L-alanine. This reaction covalently attaches poly-D,L-alanyl side chains to amino groups. The polyalanine side chains greatly increase aqueous solubility but appear to have little effect on the capacity to refold. Like other modifications, including some which extensively alter net electrical charge, polyalanylation does not appear to greatly decrease the capacity to correctly refold (Frensdorff et al., 1967). The influence of small amounts of low-molecular-weight thiols and disulfides during reoxidation promotes the recovery of activity. A definite ratio of disulfide to sulfhydryl has been shown to best promote the correct reoxidation of chicken egg white lysozyme (Wetlaufer and Saxena, 1968).

The extent of disulfide-bond reduction depends on both the reducing agent and the structure of the protein. Destabilization by denaturing agents or stabilization by various agents can greatly alter the numbers of disulfide bonds cleaved. The disulfide bonds in chicken ovotransferrin, for example, are readily cleaved by reducing agents without denaturants, but in the iron complex of ovotransferrin they are very resistant to reduction (Azari and Feeney, 1961). Reduction in the absence of denaturants selectively cleaves only the more accessible or labile disulfide bridges (Bewley and Li, 1969). Selective cleavages of small numbers of disulfide bridges

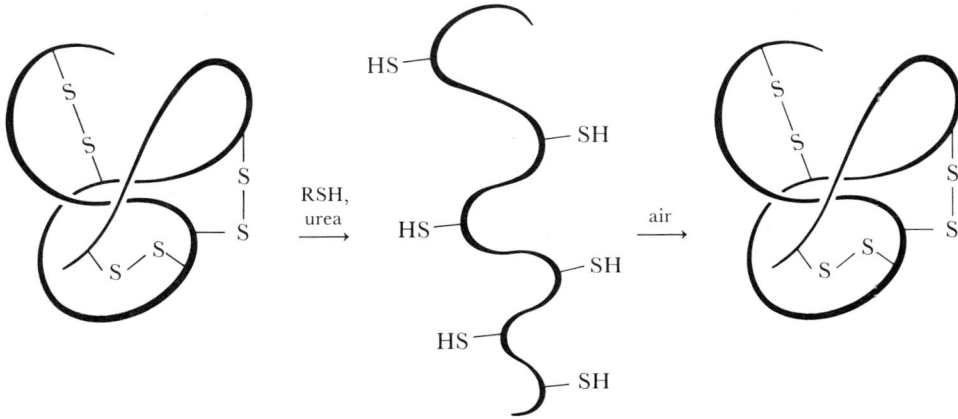

FIGURE 3-2 Schematic illustration of the fate of protein S—S bonds during reduction and unfolding followed by renaturation and reoxidation.

and its effect on biological activity can sometimes be used to ascertain their proximity or importance to the active center. The relative reactivity of individual disulfide bonds is a reflection of their environment and, as such, is indicative of their position in a protein's three-dimensional structure. Selective reduction of specific disulfide bonds in both papain (Arnon and Shapira, 1969) and pancreatic ribonuclease A (Sperling et al., 1969) has permitted the introduction of mercuric ions bridging the two resultant thiol groups in each. This general approach may have great importance for X-ray crystallography as a specific means to produce heavy-metal-containing protein derivatives.

Reagents employed for the reduction of protein disulfide bonds are described in Section 8-1.

3-1.3 CHEMICAL CLEAVAGE OF PEPTIDE BONDS

Chemical procedures for specific cleavage of peptide bonds have been highly useful as tools in studying amino acid sequences. These cleavage methods have been described elsewhere (Witkop, 1968; Spande et al., 1970). Since these methods are destructive to the structure of most proteins, they are seldom useful for studies of structure to function relationships.

The most widely used of these reagents, cyanogen bromide, cleaves at the carboxyl bond of methionine residues, converting internal methionine residues into carboxy-terminal homoserine lactone. Although reaction conditions are harsh, the reagent has been used with very stable proteins for the selective modification of methionine. It has been shown, for example, that one of the two methionine residues of turkey ovomucoid can react with cyanogen bromide without loss of the protein's activity (Feinstein and Feeney, 1967). For a further discussion of cyanogen bromide, see Section 10-3. N-bromosuccinimide, another popular reagent for cleaving peptide bonds, is discussed briefly in Section 8-9.

3-1.4 Cross-Linking Reagents

Bifunctional reagents, reagents with two reactive groups, can be used to introduce both inter- and intramolecular cross-links into proteins (Figure 3-3). They can be used to stabilize tertiary structure, to prepare cross-linked models for study of protein-protein interaction, or as probes for determining distances between reactive groups

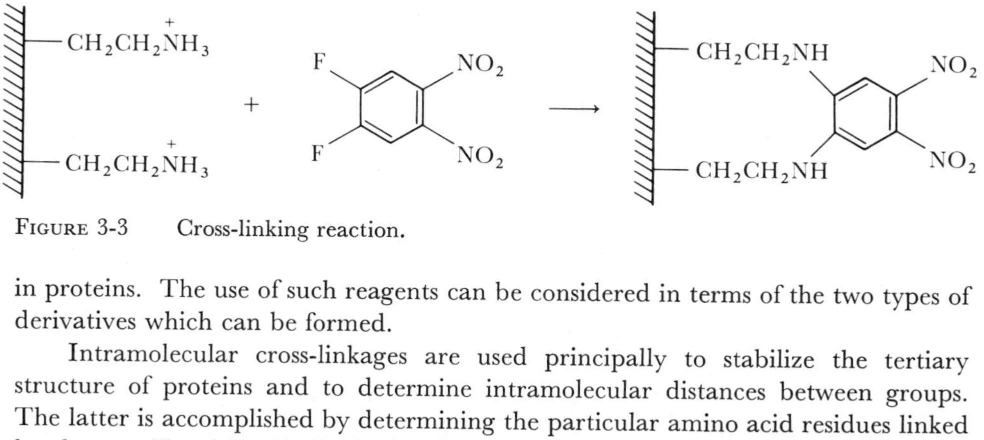

FIGURE 3-3   Cross-linking reaction.

in proteins. The use of such reagents can be considered in terms of the two types of derivatives which can be formed.

Intramolecular cross-linkages are used principally to stabilize the tertiary structure of proteins and to determine intramolecular distances between groups. The latter is accomplished by determining the particular amino acid residues linked by the two "heads" of individual bifunctional reagents. By using several reagents, or by varying the distance between the two "heads" with several homologous reagents, it is possible to determine relative distances between several groups. Comparative study of analogous monofunctional reagents is usually desirable to detect alterations due to the modification.

Intermolecular cross-linkages may join molecules of the same kind or of different kinds. The products can be used as models for study of protein-protein interactions, as high-molecular-weight analogs of one or both of the component proteins, or as substances combining the desirable attributes of both components into one molecule. Intermolecular cross-linkages can be used to join biologically active molecules to insoluble carriers or to link adjacent subunits in naturally occurring subunit proteins as a means to determine their geometrical arrangement.

The reaction conditions largely predetermine which kind of linkage, inter- or intramolecular, will predominate. At high dilutions, formation of intramolecular linked products is greatly favored. The main variables, important to the final outcome, are the protein concentration, the ratio of reagent to protein, and factors such as pH and ionic strength to the extent that they effect the protein-protein interactions during modification.

A cross-linking reagent, dimethyl suberimidate, has been used to study the subunit structure of several oligomeric proteins. After modification the proteins were examined by electrophoresis in polyacrylamide containing sodium dodecyl sulfate. For oligomers composed of identical subunits, the number of principal species observed was identical to the number of subunits. Application of the method to two proteins composed of dissimilar subunits revealed differences in the reactivities of the different subunits. The method appears capable of revealing some detail of structure and reactivity of oligomeric proteins (Davies and Stark, 1970).

A few of the many bifunctional reagents which have been used for the cross-linking of proteins are given in Table 3-1. General principles and applications of bifunctional reagents to proteins have been discussed in a recent review (Wold, 1967).

3-1.5 ATTACHMENT TO SOLID SUPPORTS

Biologically active proteins bound to water-insoluble supports or imbedded in water-insoluble matrices are of both theoretical and practical interest. A large number of techniques have been used to prepare such materials, retaining in some degree their particular biological activity. On a theoretical level, such proteins may serve as useful model systems of particulate or membrane-bound proteins. The interactions of bound enzymes with substrates are kinetically interesting for studying the effects of steric hindrance or limited accessibility on enzymatic activity. With macromolecular substrates, these preparations can be especially useful for obtaining limited enzyme action. Insolubility in water provides an excellent means for mechanical separation from substrates and permits such proteins to be packed in columns through which a soluble substrate may be passed. By varying the flow through such a column, it is possible to vary the amount of enzymatic action.

Water-insoluble protein derivatives can be prepared by covalently or non-covalently binding the protein to a natural or synthetic water-insoluble polymer or by the action of certain protein cross-linking reagents (Manecke, 1968). Cellulose and polystyrene are two of the most commonly used solid supports; copolymers of ethylene and maleic anhydride have also been widely used. Hydrophobicity, surface area, mechanical, chemical, and other characteristics of the support have important

TABLE 3-1   Bifunctional protein reagents

Reagent	Description
Mercuric ion Hg^{2+}	Reversible linkage of ∼5 Å between sulfhydryl groups. Stepwise reaction; first sulfhydryl group reacts very fast, followed by slow reaction with the second. Cross-linking occurs only when Hg^{2+}/—SH is 0.5 or less. Has been used to prepare dimers of mercaptalbumin (Hughes, 1947) and *E. coli* DNA polymerase (Jovin et al., 1969) and as an intramolecular bridge in partially reduced papain (Arnon and Shapira, 1969) and pancreatic ribonuclease (Sperling et al., 1969).
3,6-Bis(mercurimethyl)-dioxane	Similar to mercuric ion but results in 15-Å linkages. Gives a much increased rate of formation of mercaptalbumin dimers (Kay and Edsall, 1956).
N,N′-(1,3-phenylene) bismaleimide	Insoluble in water; highly specific for sulfhydryls; linkages formed irreversibly. Has been used to form dimers of bovine serum albumin (Moore and Ward, 1956); *ortho* isomer has also been used (Moore and Ward, 1956). The similar reagent bis(N-maleimidomethyl) ether has been used to form intramolecular cross-links in human and horse carbonmonoxy-hemoglobin (Simon and Konigsberg, 1966).
N,N′-Ethylene-bis-(iodoacetamide)	Relatively specific for sulfhydryl groups, reaction with aldolase gave an apparently dimeric species which was still active (Ozawa, 1967a). Similar reagents having six and eleven carbon methylene bridges were also used.
1,5-Difluoro-2,4-dinitrobenzene	Insoluble in water, forms irreversible linkages of ∼3 Å principally with amino groups but also with tyrosine. Similar to dinitrofluorobenzene (Section 6-5). Reaction with insulin gave both intra- and intermolecular bridges (Zahn and Meienhofer, 1958); an intramolecularly cross-linked derivative of pancreatic ribonuclease was catalytically active (Marfey et al., 1965).
p,p′-Difluoro-m,m′-dinitrodiphenyl sulfone	Insoluble in water, forms irreversible cross-linkages of 11–12 Å, reacts principally with amino and phenolic groups. Similar to dinitrofluorobenzene (Section 6-5) but less reactive. Has been used with bovine serum albumin (Wold, 1961), collagen (Sykes, 1958), carboxyhemoglobin (Macleod and Hill, 1968), and several mixed intramolecular conjugates (Ram, 1963; Ram et al., 1963).

TABLE 3-1   Bifunctional protein reagents (Cont)

Reagent	Description
Dimethyl adipimidate $\overset{+NH_2}{\underset{OCH_3}{C}}-CH_2-CH_2-CH_2-CH_2-\overset{NH_2^+}{\underset{OCH_3}{C}}$	Water soluble, specific for amino groups, product is positively charged; reacts similarly to ethyl acetimidate (Section 5-3). Has been used to determine inter-residue distances in pancreatic ribonuclease (Hartman and Wold, 1967). Diethyl malonimidate has been used similarly with bovine serum albumin and γ-globulin (Dutton et al., 1966).
Phenol-2,4-disulfonyl chloride $ClO_2S-\overset{SO_2Cl}{\underset{OH}{\bigcirc}}$	Water soluble, reacts principally with amino groups, forms linkages of 6.7–7 Å; cross-linked derivative of egg white lysozyme was catalytically active (Moore and Day, 1968).
Hexamethylenediiso-cyanate $O{=}C{=}N-CH_2-CH_2-CH_2-CH_2-CH_2-CH_2-N{=}C{=}O$	Insoluble in water, reacts principally with amino groups; has been used with pancreatic ribonuclease and α-chymotrypsin (Ozawa, 1967b); diisothiocyanate has also been used (Ozawa, 1967b); The similar reagent azophenyl-p-diisocyanate has been used with whale myoglobin (Fasold, 1965a, 1965b).
Glutaraldehyde $\overset{O}{\underset{H}{\overset{\|}{C}}}-CH_2-CH_2-CH_2-\overset{H}{\underset{O}{\overset{\|}{C}}}$	Complex reaction with proteins apparently involving many different side chains. Confers great structural rigidity. Reaction with crystalline carboxypeptidase A increased the mechanical stability without eliminating catalytic activity (Quiocho and Richards, 1966). Treatment of trypsin gives an insoluble product retaining much of its catalytic activity (Habeeb, 1967).
Formalin (formaldehyde) $\overset{H}{\underset{H}{\overset{\|}{C}}}{=}O \quad (H_2O)$	Highly reactive water-soluble reagent, undergoes a variety of unknown cross-linking reactions (Fraenkel-Conrat and Olcott, 1948; Blass et al., 1965).
Woodward's reagent K (structure with SO_3^-, phenyl, and N^+—CH_2—CH_3)	Water-soluble reagent; links existing carboxyl and amino groups. Has been used to prepare oligomeric α-chymotrypsin (Patel and Price, 1967).
Bisdiazobenzidine $^+N_2-\bigcirc-\bigcirc-N_2^+$	Highly reactive water-soluble reagent, reacts primarily with tyrosine and histidine residues. Used to prepare water-insoluble catalytically active derivatives of papain (Silman et al., 1966).

effects on properties of individual water-insolubilized proteins. Covalent linkage to the protein, it is clear, must involve groups other than those essential for a protein's biological activity if the activity is to be retained.

The specific adsorption characteristics of one protein to another, or between a protein and a low-molecular-weight substance, can be used for protein purification. Such a procedure has been particularly useful for the purification of specific antibodies from mixed immunoglobulin fractions. These purifications are normally quite difficult by other procedures, since most immunoglobulins have remarkably similar properties except for their combining specificities. The mixed antibodies are passed through a column containing the particular antigen (or hapten), covalently attached to an insoluble support under conditions which favor complex formation. After passage of the nonspecific immunoglobulins from the column, elution of the bound specific antibodies is done under conditions favoring dissociation of the antibody-antigen complex (Robbins et al., 1967; Campbell et al., 1951). A similar approach has been employed for the purification of certain proteolytic enzymes and protein inhibitors of these enzymes. In the first case, the inhibitor is bound to the insoluble support and used to purify the enzyme, and, in the other, enzyme bound to the support is used to purify the inhibitor. An example of the purification of an enzyme by chromatography on a column of an insolubilized inhibitor is the successful fractionation of bovine trypsin on ovomucoid-Sepharose (Feinstein, 1970) (see Figure 3-4). α-Chymotrypsin, which is not inhibited by chicken ovomucoid, is not adsorbed.

One of the first striking examples of purification of a nonantibody protein by this type of procedure was the purification of the egg white protein avidin on columns of biotin linked to cellulose. Because avidin forms a very strong complex with biotin, such columns could be used to prepare highly purified samples of avidin (McCormick, 1965). Haynes and Walsh (1969) have described a method involving the adsorption of proteins as monolayers on colloidal silica particles (duPont Ludox HS), followed by intermolecular cross-linking with glutaraldehyde. Such preparations retain high biological activity. The particles are easily sedimented and readily dispersible.

Water-insoluble derivatives of enzymes, antigens, and antibodies have recently been reviewed by Silman and Katchalski (1966). Resins designed for the insolubilization of enzymes[1] have recently become available, as have several insolubilized enzymes[2]. The various types of insolubilized proteins should continue to find many uses.

3-2 INTRODUCTION OF SPECIAL-PURPOSE GROUPS

The introduction of special groups into proteins by chemical modification is an important procedure in many physical and chemical studies. Groups may be introduced

---
[1] Koch-Light Laboratories, Buckinghamshire, England.
[2] Miles Laboratories, Elkhart, Indiana.

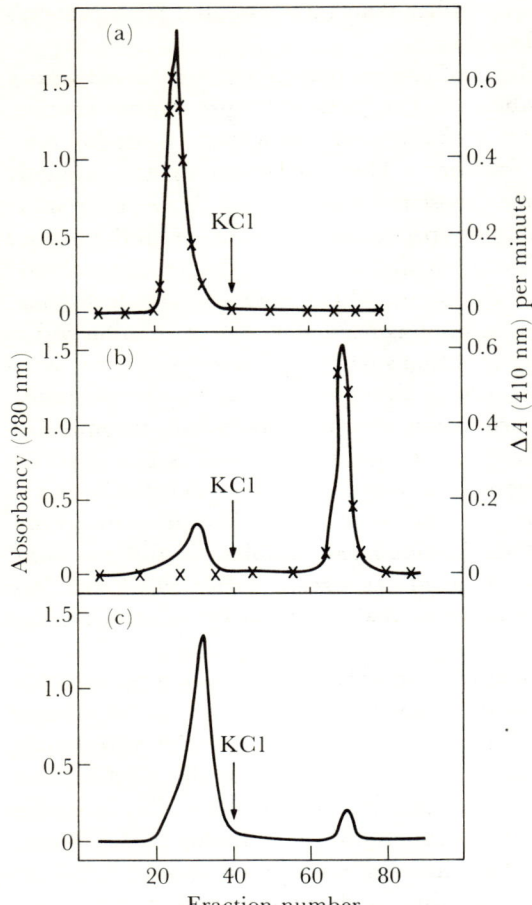

FIGURE 3-4 Column chromatography of trypsin and α-chymotrypsin on Sepharose and ovomucoid-Sepharose. First step of elution with 0.10 M triethanolamine buffer, pH 8.1, containing 0.02 M CaCl$_2$. Second step of elution with 0.20 M KCl–HCl, pH 2.0. Solid line indicates absorbancy at 280 nm; × indicates hydrolysis of BAPA. (a) Trypsin chromatography on Sepharose. (b) Trypsin chromatography on ovomucoid-Sepharose. (c) α-Chymotrypsin chromatography on ovomucoid-Sepharose. (From Feinstein, 1970.)

to confer a new property or to enhance an existing property. They can act as labels identifying particular areas of a protein or convey information concerning the environment at their binding sites. Modifications changing the susceptibility of peptide chains to proteolytic enzymes can be used to alter the course of enzymatic digestion.

3-2.1 Changing the Susceptibility to Proteolysis

The enzymatic hydrolysis of peptide bonds depends upon recognition of the susceptible bonds by the particular enzyme. Many proteolytic enzymes, particularly those commonly used for degradation in sequencing studies, have rather narrow substrate specificities and thus cleave at a limited number of sites. Bovine trypsin, for example, hydrolyzes only bonds involving the carboxyl groups of arginine and lysine. Ideally, such limited proteolysis results in the formation of a limited number of peptides which may be isolated and individually purified. In order to facilitate the sequencing of such peptides, however, it is often desirable either to increase or decrease the number of cleavage points. Many procedures can be used for this purpose, but only a few have been shown to be of particular value.

The ε-amino groups of lysine residues are susceptible to modification in ways which render them resistant to hydrolysis by trypsin. Both trifluoroacetylation and amidination (Subsections 5-1.6 and 5-3.1) are particularly useful for this purpose, since in each case it is a simple matter to remove the blocking group later, thereby regenerating the trypsin-susceptible bonds. During tryptic digestion of the blocked polypeptide chain, only arginine residues are cleaved. Fractionation and purification of the peptides formed, followed by removal of the blocking groups and retreatment with trypsin, causes the lysine bonds to be hydrolyzed. A rather simple analysis of the resulting peptides greatly facilitates the problem of sequencing. Other methods for reversibly blocking amino groups would presumably be useful for the same purpose (see maleic anhydride, Subsection 5-1.4).

Modification of guanidino groups of arginine residues, similarly, result in their conversion to trypsin-resistant residues. Treatment with 1,2-cyclohexanedione, benzil, or malonylaldehyde, for example, can be used for limiting the cleavage by trypsin to lysine residues. Unfortunately, these modifications cannot later be reversed and, insofar as is known, have not been applied to sequence analysis. Modification with phenylglyoxal appears more promising than the use of other arginine reagents, since its reported modification of arginyl residues can apparently be reversed (see Subsection 10-1.3).

Increasing the number of hydrolyzable bonds is frequently desirable in order to obtain smaller, more easily manageable peptides and to lower the amount of unhydrolyzable residual core. For this purpose, methods have been developed to convert cysteinyl residues to trypsin-susceptible S-(2-aminoethyl)cysteinyl residues using 2-bromoethylamine (Lindley, 1956) or ethylenimine (Raftery and Cole, 1963).

The substituted cysteine side chains resemble those of lysine, differing only in the substitution of a single —S— for a —CH$_2$—. Peptide bonds of such residues are cleaved by trypsin at a slightly lower rate than those of lysine (Wang and Carpenter, 1968).

3-2.2 THE AUGMENTATION OF IMMUNOGENICITY

Chemical modification of proteins for the purpose of introducing groups having a high capacity to elicit the production of antibodies when injected into experimental animals is a widely used immunochemical technique. This capacity is termed *immunogenicity*, and groups introduced into the protein to enhance this capacity are called haptens. Haptens themselves do not elicit production of antibodies unless conjugated to a protein. Haptens are useful for increasing the immunogenicity of weakly immunogenic proteins and for directing the production of antibodies to specific chemically defined determinants (the haptens).

Attachment of the hapten to the protein is usually accomplished by one of two alternate reactions. These are the reaction with diazonium ions (Section 9-3) and with aryl halides (Section 6-5). While a variety of groups have been used as haptens, aromatic groups of these types have been most successful.

Antihapten antibodies have the ability to combine with the immunizing protein-hapten conjugate (antigen) and with the separate protein and hapten components. Because complex formation between an antibody and a small molecule is considerably less complicated than between an antibody and large antigens, study of hapten-antibody interactions has greatly facilitated our understanding of antigen-antibody reactions. The formation of such complexes can be treated like other protein–small molecule interactions. Equilibrium and rate constants, and the number of binding sites, can be determined by relatively simple physical-chemical techniques. For example, the effects of modification of antibody upon the combining capacity can be easily followed. By studying the binding of a series of related haptens, effects of structure on binding can be determined. The study of hapten binding by antibodies has been an important source of information on structural requirements for antibody-antigen reactions.

Site-specific modification of proteins has recently been a key to understanding antibody heterogeneity. The combining specificity of antibodies against hapten-protein conjugates is directed against not just the hapten but also parts of the protein in the vicinity of the hapten. Injection of an active-site-labeled derivative of papain, having a single dinitrophenyl hapten bound to its active-center sulfhydryl group, resulted in the formation of antibodies directed against the hapten and also the active-site region, which contained the hapten. These antibodies were more uniform in their combining specificity than antihapten antibody produced in the usual way against haptens attached at a number of sites (Trump and Singer, 1970). Heterogeneity apparently results, in part, in the more typical cases from haptens being introduced randomly at a large number of sites.

Affinity labeling of antibodies has been used to label and identify combining-site residues. The reaction with diazonium ions which is frequently used to prepare protein-hapten immunogens can also be used to label antibody combining sites. As described in Chapter 2, this method predominantly labels tyrosine residues. The dual role of the diazonium reagent in such a sequence assures the specificity of the labeling procedure.

3-2.3 Environmentally Sensitive Labels

Groups whose spectral properties are sensitive to environment can be attached to proteins and used to indicate or monitor various aspects of the protein environment. Changes in protein conformation or any other changes affecting these labels are reflected by changes in a characteristic spectral property. Because it is usually the properties of the native protein which are of interest, it is important that the labels do not significantly alter those properties. By using various procedures to obtain site-specific labeling, it has been possible to monitor specific sites and follow changes accompanying both specific and nonspecific interactions as, for example, the binding of a substrate or an inhibitor.

Fluorescent Labels. The introduction of fluorescent labels into proteins is a well-known method used to locate proteins in tissues, to demonstrate protein interactions, and as environmental probes of protein structure. Fluorescent techniques are particularly valuable because of their sensitivities. At current levels of technology, usually approximately 100-fold less protein is required than with comparable absorption techniques. The emission maxima of most tryptophan-containing proteins are between 330 and 350 mμ. These are shifted to 350 to 355 mμ upon denaturation. Proteins which contain tyrosine, but no tryptophan, have relatively weak fluorescence maxima near 303 mμ. The native fluorescence of most proteins is insufficient for certain kinds of measurements. Procedures have been developed for this reason to bind certain fluorescent dyes to proteins as a means to increase their fluorescence.

A technique known as *fluorescence polarization* has been used to determine the rotational relaxation time of proteins conjugated to fluorescent dyes, and thereby to obtain a measure of both their sizes and shapes. This technique is useful for determining molecular size and shape parameters and as a means to monitor conformational changes in proteins. It is useful for studying association-dissociation phenomena such as the combination of antibodies with fluorescent-labeled antigens which result in significant size or shape changes. The method is dependent upon the existence of a preferred plane or direction for both the excitation and the emission of light. In the case of colloidal-sized molecules, or in very viscous solutions, normal depolarization caused by random molecular or Brownian motion is reduced such that, if excitation is with polarized light, emission may also be measurably polarized. The extent of polarization of the emitted light is dependent upon a number of factors including the size of the protein-dye conjugate and upon the lifetime τ of the excited intermediate. Dyes with different mean lifetimes (τ_0) are useful for correspondingly different molecular-weight ranges (see Table 3-2).

TABLE 3-2 Lifetimes of excited states of dye-protein conjugates and their applicability to rotational relaxation time measurements

Dye-reagent	$\tau_0 \times 10^9$ [a] (sec)	Most useful molecular-weight range for fluorescence depolarization studies[b]	References
3-Phenyl-7-isocyanatocoumarin	2.5	10^4	Chadwick et al., 1960
Fluoresceinisothiocyanate	5.0	$5 \times 10^3 - 5 \times 10^4$	Gill, 1965
Dimethylaminonaphthalene-sulfonyl chloride	12	$10^4 - 10^5$	Hartley and Massey, 1956; Young and Potts, 1963
Anthracene isocyanate	40	$10^5 - 3 \times 10^5$	Harrington et al., 1956
3-Hydroxypyrene-5,8,10-trisulfonic acid	90	$10^5 - 10^6$	Chadwick et al., 1960
Pyrenebutyric acid anhydride	100	10^5	Knopp and Weber, 1967, 1969

[a] As protein-bound conjugates. [b] Assuming spherical molecules.

Excited molecules may return to their ground state by many pathways. If return is by emission of light, the molecule is said to be fluorescent. The quantum yield of fluorescence depends upon the efficiency of processes competing with the radiation transition. If the lifetime of the excited state is short, then the opportunity for other types of energy transfer (i.e., quenching) are low, and the quantum yield will generally be high. Thus, intensity of fluorescence, either the natural fluorescence of a protein due to tryptophan and tyrosine or that of an attached fluorescent dye, is in part a function of the environment of the excited state. Amino, carboxyl, and tyrosyl residues, for example, can interact with excited indole chromophores, causing extensive quenching. This quenching capacity is strongly affected by distance. Changes in the environment of fluorescent chromophores are thus reflected by changes in the intensity of fluorescence. These changes in fluorescence are sensitive criteria of conformation changes.

The emission maxima of fluorescent chromophores depend upon the dielectric constant of their environment. In general, a red-shift in the emission is observed upon transfer to solvents of higher dielectric constant, as from an organic solvent into water. The emission maxima of fluorescent dyes attached to proteins are, in general, at a shorter wavelength than when dissolved in water. This is usually interpreted in terms of a limited accessibility of the dye to water. Changes in location of these maxima result from changes in conformation of the dye-protein conjugate. The wavelength of emission is thus sensitive to structural changes in the vicinity of such chromophores.

Application of fluorescent techniques to proteins and protein–fluorescent dye conjugates have been recently reviewed (Weber and Teale, 1965; Steiner and Edelhoch, 1962; Brand and Witholt, 1967). Much of the foregoing discussion also applies to noncovalently bound fluorescent probes. Some of these probes have been discussed recently by Edelman and McClure (1968).

Environmentally Sensitive Chromophoric Groups. Reagents which bind covalently to proteins and which have a chromophore whose spectral properties are sensitive to environment have been called "reporter" groups (Burr and Koshland, 1964; Koshland et al., 1964). Spectra of the bound reagents change with changes in the pH or polarity of their environment and in this way can be used as internal monitors of protein structure. The term is also applied to reagents whose fluorescence spectra are sensitive to environment like some of those described in the preceding section. When α-chymotrypsin was treated with 2-bromoacetamido-4-nitrophenol, 0.6 moles were bound to a methionine residue three positions removed from the active-site serine. Combination of this chymotrypsin derivative with various substrates or substratelike materials generated different spectra indicating alterations in the chromophore's environment (Figure 3-5).

2-Methoxy-5-nitrobenzyl bromide (Horton et al., 1965), 4-bromo-5-nitroimidazole (Cohen, 1967), and N-benzoxazolylphenylmaleimide (Sekine et al., 1967) have environmentally sensitive spectra and have been used for their reporter capabilities. Such use of 2-bromoacetamido-4-nitrophenol (Kirtley and Koshland, 1967) and of 2-hydroxy-5-nitrobenzyl bromide has recently been reviewed (Horton and Koshland, 1967).

Spin Labeling. Techniques have been developed to attach small paramagnetic groups, so-called "spin labels," to proteins and other biological macromolecules. These labels or probes allow one to examine by electron paramagnetic resonance, proteins which are not naturally paramagnetic. The electron paramagnetic resonance spectra of such conjugates can be used to study environmental properties of proteins in the vicinity of the spin label. Changes in environment caused by changes in conformation of such a protein are indicated by changes in its resonance spectra and may be interpreted in terms of the relative conformational restraint upon the spin

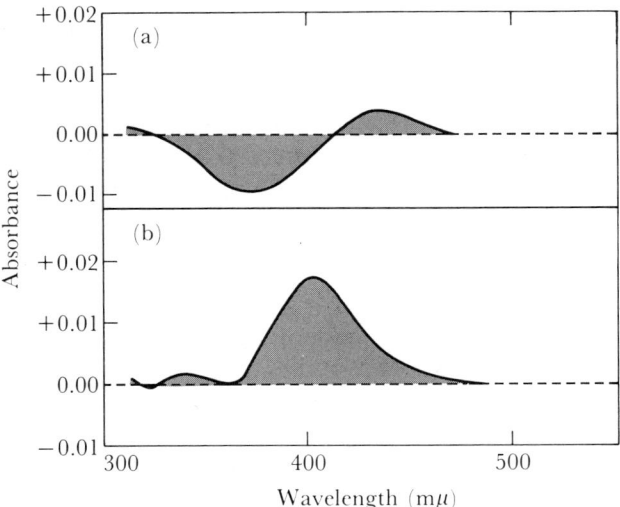

FIGURE 3-5 Difference spectra of chymotrypsin labeled with the reporter group acetamido-4-nitrophenol on a methionine three residues from the active-site serine upon combination with the substratelike molecules (a) benzoyl-L-phenylalanine, (b) benzoyl-D-phenylalanine. Conditions: $7.4 \times 10^{-5}\,M$ enzyme, $0.05\,M$ phosphate buffer, $1.5 \times 10^{-3}\,M$ substrate (pH 6.49). (From Burr and Koshland, 1964.)

label. Such probes are also subject to viscosity and motional effects of their environment and can be used to study these phenomena. The type of labeling molecule and the nature of the attachment site can be varied to some extent, which can make the technique quite versatile. Because the ESR technique is intrinsically very sensitive, the method of spin labeling is useful even when only a small proportion of sites can be labeled.

The most widely used spin labels have been nitroxide radicals which are usually

prepared from a corresponding secondary amine by oxidation with hydrogen peroxide in the presence of a catalyst. The unpaired electrons responsible for the ESR signal are associated with the N—O bond. R is a particular functional group used to attach the label to the protein. The ESR spectrum is composed basically of a three-line hyperfine pattern; its precise character is dependent upon both the spin label and its environment. It reflects the relative rotational freedom of the nitroxide radical, becoming quite asymmetric as rotation becomes more restricted. This restricted rotation may occur if the solution viscosity increases or, for example, when the label is bound to the protein. The relative rotational freedom is a composite of the tightness of fit between the macromolecule and spin label, and of the rotational time of the macromolecule. Changes in these can be used to monitor conformational changes. The polarity of the environment affects the spectrum and so enables one to determine something of this aspect of the attachment site.

More detailed descriptions of the principles of ESR and spin labeling, as applied to the study of proteins, are given in two recent reviews (Hamilton and McConnell, 1968; Griffith and Waggoner, 1969).

3-2.4 ISOMORPHIC REPLACEMENT

Interest in X-ray crystallographic studies of proteins has promoted active investigation of chemical methods to produce derivatives containing the necessary heavy atom. Isomorphic replacement of a small atom for a large atom is necessary for most X-ray studies. True isomorphic replacement occurs without concurrent structural changes in the protein. The heavy atom serves as a reference point around which the other atoms may be located. Complete assurance that conformational changes have not occurred is of great concern. With many enzymes, the retention of enzymatic activity by heavy-atom-containing crystals has served to demonstrate some degree of the native structure.

The methodology used to prepare heavy-atom-labeled crystals has been developed for the most part by trial and error. In many cases, suitable derivatives are prepared by soaking protein crystals in solutions containing heavy-metal ions or by crystallization in the presence of such ions. These procedures are tedious and unsure. Often, binding of the metal leads to distortion of the native structure, even sometimes to destruction of the crystal lattice. In other cases, no significant binding of the metal ion is observed. Nevertheless, many successful substitutions have been achieved.

A relatively new approach to the introduction of heavy atoms into protein crystals utilizes covalent attachment to the protein of a heavy-atom-containing or -chelating group (Figure 3-6). The first problem with such a procedure is the necessity of introducing a single group at a specific site. The techniques of affinity labeling and pseudosubstrate labeling are possible approaches to this problem, but they have not been successfully exploited experimentally. If such a group can be introduced specifically and, in the case of the chelating group, if it binds a heavy ion with an affinity much greater than that of other sites in the protein, then it should be possible to prepare in a somewhat systematic way heavy-atom-containing crystals suitable for X-ray studies. Such a study of active-site-labeled p-iodobenzenesulfonyl-chymotrypsin, however, showed it not to be isomorphous with native α-chymotrypsin, illustrating difficulties present even with specific labeling procedures (Sigler et al., 1964).

Metal-ion-chelating groups such as 3-aminotyrosine (Sokolovsky et al., 1967) and picolinamidine groups (Benisek and Richards, 1968) have been proposed to be potentially useful for preparing heavy-metal-containing protein derivatives. There

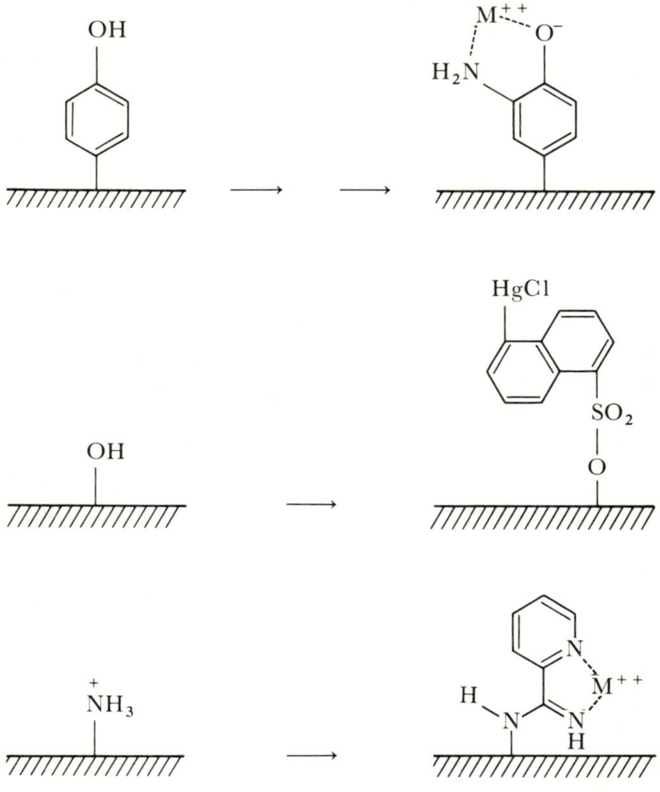

FIGURE 3-6   Heavy-metal-containing protein derivatives suitable for X-ray diffraction studies.

appears to be no great difficulty with this approach, except perhaps, in preparing the specifically modified derivatives and showing them to be isomorphous with the native protein. Several specific aminotyrosine enzyme derivatives have already been described (Cautrecasas et al., 1968, 1969). Reduction of disulfide bonds and their bridging with mercuric ion has already been mentioned. The potential of this procedure for crystallographic studies will depend upon whether such derivatives are found to be isomorphous with the unmodified proteins.

REFERENCES

Arnon, R., and E. Shapira (1969): *J. Biol. Chem.*, **244,** 1033.
Azari, P. R., and R. E. Feeney (1961): *Arch. Biochem. Biophys.*, **92,** 44.
Benisek, W. F., and F. M. Richards (1968): *J. Biol. Chem.*, **243,** 4267.
Bewley, T. A., and C. H. Li (1969): *Int. J. Protein Res.*, **1,** 117.
Blass, J., B. Bizzini, and M. Raynaud (1965): *Compt. Rend. Acad. Sci. (Paris)*, **261,** 1448.
Brand, L., and B. Witholt (1967): *Methods Enzymol.*, **11,** 776.
Burr, M., and D. E. Koshland (1964): *Proc. Nat. Acad. Sci.*, **52,** 1017.
Campbell, D. H., E. Luescher, and L. S. Lerman (1951): *Proc. Nat. Acad. Sci.*, **37,** 575.
Cautrecasas, P., S. Fuchs, and C. B. Anfinsen (1968): *J. Biol. Chem.*, **243,** 4787.
Cautrecasas, P., S. Fuchs, and C. B. Anfinsen (1969): *J. Biol. Chem.*, **244,** 406.
Chadwick, C. S., P. Johnson, and E. G. Richards (1960): *Nature*, **186,** 239.
Cohen, L. A. (1968): *Ann. Rev. Biochem.*, **37,** 695.
Davies, G. E., and G. R. Stark (1970): *Proc. Nat. Acad. Sci.*, **66,** 651.
Dutton, A., M. Adams, and S. J. Singer (1966): *Biochem. Biophys. Res. Commun.*, **23,** 730.
Edelman, G. M., and W. O. McClure (1968): *Accounts Chem. Res.*, **1,** 65.
Fasold, H. (1965a): *Biochem. Z.*, **342,** 288.
Fasold, H. (1965b): *Biochem. Z.*, **342,** 295.
Feinstein, G. (1970): *FEBS Letters*, **7,** 353.
Feinstein, G., and R. E. Feeney (1967): *Biochim. Biophys. Acta*, **140,** 55.
Fraenkel-Conrat, H., and H. S. Olcott (1948): *J. Am. Chem. Soc.*, **70,** 2673.
Freisheim, J. H., K. A. Walsh, and H. Neurath (1967): *Biochemistry*, **6,** 3010.
Frensdorff, A., M. Wilchek, and M. Sela (1967): *Eur. J. Biochem.*, **1,** 281.
Frist, R. H., I. J. Bendet, K. M. Smith, and M. A. Lauffer (1965): *Virology*, **26,** 558.
Gerhart, J. C., and H. K. Schachman (1965): *Biochemistry*, **4,** 1054.
Gerhart, J. C., and H. K. Schachman (1968): *Biochemistry*, **7,** 538.
Gill, T. J. (1965): *Biopolymers*, **3,** 43.
Gounaris, A. D., and G. E. Perlmann (1967): *J. Biol. Chem.*, **242,** 2739.
Griffith, O. H., and A. S. Waggoner (1969): *Accounts Chem. Res.*, **2,** 17.
Habeeb, A. F. S. A. (1967): *Arch. Biochem. Biophys.*, **121,** 652.
Hamilton, C. L., and H. M. McConnell (1968): in *Structural Chemistry and Molecular Biology*, edited by A. Rich and N. Davidson, W. H. Freeman, San Francisco.
Harrington, W. F., P. Johnson, and R. H. Ottewill (1956): *Biochem. J.*, **62,** 569.
Hartley, B. S., and V. Massey (1956): *Biochim. Biophys. Acta*, **21,** 58.
Hartman, F. C., and F. Wold (1967): *Biochemistry*, **6,** 2439.
Hass, L. F. (1964): *Biochemistry*, **3,** 535.
Haynes, R., and K. A. Walsh (1969): *Biochem. Biophys. Res. Commun.*, **36,** 235.
Horton, H. R., H. Kelly, and D. E. Koshland (1965): *J. Biol. Chem.*, **240,** 722.
Horton, H. R., and D. E. Koshland (1967): *Methods Enzymol.*, **11,** 556.
Hughes, W. L. (1947): *J. Am. Chem. Soc.*, **69,** 1836.
Jovin, T. M., P. T. Englund, and A. Kornberg (1969): *J. Biol. Chem.*, **244,** 3009.
Kay, C. M., and J. T. Edsall (1956): *Arch. Biochem. Biophys.*, **65,** 354.
Keresztes-Nagy, S., and I. M. Klotz (1965): *Biochemistry*, **4,** 919.
Kirtley, M. E., and D. E. Koshland (1967): *Methods Enzymol.*, **11,** 866.
Klotz, I. M., and S. Keresztes-Nagy (1963): *Nature*, **195,** 900.
Knopp, J., and G. Weber (1967): *J. Biol. Chem.*, **242,** 1353.
Knopp, J., and G. Weber (1969): *J. Biol. Chem.*, **244,** 6309.
Koshland, D. E., Y. D. Karkhanis, and H. G. Latham (1964): *J. Am. Chem. Soc.*, **86,** 1448.
Lindley, H. (1956): *Nature*, **178,** 647.

Macleod, R. M., and R. J. Hill (1968): *Fed. Proc.*, **27**, 521.
Manecke, G. (1968): *Biochem. J.*, **107**, 2P.
Marfey, P. S., M. Uziel, and J. Little (1965): *J. Biol. Chem.*, **240**, 3270.
McCormick, D. B. (1965): *Anal. Biochemistry*, **13**, 194.
Moore, G. L., and R. A. Day (1968): *Science*, **159**, 210.
Moore, J. E., and W. H. Ward (1956): *J. Am. Chem. Soc.*, **78**, 2414.
Ozawa, H. (1967a): *J. Biochem. (Tokyo)*, **62**, 531.
Ozawa, H. (1967b): *J. Biochem. (Tokyo)*, **62**, 419.
Patel, R. P., and S. Price (1967): *Biopolymers*, **5**, 583.
Polyanovsky, O. L. (1965): *Biochem. Biophys. Res. Commun.*, **19**, 364.
Quiocho, F. A., and F. M. Richards (1966): *Biochemistry*, **5**, 4062.
Raftery, M. A., and R. D. Cole (1963): *Biochem. Biophys. Res. Commun.*, **10**, 467.
Ram, J. S. (1963): *Biochim. Biophys. Acta*, **78**, 228.
Ram, J. S., S. S. Tawde, G. B. Pierce, and A. R. Midgley (1963): *J. Cell. Biol.* **17**, 673.
Robbins, J. B., J. Haimouich, and M. Sela (1967): *Immunochemistry*, **4**, 11.
Sekine, T., Y. Kanaska, T. Kameyama, M. Machida, and A. Takeda (1967): *Abstr. Intern. Congr. Biochem. (Tokyo)*, **7**, 773.
Sigler, P. B., H. C. W. Skinner, C. L. Coulter, J. Kallos, H. Braxton, and D. R. Davies (1964): *Proc. Nat. Acad. Sci.*, **51**, 1146.
Silman, H. I., M. Albu-Weissenberg, and E. Katchalski (1966): *Biopolymers*, **4**, 441.
Silman, H. I., and E. Katchalski (1966): *Ann. Rev. Biochem.*, **35**, 873.
Simon, S. R., and W. H. Konigsberg (1966): *Proc. Nat. Acad. Sci.*, **56**, 749.
Sokolovsky, M., J. F. Riordan, and B. L. Vallee (1967): *Biochem. Biophys. Res. Commun.*, **27**, 20.
Spande, T. F., B. Witkop, Y. Degani, and A. Patchornik (1970): *Adv. Protein Chem.*, **24**, 97.
Sperling, R., Y. Burstein, and I. Z. Steinberg (1969): *Biochemistry*, **8**, 3810.
Steiner, R. F., and H. Edelhoch (1962): *Chem. Rev.*, **62**, 457.
Sykes, R. L. (1958): *Makromol. Chem.*, **27**, 157.
Trump, G. N., and S. J. Singer (1970): *Proc. Nat. Acad. Sci.*, **66**, 411.
Wang, S. S., and F. H. Carpenter (1968): *J. Biol. Chem.*, **243**, 3702.
Weber, G., and F. W. J. Teale (1965): *Proteins*, **3**, 445.
Wetlaufer, D. B., and V. P. Saxena (1968): 156th Meeting of the American Chemical Society, Biological Chemistry Division, Abstract #164.
Witkop, B. (1968): *Science*, **162**, 318.
Wold, F. (1961): *J. Biol. Chem.*, **236**, 106.
Wold, F. (1967): *Methods Enzymol.*, **11**, 617.
Young, D. M., and J. T. Potts (1963): *J. Biol. Chem.*, **238**, 1995.
Zahn, H., and J. Meienhofer (1958): *Makromol. Chem.*, **26**, 153.

4

special problems in analysis of chemically modified proteins

Advancements in protein modification have resulted from the development of better analytical methods as well as better modification methods. Determining, as exactly as possible, which groups have been modified and the extent to which they have been modified is a necessary first step to relating changed properties to specific chemical changes. Proof that unwanted side reactions did not occur is also a critical necessity. Much of the recent success in protein modification is due to development of better analytical methods for such characterization.

While it is not our purpose to describe comprehensively the analytical aspects of protein chemistry, a few brief remarks seem appropriate because of the importance of these aspects to other topics covered. A number of general references are available (Alexander and Block, 1960a, 1960b, 1961; Alexander and Lundgren, 1966; Schroeder, 1968; Bailey, 1967; Hirs, 1967; Leach, 1969).

4-1 EXTENT OF MODIFICATION

The effects of protein modification on biological activity can be clearly interpreted only when both the qualitative and quantitative nature of the chemical changes are known. Because stoichiometric modification of proteins is usually difficult and not always necessary, determination of the extent of modification is frequently one of the most important aspects of their characterization. Quantitation of site-specific modification and other selective modifications is often based on the extent of loss of biological activity. This has little utility as a general approach, however, until other quantitative data have been obtained to show the reaction selective. This usually requires careful correlation of the change in activity with the extent of modification. This crucial test of specific modification requires few if any special techniques.

Such protein modifications typically show a one-to-one correspondence between the number of residues modified and the change in biological activity. The properties associated with specific modification can also be frequently demonstrated in other ways. Such reactions usually have a strong requirement for the protein's native conformation and, for example, do not occur in 8 M urea or after strong heat treatment. The binding of specific ligands (i.e., enzyme substrates, reversible inhibitors, etc.) frequently eliminates such reactions. Isolation from a modified protein of a single peptide containing the unique modified residue is often the clearest proof of specific reaction.

The most frequently used approach for the identification of essential groups involves reaction with nonspecific reagents and determination of the effect of such on properties. The extensive modifications which are often required sometimes promote changes in structure and accompanying nonspecific changes in properties. By correlating the extent of modification with activity, it is sometimes possible to attribute changed properties to specific chemical events and, by so doing, to discount somewhat the possible role of conformational changes. Loss of activity upon modification is always subject to this possibility.

Determinations of the extent of modification can be done by analyses on intact proteins or on hydrolysates. Analyses of intact proteins are generally more convenient but usually less sensitive and somewhat less accurate. An inherent limitation to determinations on intact proteins results from the unpredictable effects of protein structure. The analysis of hydrolysates is restricted to those modified residues which do not revert to the original amino acid upon hydrolysis.

Analyses of intact proteins are the preferred means to determine the extent of those modifications which are accompanied by well-defined changes in absorbance. Most modifications of tyrosine and tryptophan, for example, bring about well-characterized spectral changes (these are described later in this book in connection with individual reagents). Modification of other groups, wherein chromophoric groups become linked to the protein, are also in many cases easily quantitated from changes in absorbance; these too are described later.

Procedures for the quantitative estimation of specific amino acids in proteins can be used to follow modification. These relatively simple procedures can be used in many cases to determine the extent of modification from the decreased amount of the particular unmodified side-chain groups. In Table 4-1 we have listed some of the most useful of these procedures. Radioactively labeled reagents can sometimes be used where other means are lacking or to improve the sensitivity attainable with others. The need for a large excess of reagent frequently precludes the use of radioactive reagent due to the cost.

Standard conditions for the nonenzymatic hydrolysis of proteins were developed to give optimum recovery of the normal amino acids. The most commonly used procedure employs 6 M HCl as the hydrolytic agent at 110°C for 18 to 24 hours (Moore and Stein, 1963). The same hydrolysis procedure is suitable for many

Table 4-1 Analytical methods for quantitation for amino acid side chains in proteins

Group	Method or reagent(s)	References
Amino	Trinitrobenzene sulfonate (Section 6-5)	Habeeb, 1966
	Ninhydrin	Moore and Stein, 1954; Slobodian et al., 1962
Carboxyl	Water-soluble carbodiimide and an amine (Section 7-2)	Hoare and Koshland, 1967
Guanidino	α-Naphthol and sodium hypochlorite	Sakaguchi, 1925
Imidazole	Diazonium-1-H-tetrazole (Section 9-3)	Sokolovsky and Vallee, 1966; Takenaka et al., 1969
Indole	p-Dimethylaminobenzaldehyde and H_2SO_4	Spies and Chambers, 1949; Spies, 1967
	Hydrolysis in 6 M HCl plus thioglycolic acid	Matsubara and Sasaki, 1969
	N-Bromosuccinimide (Section 8-9)	Spande and Witkop, 1967
	2-Hydroxy-5-nitrobenzyl bromide (Section 6-6)	Barman and Koshland, 1967
Sulfhydryl	p-Mercuribenzoate (Section 10-2)	Boyer, 1954; Boyer and Segal, 1954
	5,5'-Dithiobis(2-nitrobenzoic acid) (Section 8-3)	Ellman, 1959

modified proteins. Hydrolytic procedures are potentially suitable for the analysis of all modified residues which do not revert, during hydrolysis, to the original amino acid. Quantitation can be based upon the amount of new product found or upon the decreased amount of the original. Reversion to the original amino acid upon hydrolysis is a handicap in the analysis of most acylated proteins. Alkaline hydrolysis is rarely employed due to its destruction of several amino acids (i.e., arginine, serine, threonine, cysteine, and cystine), but occasionally is used for analysis of tryptophan. Alkaline hydrolysis is also useful for the detection of methionine sulfoxide in proteins, since this derivative largely reverts to methionine during acid hydrolysis (Dreze, 1956; Ray and Koshland, 1962).

4-2 SIDE REACTIONS

Like most reactions between organic compounds, modifications of proteins are frequently accompanied by side reactions. Proteins are particularly susceptible to unwanted reactions, because of their great complexity. The detection of side reactions can be essential to the development of a correct understanding of structure-function relationships. In the present discussion, we shall consider unwanted chemical changes which accompany modification; a few generally important reactions are indicated in Table 4-2. Physical changes in protein structure are discussed separately.

TABLE 4-2   Possible side reactions of protein modification

Groups	Treatment	Effects	Detection
Peptide bonds	Alkaline pH's	Hydrolysis	Molecular weight and terminal amino acid determinations
	Acidic pH's	N → O acyl shift	Peptide mapping; determination of amino-terminals
Thiol groups	Oxidation	See Sections 8-3 to 8-9	Determine —SH groups (Table 4-1)
	Disulfide interchange[a]	Mispairing of disulfide and thiol groups	Peptide mapping; determination of molecular weight
Disulfide bonds	Reduction	See Section 8-1	Determine —SH groups (Table 4-1)
	Thiol interchange[a]	Mispairing of disulfide and thiol groups	Peptide mapping; determination of molecular weight
	Alkaline pH's	Hydrolysis and cleavage	Peptide mapping; determination of molecular weight
Methionyl groups	Oxidation	—CH$_2$—S—CH$_3$ with =O, and —CH$_2$—SO$_2$—CH$_3$	See Section 8-7
Amide groups	Alkaline pH's	Hydrolysis to carboxyl groups	Electrophoresis; peptide mapping

[a] Enhanced by low-molecular-weight thiols and disulfides.

Group-specific protein reagents represent an ideal type. In practice, however, few of the presently available reagents achieve this degree of selectivity. With some reagents, several different groups may be modified. Simultaneous modification of more than one kind of group makes it difficult to attribute changes in properties specifically to the modification of a particular group. To do so usually requires additional information. Such information may already be existent or may have to be obtained through the use of a different modification procedure.

Lack of specificity is not always a disadvantage. Certainly, when modification fails to bring about a change in properties, the conclusion that the modified residues are not essential to those properties is equally as strong as that which would have resulted from the use of a more specific reagent. Additionally, it may be possible with the less specific reagent to rule out involvement of other groups which were also modified. The advantages of a specific reagent are relatively less important for proteins lacking one or more of the groups that are potentially reactive with the reagent.

Super-reactivity of a group in a protein can lead to its modification under conditions, or by a reagent, which would not normally affect groups of that type. As described earlier, such enhanced reactivity is frequently due to a group's presence in an active center, and is thus of particular consequence to biological activity. The unexpected reaction of iodoacetic acid with a carboxyl group in the active site of ribonuclease T1, under conditions where only sulfhydryl groups normally react, is an example of what we refer to as super-reactivity (Takahashi et al., 1967).

Determination of the amount of reagent bound as compared to the number of groups available for modification can be used to determine the extent of reaction and sometimes the specificity. The incorporation of ^{14}C-cyanate into several proteins, for example, was shown to be greater than their content of amino groups and was one of the first indications of its reaction with other side chains (Stevens and Feeney, 1963; Shaw et al., 1964).

Oxidation of sulfhydryl groups is one of the most important potential side reactions in many protein modifications; their loss can be detected by several convenient procedures (Table 4-1). The N → O acyl shift is a well-recognized reaction of proteins in strongly acidic solutions. In this reaction an acyl bond to an amino group of serine or threonine is transferred to the adjacent hydroxyl group (Equation 1). The peptide bond is thus broken and an ester linkage formed in its place (Bergmann et al., 1921). The reaction is indicated by the appearance of a new amino-terminal serine or threonine. The N → O acyl shift has been employed for the selective cleavage of peptide chains (Iwai and Ando, 1967).

$$\left[\begin{array}{c} R-CH-OH \\ \diagup \\ -C-CH \\ \| \quad \diagdown \\ O \quad NH-C- \\ \| \\ O \end{array} \right] \underset{}{\overset{H^+}{\rightleftarrows}} \left[\begin{array}{c} R-CH-O \\ \diagup \quad \diagdown \\ -C-CH \quad C- \\ \| \quad \diagdown \diagup \\ O \quad \overset{+}{NH_3} \; O \end{array} \right] \tag{1}$$

Alterations of groups, other than amino acid side chains, attached to proteins are some of the least studied side reactions of protein modification. The hydrolysis of phosphate ester moieties or reactions involving attached carbohydrate chains are frequently overlooked as possibilities of many modification procedures. Many acylating agents, for example, are able to react with carbohydrates to produce corresponding saccharide esters in addition to their effects on other protein groups. Such reactions have received little attention and their importance is, for practical purposes, unknown.

Detection of the unwanted or so-called side reactions can be very difficult in some cases. When the nature of such potential reactions is known, methods for detection of the products are also frequently known. The real difficulty, which fortunately is uncommon, is in detecting a product of unknown structure and with unknown properties. Potential side reactions are described in connection with the use of each reagent in Part II.

4-3 CHANGES IN CONFORMATION

Changes in noncovalent structure are among the most important side effects of protein modification. As discussed previously, changes in conformation frequently occur, either due to the modified protein's new properties, or as a consequence of too harsh reaction conditions. Their possible occurrence should be investigated whenever changes in activity are obtained and, if found, must a priori be considered important in evaluation of the changed activity. It is unfortunate that there are not yet any completely reliable techniques to detect all conformational changes. While there are a great number of ways to detect some changes in conformation, the possible occurrence of limited undetected conformational changes can in virtually no case be excluded. The failure to detect such a change should encourage further investigation rather than complacency. Many of the techniques used to detect conformational changes of unmodified proteins are also sensitive to many modifications. Inability to distinguish between changes due to the modification and due to changes in conformation renders many techniques of dubious value.

Both physical and chemical techniques can be used to detect changes in protein conformation. An obvious manifestation of conformational change is that which sometimes results in insolubility. Denaturation, as discussed in Chapter 2, is the general term referring to changes having such gross consequences. Chemical methods, and there are many, detect conformational differences based on measurement of changes in the reactivity of protein groups. Spectrophotometric titrations, for example, give a measure of the dissociation equilibrium of tyrosine phenolic groups. Alterations in the dissociation characteristics reflect changes in the accessibility of tyrosine groups and, as such, in protein conformation. In the same way, reaction with reagents to determine the number of buried and exposed groups can be used to detect changed accessibility and, therefore, to detect changes in conformation.

Conformational changes bringing about great changes in shape or in molecular weight are easily detected. For example, increased association or dissociation of protein subunits can be easily detected by one of the many methods for determining molecular weight (ultracentrifugation, gel filtration, etc.). Methods for determining size and shape parameters of proteins are given in the list of physical techniques in Table 4-3.

Spectrapolarimetry is one of the more sensitive tools for the detection of conformational changes in proteins. Optical rotatory dispersion (ORD) is a sensitive tool to detect changes in the conformation of polypeptide chains. Characteristic ORD spectra of proteins can be converted into linear plots and the percentage helicity of the polypeptide estimated. Changes in conformation of the peptide backbone are usually reflected by changes in helix content. A less exact procedure, requiring a less expensive instrument, is based on measurement of the optical rotation at a single wavelength.

A more directly interpretable optical technique for the study of conformational properties of proteins is circular dichroism. Circular dichroism of a typical protein in the ultraviolet exhibits a characteristic pattern of positive and negative bands that result from peptide bond transitions and transitions of side-chain chromophores. Changes in the circular dichroism spectrum result from changes in side-chain or polypeptide conformation.

Nuclear magnetic resonance spectroscopy is probably the most revealing means to examine proteins in solution. Nuclear magnetic resonance (NMR) is a form of spectroscopy based on the absorption of radiofrequency electromagnetic radiation by nuclei in strong external magnetic fields. For theoretical considerations, techniques, and a survey of the many applications of NMR spectroscopy to studies of amino acids and proteins, the reader is referred to the recent review by Roberts and Jardetzky (1970).

NMR allows the structure of proteins in solution to be considered in some detail and to be compared for evidences of structural differences, whether arising from chemical differences between particular residues or from subtle conformational differences. Changes in NMR spectra can be used to monitor changes in protein structure over a period of time during which a variety of processes and tests can be conducted. In addition to problems of protein structure, NMR can be used to examine the structure of binding sites and modes of binding of small molecules. Studies of proteins in solution by NMR can greatly supplement studies of protein crystals by X-ray crystallography.

4-4 OTHER CONSIDERATIONS

Incomplete modification of groups in a protein can result either in the formation of a homogeneous population of partially modified molecules or in a heterogeneous population containing molecules modified to differing extents. The latter situation poses

TABLE 4-3 Physical techniques to determine size, shape, and ionic charge parameters of modified proteins

Techniques	Uses
Electrophoresis	To determine homogeneity or purity and to detect changes in ionic state (Whitaker, 1967; Cann, 1969). To compare size (molecular weight) and ionic charge under nondissociating conditions compatible with biological activity (Hedrick and Smith, 1968; Hedrick et al., 1969) or to detect and determine molecular weights of polypeptide components under dissociating (sodium dodecyl sulfate) and reducing (β-mercaptoethanol) conditions (Shapiro et al., 1967; Weber and Osborn, 1969; Dunker and Rueckert, 1969).
Gel filtration	To determine molecular weight under conditions compatible with biological activity (Whitaker, 1963; Andrews, 1964; DeVincenzi and Hedrick, 1967) or in dissociating and/or reducing media (sodium dodecyl sulfate, urea, guanidine/HCl, or β-mercaptoethanol); to determine the size and number of polypeptide chains (Davison, 1968; Fish et al., 1969; Burgess, 1969). To desalt and to fractionate different-sized components (Porath and Flodin, 1959).
Ultracentrifugation	To determine sedimentation coefficients (Schachman, 1959) or molecular weights (Schachman, 1959; Van Holde, 1967; Yphantis, 1964) and homogeneity in each respect. To measure small differences in sedimentation (Richards and Schachman, 1968; Schumaker and Adams, 1968; Gerhart and Schachman, 1968). To determine sedimentation coefficients or molecular weights under fully dissociating conditions (Greene and Feeney, 1968; Mann et al., 1970).
Optical rotatory dispersion	To determine helical content of polypeptide chains and to detect changes in conformation (Urnes and Doty, 1961; Beychok, 1968; Jirgensons, 1969).
Circular dichroism	To determine helical content of polypeptide chains and to detect changes in conformation (Beychok, 1968; Gorbunoff, 1969; Pflumm and Beychok, 1969; Jirgensons, 1970).
Hydrogen exchange	To determine the kinetics of proton exchange between peptide chains and solvent. To determine the number of slow-, medium-, and fast-exchanging hydrogens and, from changes in these, to detect changes in protein conformation (Englander, 1967; Hvidt and Nielsen, 1966; Sabato and Ottesen, 1967).
Viscosimetry	To determine relative, specific, and intrinsic viscosity and, from changes in these, to detect changes in size or shape (Yang, 1961).
X-ray diffraction	To determine three-dimensional conformation in atomic detail (Stryer, 1968; Blake, 1968; Fraser and MacRae, 1969).
Nuclear magnetic resonance	To define conformations in solution, study mechanisms of folding and unfolding, and to follow changes in conformation or structure of individual residues and larger regions of polypeptide chains (Roberts and Jardetzky, 1970).

a major obstacle to any interpretation wherein a partial or incomplete change in activity is also observed. As a test of the constancy of composition, electrophoresis is the procedure most often utilized; but this is usually useful only when charged groups have been affected. A laborious but sometimes successful procedure involves proteolytic degradation and an examination of the resulting peptides. Determining the number of groups modified with increasing severity of treatment can also sometimes help differentiate between the two possibilities. If the extent of modification plateaus at less than complete modification and increasing amounts of reagent fail to significantly exceed this level, homogeneity is suggested.

Partial loss of enzymatic activity can be explained in terms of either of two models, partial inactivation of all the enzyme molecules or complete inactivation of part of the molecules. The two situations are usually difficult to distinguish. Partial inactivation of the entire population is normally accompanied by changes in kinetic parameters, pH profiles, and the like, whereas a decrease in the turnover number without a corresponding change in K_m is frequently associated with the existence of a fraction of molecules whose activities have not been changed. A distinction can be made in some cases by using low-molecular-weight substratelike functional reagents to directly titrate the amount of active enzyme (Schonbaum et al., 1961). These methods are dependent upon the amount of active enzyme and not upon its relative catalytic activity. Unfortunately, these specific substrates are available for only a small number of enzymes.

4-5 THE USE OF VARIATIONS IN PROTEIN STRUCTURE TO STUDY PROTEIN FUNCTION

A simple way to obtain proteins with varied structure other than by chemical modification is to take advantage of those variations which exist naturally between homologous proteins from different species. Such proteins may have a few amino acid differences or may be so different as to obscure their relationship. In either case, they can be useful for studying the relationship of structure to function (Feeney and Allison, 1969).

A histidine residue in chicken egg white lysozyme, for example, was once thought to be essential for enzymatic activity, but this idea was abandoned when histidine was shown to be absent from duck lysozyme (Jolles et al., 1965). Similarly, homologous ovomucoids from different avian species vary in their content of amino acid residues known to be important to their trypsin-inhibitory activities. Arginine, for example, which is required by chicken ovomucoid for trypsin-inhibitory activity, is absent in the homologous proteins from several other egg whites. The function of the missing arginine for the inhibition of trypsin is supplied, in those cases which have been studied, by a lysine residue (Feeney and Allison, 1969; Liu et al., 1968).

As knowledge of amino acid sequences becomes more extensive, relationships between structure and function will become more clear. It is hypothesized that only those parts of a protein that are highly conserved play a fundamental role in catalysis or in the maintenance of structure. Such knowledge has already been obtained for several protein families—for example, the respiratory enzyme cytochrome c. It is interesting, however, that in none of these cases has the relationship between chemical structure and biochemical mechanism been greatly facilitated. Unveiling the chemical mysteries contained in these complex molecules will require much more information than can be obtained from knowledge of the linear arrangement of amino acids.

REFERENCES

Alexander, P., and R. J. Block (1960a): *Analytical Methods of Protein Chemistry*, vol. 1, Pergamon Press, New York.
Alexander, P., and R. J. Block (1960b): *Analytical Methods of Protein Chemistry*, vol. 2, Pergamon Press, New York.
Alexander, P., and R. J. Block (1961): *Analytical Methods of Protein Chemistry*, vol. 3, Pergamon Press, New York.
Alexander, P., and H. R. Lundgren (1966): *Analytical Methods of Protein Chemistry*, vol. 4, Pergamon Press, New York.
Andrews, P. (1964): *Biochem. J.*, **91**, 222.
Bailey, J. L. (1967): *Techniques in Protein Chemistry*, 2d ed., Elsevier Publishing Co., Amsterdam.
Barman, T. E., and D. E. Koshland (1967): *J. Biol. Chem.*, **242**, 5771.
Bergmann, M., E. Brand, and R. Dreyer (1921): *Ber.*, **54**, 936.
Beychok, S. (1968): *Ann. Rev. Biochem.*, **37**, 437.
Blake, C. C. F. (1968): *Adv. Protein Chem.*, **23**, 59.
Boyer, P. D. (1954): *J. Am. Chem. Soc.*, **76**, 4331.
Boyer, P. D., and H. L. Segal (1954): *The Mechanism of Enzyme Action*, edited by W. D. McElroy and B. Glass, Johns Hopkins Press, Baltimore.
Burgess, R. R. (1969): *J. Biol. Chem.*, **244**, 6168.
Cann, J. R. (1969): *Physical Principles and Techniques of Protein Chemistry*, Academic Press, New York.
Davison, P. F. (1968): *Science*, **161**, 906.
DeVincenzi, D. L., and J. L. Hedrick (1967): *Biochemistry*, **6**, 3489.
Dreze, A. (1956): *Biochem. J.*, **62**, 3P.
Dunker, A. K., and R. R. Rueckert (1969): *J. Biol. Chem.*, **244**, 5074.
Ellman, G. L. (1959): *Arch. Biochem. Biophys.*, **82**, 70.
Englander, S. W. (1967): *Poly-α-Amino Acids: Protein Models for Conformational Studies*, edited by G. Fasman, M. Dekker, New York.
Feeney, R. E., and R. G. Allison (1969): *Evolutionary Biochemistry of Proteins. Homologous and Analogous Proteins from Avian Egg Whites, Blood Sera, Milk, and Other Substances*, Wiley-Interscience, New York.
Fish, W. W., K. G. Mann, and C. Tanford (1969): *J. Biol. Chem.*, **244**, 4989.
Fraser, R. D. B., and T. P. MacRae (1969): *Physical Principles and Techniques of Protein Chemistry*, part A, Academic Press, New York.
Gerhart, J. C., and H. K. Schachman (1968): *Biochemistry*, **7**, 538.
Gorbunoff, M. J. (1969): *Biochemistry*, **8**, 2591.

Greene, F. C., and R. E. Feeney (1968): *Biochemistry*, **7,** 1366.
Habeeb, A. F. S. A. (1966): *Anal. Biochem.*, **14,** 328.
Hedrick, J. L., and A. J. Smith (1968): *Arch. Biochem. Biophys.*, **126,** 155.
Hedrick, J. L., A. J. Smith, and G. E. Bruening (1969): *Biochemistry*, **8,** 4012.
Hirs, C. H. W. (1967): *Methods Enzymol.*, **11,** 27.
Hoare, D. G., and D. E. Koshland (1967): *J. Biol. Chem.*, **242,** 2447.
Hvidt, A., and S. O. Nielsen (1966): *Adv. Protein Chem.*, **21,** 287.
Iwai, K., and T. Ando (1967): *Methods Enzymol.*, **11,** 263.
Jirgensons, B. (1969): *Optical Rotatory Dispersion of Proteins and Other Macromolecules*, Springer-Verlag, New York.
Jirgensons, B. (1970): *Biochim. Biophys. Acta*, **200,** 9.
Jolles, J., G. Sportorno, and P. Jolles (1965): *Nature*, **208,** 1204.
Leach, S. J. (1969): *Physical Principles and Techniques of Protein Chemistry*, part B, Academic Press, New York.
Liu, W. H., G. Feinstein, D. T. Osuga, R. Haynes, and R. E. Feeney (1968): *Biochemistry*, **7,** 2886.
Mann, K. G., W. W. Fish, A. C. Cox, and C. Tanford (1970): *Biochemistry*, **9,** 1348.
Matsubara, H., and R. M. Sasaki (1969): *Biochem. Biophys. Res. Commun.*, **35,** 175.
Moore, S., and W. H. Stein (1954): *J. Biol. Chem.*, **211,** 907.
Moore, S., and W. H. Stein (1963): *Methods Enzymol.*, **6,** 819.
Pflumm, M. N., and S. Beychok (1969): *J. Biol. Chem.*, **244,** 3973.
Porath, J., and P. Flodin (1959): *Nature*, **183,** 1657.
Ray, W. J., and D. E. Koshland (1962): *J. Biol. Chem.*, **237,** 2493.
Richards, E. G., and H. K. Schachman (1968): *J. Phys. Chem.*, **63,** 1578.
Roberts, G. C. K., and O. Jardetzky (1970): *Adv. Protein Chem.*, **24,** 448.
Sabato, G. D., and M. Ottesen (1967): *Methods Enzymol.*, **11,** 734.
Sakaguchi, S. (1925): *J. Biochem. (Tokyo)*, **5,** 25.
Schachman, H. K. (1959): *Ultracentrifugation in Biochemistry*, Academic Press, New York.
Schonbaum, G. R., B. Zerner, and M. L. Bender (1969): *J. Biol. Chem.*, **236,** 2930.
Schroeder, W. A. (1968): *The Primary Structure of Proteins*, Harper & Row, New York.
Schumaker, V., and P. Adams (1968): *Biochemistry*, **7,** 3422.
Shapiro, A. L., E. Vinuela, and J. V. Maizel (1967): *Biochem. Biophys. Res. Commun.*, **28,** 815.
Shaw, D. C., W. H. Stein, and S. Moore (1964): *J. Biol. Chem.*, **239,** PC671.
Slobodian, E., G. Mechanic, and M. Levy (1962): *Science*, **135,** 441.
Sokolovsky, M., and B. L. Vallee (1966): *Biochemistry*, **5,** 3574.
Spande, T. F., and B. Witkop (1967): *Methods Enzymol.*, **11,** 498.
Spies, J. R. (1967): *Anal. Chem.*, **39,** 1412.
Spies, J. R., and D. C. Chambers (1949): *Anal. Chem.*, **21,** 1249.
Stevens, F. C., and R. E. Feeney (1963): *Biochemistry*, **2,** 1346.
Stryer, L. (1968): *Ann. Rev. Biochem.*, **37,** 25.
Takahashi, K., W. H. Stein, and S. Moore (1967): *J. Biol. Chem.*, **242,** 4682.
Takenaka, A., T. Suzuki, O. Takenaka, H. Horinishi, and K. Shibata (1969): *Biochim. Biophys. Acta*, **194,** 293.
Urnes, P., and P. Doty (1961): *Adv. Protein Chem.*, **16,** 401.
Van Holde, K. E. (1967): *Fractions*, No. 1, Beckman Instruments, Inc., Palo Alto, p. 1.
Weber, K., and M. Osborn (1969): *J. Biol. Chem.*, **244,** 4406.
Whitaker, J. R. (1963): *Anal. Chem.*, **35,** 1950.
Whitaker, J. R. (1967): *Paper Chromatography and Electrophoresis*, vol. 1, Academic Press, New York.
Yang, J. T. (1961): *Adv. Protein Chem.*, **16,** 323.
Yphantis, D. A. (1964): *Biochemistry*, **3,** 297.

II
the chemistry and chemical reactions of the reagents

5
acylating and similar reagents

5-1 ACYLATING AGENTS

$$\text{P}-NH_2 + \underset{O}{\overset{X}{C}}-R \longrightarrow \text{P}-NH-\underset{O}{\overset{\|}{C}}-R + HX \qquad (1)$$

$$\text{P}-\!\!\!\bigcirc\!\!\!-O^- + \underset{O}{\overset{X}{C}}-R \longrightarrow \text{P}-\!\!\!\bigcirc\!\!\!-O-\underset{O}{\overset{\|}{C}}-R + X^- \qquad (2)$$

Amino and tyrosyl groups of proteins can be readily acylated under conditions wherein other side chains either do not react at significant rates or the products so formed are labile. Acylation of histidine and cysteine residues, for example, is rarely observed since the resulting products readily hydrolyze in aqueous solutions. Serine and threonine hydroxyl groups, on the other hand, are not easily acylated in aqueous solutions because they are such weak nucleophiles.

The many reagents employed to acylate proteins differ considerably in structure and reactivity, but all presumably follow the carbonyl-addition pathway (Equation 3). Reaction rates depend upon the rate of attack of nucleophile and upon the fraction of tetrahedral intermediates which undergo fission at the carbonyl—X bond. Because the nucleophilic groups to be acylated ionize at characteristic pH's, acylation rates are affected greatly by pH. The pH at which acylation of a group can reasonably be achieved thus depends upon its pK. Acylation rates for homologous nucleophiles are inversely related to their pK's except at high pH. In a protein, high reactivity of a group is usually attributed to a low pK.

$$R-\overset{O}{\underset{X}{C}}:NH_2 \rightleftharpoons \left[R-\overset{O^-}{\underset{X}{C}}-\overset{+}{N}H_2\right] \overset{-H^+}{\rightleftharpoons} \left[R-\overset{O^-}{\underset{X}{C}}-NH\right] \rightleftharpoons R-\overset{O}{C}\underset{NH}{\diagdown} + X^-$$

(3)

Steric and proximity effects of topologically adjacent residues frequently make individual residues resistant to reaction. In such cases, hydrolysis of the acylating agent becomes the favored reaction. Reaction with water, because it is present in large amounts, usually accounts for the consumption of considerable reagent. Even when using a large excess of reagent, complete reaction is commonly very difficult to obtain unless denaturing conditions are employed. Treatment of most native proteins with acetic anhydride, for example, seldom leads to the acetylation of more than 60% and, only infrequently, 90% of the amino groups.

The lower reactivity with most acylating agents of tyrosine phenolic groups as compared to amino groups in proteins is only partially due to their generally higher pK's. If tyrosines were not usually more protected from reaction in proteins, relatively little selectivity would be obtained. In the reaction of acetic anhydride with free tyrosine, for example, wherein steric factors are relatively less important, acylation of the amino and phenolic groups proceed at similar rates (Fraenkel-Conrat and Colloms, 1967). A tyrosine residue in tobacco mosaic virus coat protein reacts with acetic anhydride more rapidly than either of that protein's two ε-amino groups (Fraenkel-Conrat and Colloms, 1967). Reagents differ in their relative specificity for amino and tyrosyl groups. Under mild conditions, N-acetylimidazole, for example, has been shown to preferentially react with tyrosyl groups in several proteins (Riordan et al., 1965). Most acylating agents, however, react more readily with amino groups.

Acylation of tyrosine phenolic groups is distinguished from that of amino groups by a decrease in absorption at 278 mμ accompanying the former. Such reaction is easily followed spectrophotometrically. The acylation of tyrosine residues is further distinguished from that of amino groups by the ease with which the former can be reversed. Hydroxylamine at neutral pH, or simply mild alkali treatment, are sufficient to deacylate tyrosyl residues, but not amino groups. Strong, hot acid or base is required for the latter.

5-1.1 ACETIC ANHYDRIDE

$$\boxed{P}-NH_2 + \overset{O}{\underset{O}{\overset{\parallel}{C}-CH_3}}\underset{\underset{\parallel}{O}}{\overset{}{C}-CH_3} \xrightarrow{pH > 7} \boxed{P}-NH-\overset{O}{\overset{\parallel}{C}}-CH_3 + CH_3COO^- + H^+$$

(4)

Acetic anhydride is one of the most useful reagents for protein modification. Its reaction with proteins in half-saturated sodium acetate has been shown to be relatively specific for amino groups (Fraenkel-Conrat, 1959). Tyrosine phenolic hydroxyl groups are also acetylated but under these alkaline conditions the O-acetyltyrosines quickly hydrolyze (Uraki et al., 1957; Fraenkel-Conrat, 1959; Smyth, 1967).

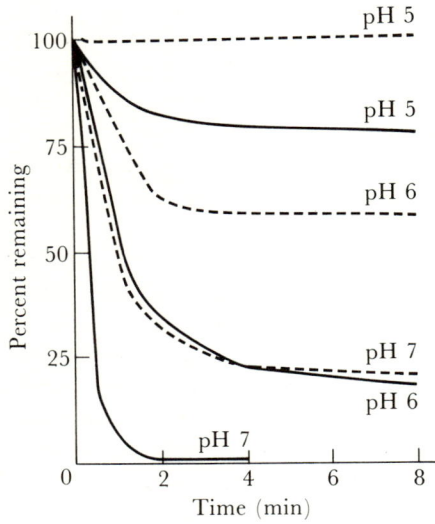

FIGURE 5-1   Rates of acetylation of alanine (dashed curves) and triglycine (solid curves) by acetic anhydride. Reactions were carried out at 30°C on 0.5 μmole/ml of NH$_2$ group using a twofold excess of acetic anhydride. (Adapted from Smyth, 1967.)

At moderately alkaline pH's, relative reaction rates of homologous amines largely reflect the extent of their protonation. High pK's thus result in correspondingly lower reaction rates. The acetylation of alanine (pK 9.0) by acetic anhydride, for example, proceeds relatively much less readily at low pH than that of triglycine (pK 7.9) (Figure 5-1). At higher pH's less differentiation is found. Lower pH's favor reaction with groups having low pK's and give less extensive reaction. The extent of reaction, of course, is determined by the amount of reagent and can be increased by using an excess. Acetylation of α-chymotrypsin, for example, goes to completion at pH 6.7 (including the acylation of its poorly reactive NH$_2$-terminal isoleucine) upon treatment with 1% acetic anhydride (Oppenheimer et al., 1966). Tyrosine as well as amino groups are acylated by such treatment.

Acetoacetate decarboxylase is rapidly inactivated by acetic anhydride upon specific acetylation of a single lysine residue (O'Leary and Westheimer, 1968). This lysine appears unusually nucleophilic and hence highly susceptible to modification. The same residue is modified when the enzyme is reduced in the presence of its substrate and is therefore presumably in the active site (Fridovich and Westheimer, 1962; Warren et al., 1966).

Retention of enzymatic activity after acetylation of all amino groups of chicken egg white lysozyme has been taken as evidence against their participation in its catalytic activity (Yamasaki et al., 1968; Davies and Neuberger, 1969). Similarly, acetylation of the ε-amino groups in trypsin has little effect on its enzymatic activity, showing them to not be essential to that activity (Labouesse and Gervais, 1967). Acetylation of its NH_2-terminal isoleucine, however, is inactivating (Chevallier et al., 1969). A similar result was found for α-chymotrypsin (Oppenheimer et al., 1966). Extensive acetylation often brings about significant changes in conformation and marked reductions in solubility. Extensive acetylation is thus useful for determining the essentiality of amino groups only in certain proteins (Bethune et al., 1964; Lee et al., 1963). Only the retention of activity following such treatment permits an unambiguous conclusion.

$$\text{P}-\text{C}_6\text{H}_4-\text{O}-\overset{\overset{\text{O}}{\|}}{\text{C}}-\text{R} + \text{NH}_2\text{OH} \xrightarrow{\text{pH} > 6}$$

$$\text{P}-\text{C}_6\text{H}_4-\text{OH} + \text{HONH}-\overset{\overset{\text{O}}{\|}}{\text{C}}-\text{CH}_3 \quad (5)$$

$$\downarrow Fe^{3+}$$

colored complex

O-Acetyltyrosine hydrolyzes under moderately alkaline conditions or can be cleaved at neutral pH by brief exposure to hydroxylamine (Figure 5-2). In each case, deacylation is accompanied by an increase in absorbance at 278 mμ ($\Delta\varepsilon_{278} = 1160\ M^{-1}\ cm^{-1}$). One mole of acetylhydroxamate is formed in the reaction with hydroxylamine. With ferric ion, this compound forms a purple-colored complex, which can be quantitatively determined at 540 mμ (Balls and Wood, 1956; Riordan and Vallee, 1963; Smyth, 1967).

Ketene as an acetylating reagent is difficult to use because of its low boiling point and high reactivity. Acetyl chloride, and acyl halides in general, are also highly reactive and offer no clear advantages over the corresponding acid anhydrides. Long-chain carboxylic acid anhydrides are not as reactive as acetic anhydride, are less soluble, and only rarely used (Dixon and Neurath, 1957; Riordan and Vallee, 1963).

5-1.2 N-Acetylimidazole

$$\text{P}-\text{C}_6\text{H}_4-\text{O}^- + \text{H}^+ + \underset{\text{N}}{\overset{\text{N}}{\bigcirc}}-\overset{\text{O}}{\underset{\|}{\text{C}}}-\text{CH}_3 \xrightarrow{\text{pH 7.5}}$$

$$\text{P}-\text{C}_6\text{H}_4-\text{O}\overset{\text{O}}{\underset{\|}{\text{C}}}-\text{CH}_3 + \underset{\overset{|}{\text{H}}}{\overset{\text{N}}{\bigcirc}}\text{N} \qquad (6)$$

$$\text{P}-\text{NH}_2 + \underset{\text{N}}{\overset{\text{N}}{\bigcirc}}-\overset{\text{O}}{\underset{\|}{\text{C}}}-\text{CH}_3 \xrightarrow{\text{pH 7.5}} \text{P}-\text{NH}-\overset{\text{O}}{\underset{\|}{\text{C}}}-\text{CH}_3 + \underset{\overset{|}{\text{H}}}{\overset{\text{N}}{\bigcirc}}\text{N} \qquad (7)$$

Like acetic anhydride, N-acetylimidazole reacts with both amino and tyrosyl groups of proteins, but is more selective for tyrosine (Riordan and Vallee, 1963; Riordan et al., 1965). Under mild conditions, for example, it was found to selectively acetylate six tyrosyl residues in bovine carboxypeptidase A without affecting amino groups (Simpson et al., 1963; Riordan et al., 1965). When extensive tyrosine modification is desired, however, considerable modification of amino groups should also be expected (Komatsu and Feeney, 1967; Riordan et al., 1965). N-Acetylimidazole in aqueous

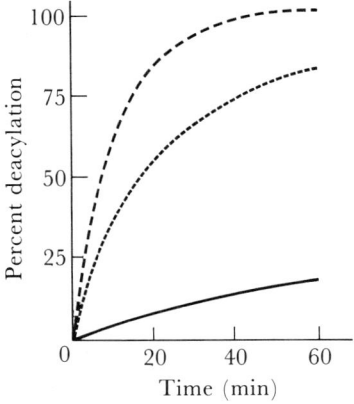

FIGURE 5-2   Deacetylation of O-acetylphenol (0.5 μmole/ml) at pH 10 (solid curve) and 11 (dashed curve) and in 0.04 M hydroxylamine at pH 7 (dotted curve), all at 30°C. (Adapted from Smyth, 1967.)

solution is subject to both acid- and base-catalyzed hydrolysis (Jencks and Carriuolo, 1959). It is normally used near pH 7.5 where it is most stable. General and specific effects of various ions, imidazole itself, pH, and other factors on the stability of acetylimidazole have been described by Jencks and Carriuolo (1959) (see Figure 5-3).

Treatment with acetylimidazole brings about a rapid destruction of the allosteric properties of rabbit muscle phosphofructokinase (Chapman et al., 1969). If, however, ATP is included during the acetylation this loss is prevented. Reaction of

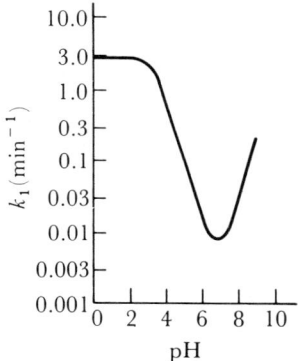

FIGURE 5-3   Logarithmic plot of the rate of N-acetylimidazole hydrolysis, against pH at 25° and 0.2 ionic strength. Ionic strength maintained with NaCl. (From Jencks and Carriuolo, 1959.)

N-acetylimidazole with rabbit muscle aldolase greatly reduces that enzyme's ability to cleave fructose-1,6-diphosphate but has little effect on its ability to cleave fructose-1-phosphate (Schmid et al., 1966; Pugh and Horecker, 1967). Loss of activity toward the former has been attributed to acylation of tyrosine in the 6-phosphate binding site.

Acetylation of seven tyrosyl residues in bovine pancreatic carboxypeptidase A lowers its peptidase activity while significantly increasing its esterase activity (Simpson et al., 1963). These changes correlate with the acetylation of two residues and are prevented by the competitive inhibitor, β-phenylpropionate. Deacylation with hydroxylamine restores the initial levels of peptidase and esterase activities, showing that the changes in enzymatic properties were due specifically to the modification of tyrosyl residues. In similar experiments with pepsin, acetylation of two tyrosyl residues by N-acetylimidazole enhanced both peptidase and esterase activities (Hollands and Fruton, 1968).

The relative specificity of N-acetylimidazole, and the mild reaction conditions, make it a suitable reagent for determining the distribution of free and buried tyrosyl

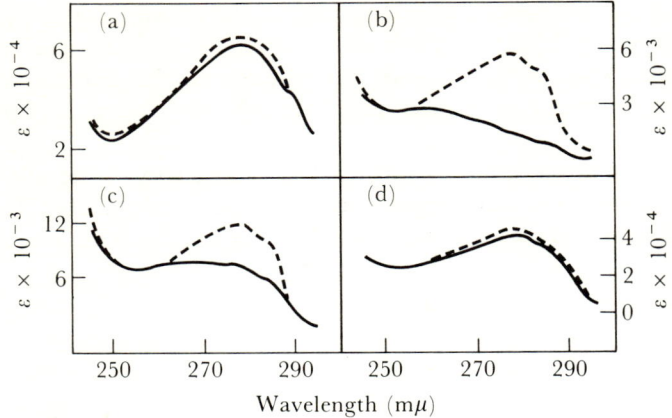

FIGURE 5-4 Absorption spectra of proteins before (dashed curves) and after (solid curves) acetylation with a 60-fold excess of N-acetylimidazole. (a) Carboxypeptidase A, (b) insulin, (c) chicken ovomucoid, (d) bovine serum albumin. (From Riordan et al., 1965.)

residues. The spectral changes accompanying the reaction afford a ready means to follow and quantitate the reaction ($\Delta\varepsilon_{278} = 1160\ M^{-1}\ cm^{-1}$) (Figure 5-4).

5-1.3 Succinic Anhydride

$$\text{P}-NH_2 + \begin{array}{c} O \\ \| \\ C-CH_2 \\ O \\ | \\ C-CH_2 \\ \| \\ O \end{array} \xrightarrow{pH>7} \text{P}-NH-\overset{O}{\underset{\|}{C}}-CH_2CH_2\overset{O}{\underset{\|}{C}}-O^- + H^+ \qquad (8)$$

Succinic anhydride resembles acetic anhydride; however, succinylation of a protein yields a quite different product. Whereas acetylation of cationic amino groups renders them electrically neutral, succinylation converts them to anionic residues. The resulting change from positive to negative charge leads to greater changes in electrostatic relationships and frequently brings about the dissociation of aggregated or subunit proteins and/or rather major conformational changes (see Section 3-1). Succinylation is sometimes preferable to acetylation for the modification of amino groups because, in some cases, products of the former are likely to be more soluble.

Succinylation of the oxygen-carrying protein hemerythrin (mol wt $\simeq$ 105,000) brings about its dissociation into eight identical subunits (mol wt $\simeq$ 14,000), as indicated by a reduction of its sedimentation coefficient from approximately 6 to 2 (Klotz and Keresztes-Nagy, 1962). Similar treatment brings about dissociation of the bovine procarboxypeptidase zymogen complex into three chemically distinct subunits, only one of which is the succinyl precursor of carboxypeptidase A (Freisheim et al., 1967).

Succinylation of the catalytic subunits of *E. coli* aspartate transcarbamylase yields a relatively homogeneous, inactive electrophoretic variant which, upon mixing with regulatory subunits, forms a complex similar in molecular weight to the native enzyme. The number and properties of the formed complexes showed each enzyme molecule to contain two catalytic subunits, each composed of three polypeptide chains (Meighen et al., 1970). Studies of these hybrid molecules indicated the expression of catalytic activity to be contributed independently by each subunit. The same hybridization technique has been shown to have general potential for the study of subunit structure in oligomeric proteins (Meighen and Schachman, 1970).

Succinylated proteins frequently exhibit increases in intrinsic viscosity and concomitant decreases in sedimentation coefficient at neutral or alkaline pH's resulting from a general unfolding and molecular expansion (Habeeb et al., 1958). Extensive succinylation of bovine serum albumin, for example, increases its Stokes radius, the susceptibility of its disulfide bonds to reduction, and reduces its ability to precipitate anti–bovine serum albumin immunoglobulins (Habeeb, 1967). Similar changes following succinylation have been observed with orosomucoid, conalbumin, and serum transferrin (Bezkorovainy et al., 1969).

$$\text{(9)}$$

O-Succinyltyrosine residues formed upon treatment of proteins with succinic anhydride are subject to a rapid intramolecularly catalyzed hydrolysis above pH 5 which results in a rapid regeneration of the original tyrosyl residues (Equation 9). Other dicarboxylic acid O-acyltyrosines are subject to similar intramolecularly catalyzed hydrolysis, although in general at much slower rates (Riordan and Vallee, 1964). (See discussion of maleic anhydride for an exception.) Succinyl derivatives of aliphatic hydroxy amino acids do not spontaneously hydrolyze in this way but can be cleaved by treatment with hydroxylamine (Gounaris and Perlman, 1967).

Succinylation, like acetylation, nearly abolishes the peptidase activity of bovine carboxypeptidase A while significantly increasing its esterase activity (Riordan and Vallee, 1964). These changes in activity have been attributed to the succinylation of tyrosine residues. Unlike its acetylated counterpart, however, succinylcarboxypeptidase A, at pH's greater than 5, undergoes a time-dependent recovery to the original levels of peptidase and esterase activities concomitant with a spontaneous deacylation of tyrosyl residues. At 0°–4°C and pH 7.5, complete reactivation takes about 36 hours. In 1.0 M hydroxylamine at the same pH only 18 minutes are required for reactivation.

5-1.4 Maleic Anhydride

$$\text{P}-NH_2 + \begin{array}{c}O=C-CH\\ O\\ O=C-CH\end{array} \xrightarrow{pH\ 8} \begin{array}{c}O\\ \text{P}-NH-C-CH\\ O-C-CH\\ O\end{array} + H^+ \qquad (10)$$

Maleic anhydride reacts with proteins in a manner similar to succinic anhydride. Products, however, differ from corresponding succinyl derivatives in that they are much more labile to hydrolysis (Butler et al., 1967). The reaction products of maleic anhydride and amino groups are stable at neutral pH but rapidly hydrolyze when acidified to pH 3.5. The half-life of ε-maleyllysine is approximately 11 hours at this pH and 37°C. Reactions of maleic anhydride with most other side chains appear to be reversed even more readily (Butler et al., 1969). Alkylation of sulfhydryl groups similar to that by N-ethylmaleimide is a potential irreversible side reaction.

Maleylation is particularly useful as a method to reversibly modify amino groups (Freedman et al., 1968). Maleyl proteins, like succinylated proteins, tend to be soluble and disaggregated at neutral pH's. Maleylation of rabbit muscle aldolase (Sia and Horecker, 1968) and phosphofructokinase (Uyeda, 1969) caused them to dissociate into subunits. Enzymatic activities were lost concomitant with the observed dissociations at relatively low levels of maleylation. Deacylation of maleylaldolase at pH 4.5 led to partial recovery of catalytic activity. Because ε-maleyllysine residues are resistant to hydrolysis by trypsin and because of the reversibility of this modification, maleylation should have great use for the degradation and sequencing of peptides.

$$\text{(11)}$$

The reaction of maleic anhydride with proteins or with lysine is accompanied by an increase in absorption at wavelengths below 280 mμ. Spectra of ε-maleyllysine, lysine plus maleic acid, β-melanocyte stimulating hormone (MSH), and maleyl MSH in 0.1 M NaOH are shown in Figure 5-5. The increased absorption at 250 mμ ($\Delta\varepsilon_{250} = 3360\ M^{-1}\ cm^{-1}$) at pH 8.0 has been used to calculate the extent of maleylation of rabbit antibodies (Freedman et al., 1968).

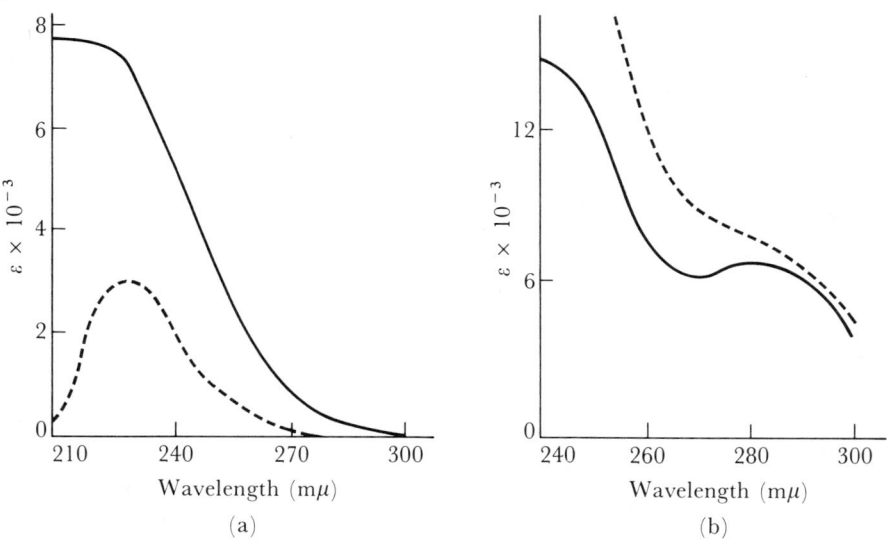

FIGURE 5-5 Spectra in 0.1 M NaOH of (a) ε-maleyllysine (solid curve) and lysine plus maleic acid (dashed curve); (b) MSH (solid curve) and maleyl MSH (dashed curve). (Adapted from Butler et al., 1969.)

Citraconic anhydride (2-methylmaleic anhydride) and 2,3-dimethylmaleic anhydride react similarly to maleic anhydride, but the acyl derivatives are even more easily hydrolyzed (Dixon and Perham, 1968). Products from reaction of the latter with amines are so unstable that deacylation occurs even at slightly alkaline pH. In contrast to the maleyl groups which are somewhat inconvenient to remove, 2,3-dimethylmaleyl groups have been found inconveniently hard to keep on. At pH 3.5 and 20°C, dimethylmaleylarginine is completely converted to arginine in 5 minutes. The lability of citraconylamides is intermediate between the amides of maleic acid and 2,3-dimethylmaleic acids. They are stable enough to withstand many hours at neutral or alkaline pH's but are sufficiently labile at low pH's to be deacylated under mild conditions. Deacylation of the citraconyl derivative of performate-oxidized insulin, for example, was complete after overnight incubation at pH 3.5 and 20°C (Dixon and Perham, 1968).

5-1.5 N-Carboxyanhydrides

$$\text{P}-NH_2 + \underset{\underset{H}{N}}{\overset{O}{\underset{\|}{C}}}\!\!\!\!\overset{O\ \ \ R}{\underset{\ }{\diagdown}}\!\!\xrightarrow[0°-4°C]{pH\sim 7} \text{P}-NH-\overset{O}{\underset{\|}{C}}-\overset{R}{\underset{|}{CH}}-NH_2 + CO_2 \longrightarrow$$

$$\text{P}-NH-\left(-\overset{O}{\underset{\|}{C}}-\overset{R}{\underset{|}{CH}}-NH-\right)_n -\overset{O}{\underset{\|}{C}}-\overset{R}{\underset{|}{CH}}-NH_2 + n\,CO_2 \qquad (12)$$

N-Carboxyanhydrides of α-amino acids react with proteins very much as do other acid anhydrides (Katchalski et al., 1964). Reactions are normally done in aqueous solution at low temperature and neutral pH. Apparently, only amino groups are modified. A molecule of carbon dioxide is evolved for each amino acid incorporated but, because of side reactions, this is not useful to follow the reaction. The new amino groups generated in the process provide focal points for further reaction such that polypeptidyl proteins can eventually be formed (Equation 12). Polypeptidyl enzymes, containing large numbers of such extraneously attached amino acid residues largely retain their catalytic properties in many cases, but they differ markedly from the native enzymes in certain physical-chemical properties (Cooke et al., 1963; Freedman and Sela, 1966; Nishikawa et al., 1968).

The N-carboxyanhydrides of D,L-alanine and glycine have been used to polypeptidylate and thereby increase the aqueous solubility of certain proteins which otherwise become insoluble upon reduction of their disulfide bonds. Upon reduction of rabbit immunoglobulin G with β-mercaptoethanol, for example, a largely insoluble material was obtained from which subsequent recovery of antigen-binding activity was extremely low. Similar reduction of the polyalanylimmunoglobulin, however, gave a soluble product which, when reoxidized by air under optimum conditions, gave up to 25% of the antigen-combining capacity present prior to reduction (Freedman and Sela, 1966).

Polyvalylribonuclease, having approximately 23 moles of added valine, differs from both native and polyalanylribonuclease in being highly subject to a reversible aggregation above 30°C. Polyvalylribonuclease has been used as a model to study hydrophobic interactions in proteins (Nishikawa et al., 1968). Polypeptidyl proteins prepared from N-carboxyanhydrides of other hydrophobic amino acids undergo similar aggregation phenomena in aqueous solutions (Glazer et al., 1962; Anfinsen et al., 1962).

Upon polypeptidylation, positively charged amino groups of a protein are displaced to loci some distance from the original. The extent of this displacement of charge appears to have fundamental importance for the Kunitz pancreatic trypsin inhibitor. Amidination or guanidination (see Section 5-3), both of which displace the individual cationic loci by approximately 1.5 Å do not destroy this protein's ability to inhibit trypsin, whereas polyalanylation, with a considerably greater displacement, gives complete inactivation (Acher et al., 1968). Polypeptidyl proteins have a net charge similar to the native protein at most pH values, but due to the slightly lower pK's of α-amino as compared to ε-amino groups, differences occur in moderate to strong alkaline solutions. The isoelectric point of RNase (∼9.6–9.8), for example, is lowered to 7.8 upon reaction with N-carboxyvaline anhydride (Nishikawa et al., 1968).

N-Carboxyanhydrides of α-amino acids are relatively unstable crystalline compounds best prepared from the corresponding amino acids by treatment with phosgene (see Katchalski et al., 1964). They are also commercially available. They can be stored in an anhydrous state at −20° for several months, but storage at higher temperatures or contact with small amounts of water leads to their polymerization. They can be quantitatively estimated from the CO_2 evolved upon treatment with dilute acid (Patchornik and Shalitin, 1961) or by several other procedures (Katchalski et al., 1964). They are widely used intermediates for peptide and polyamino acid syntheses (see Katchalski et al., 1964, or Sela and Arnon, 1967, for recent reviews).

5-1.6 Ethyl thiotrifluoroacetate

$$\boxed{P}-NH_2 + CH_3CH_2S-\overset{\overset{O}{\|}}{C}-CF_3 \xrightarrow[25°C]{pH\sim10} \boxed{P}-NH-\overset{\overset{O}{\|}}{C}-CF_3 + CH_3CH_2SH$$

(13)

Ethyl thiotrifluoroacetate and trifluoroacetic anhydride have been used to reversibly acylate protein amino groups (Schallenberg and Calvin, 1955; Goldberger, 1967). The former compound is less reactive and hence more specific in its reactions than the anhydride.

Trifluoroacetylation is of considerable interest, due to the ease with which the trifluoroacetamide groups may be removed from proteins. The strongly electron-withdrawing fluorine groups, by increasing the electrophilicity of the carbonyl carbon, greatly labilize the acyl-protein bond making it susceptible to hydrolysis under relatively mild conditions. Like many sulfur-containing compounds, ethyl thiotrifluoroacetate and the ethyl mercaptan formed during its reaction with proteins have strong odors and should be used with good ventilation.

Trifluoroacetylated proteins are rapidly deacylated in 1 M piperidine at 0°C. After 2 hours under these conditions trifluoroacetylated pancreatic ribonuclease gave a product resembling the native enzyme in most physical-chemical properties but catalytically inactive (Goldberger and Anfinsen, 1962). Nearly full catalytic activity was subsequently recovered, however, upon reduction of the disulfide bonds with thiols and slow air reoxidation. Ethyl mercaptan, formed during the trifluoroacetylation, evidently brought about some disulfide interchange to give some incorrect pairing of disulfide bonds.

$$\text{P}-\text{NH}-\overset{\overset{\text{O}}{\|}}{\text{C}}-\text{CF}_3 + \text{H}_2\text{O} \xrightarrow[25°\text{C, 30 hr}]{\text{pH 10.7, carbonate}} \text{P}-\text{NH}_2 + \overset{\overset{\text{O}}{\diagup\!\!\diagdown}}{\underset{-\text{O}}{\text{C}}}-\text{CF}_3 + \text{H}^+ \tag{14}$$

Trifluoroacetyl groups may be removed more slowly but apparently as effectively in carbonate buffer at pH 10.7 and 25°C. After 30 hours of this treatment, trifluoroacetylated cytochrome c was recovered in apparently native form with its full electron-transfer activity (Fanger and Harbury, 1965). Deacylation of trifluoroacetylated peptides absorbed upon paper chromatograms can be achieved by exposure to ammonia vapor. The procedure has been used in combination with diagonal electrophoresis to identify lysine-containing peptides (Perham and Jones, 1967).

Trifluoroacetylation has proven useful in sequencing studies for its ability to limit tryptic digestion to arginyl peptide bonds. After tryptic digestion, separation of the newly generated peptides, and removal of the trifluoroacetyl groups, a second tryptic digestion of the isolated peptides has been used to facilitate sequencing (Goldberger and Anfinsen, 1962).

The reaction of tetrafluorosuccinic anhydride with proteins is like that of succinic anhydride, but the tetrafluorosuccinamides formed are relatively much more labile, even more so than trifluoroacetamides (Braunitzer et al., 1968).

5-1.7 Diketene

$$\text{P}-\text{NH}_2 + \text{diketene} \xrightarrow{\text{pH 8.5}} \text{P}-\text{NH}-\overset{\overset{\text{O}}{\|}}{\text{C}}-\text{CH}_2-\overset{\overset{\text{O}}{\|}}{\text{C}}-\text{CH}_3 \tag{15}$$

Diketene reacts with amino groups of proteins (Equation 15). It reacts somewhat less readily with hydroxyl groups and the products can be selectively cleaved by 0.2 M NaHCO$_3$–Na$_2$CO$_3$ buffer (pH 9.5). Complete acetoacetylation of the amino groups of pancreatic ribonuclease with diketene brings about the loss of its catalytic activity (Marzotto et al., 1967; Marzotto et al., 1968). Treatment for 12 hours with a threefold excess of hydroxylamine at pH 7 and 25°C effects the complete recovery of catalytic activity, concomitant with complete removal of the acetoacetyl groups. A similar loss of catalytic activity and its recovery following treatment with hydroxylamine have been obtained with chicken egg white lysozyme (Marzotto et al., 1968). With insulin at pH 6.9, diketene reacted only with the two terminal amino groups (Lindsay and Shall, 1969). Both of the resulting monosubstituted derivatives possessed nearly full mouse convulsion activity.

$$\text{(P)}-NH-\overset{O}{\underset{\|}{C}}-CH_2-\overset{O}{\underset{\|}{C}}-CH_3 \;\; \overset{NH_2OH}{\underset{pH\,7}{\longrightarrow}}$$

$$\text{(P)}-NH-\overset{O}{\underset{\|}{C}}-CH_2-\overset{O}{\underset{\|}{C}}-CH_3 \longrightarrow \text{(P)}-NH_2 + \underset{CH_2-\overset{\|}{C}-CH_3}{\overset{O\!\!\diagdown_{C}\!\diagup^{O}\!\diagdown_{N}}{}} \quad (16)$$

Quantitation of the reaction can be on the basis of the increased absorbance at 278 mμ or by colorimetric determination at 540 mμ of the ferric-acetoacetamide complex formed in 3% FeCl$_3$ (Marzotto et al., 1967; Marzotto, 1969).

5-1.8 ETHOXYFORMIC ANHYDRIDE (DIETHYLPYROCARBONATE)

$$\text{(P)}-NH_2 + \begin{array}{c} O \\ \| \\ C-O-CH_2-CH_3 \\ O \\ C-O-CH_2-CH_3 \\ \| \\ O \end{array} \xrightarrow{pH\,4-7}$$

$$\text{(P)}-NH-\overset{O}{\underset{\|}{C}}-OCH_2-CH_3 + CH_3CH_2OH + CO_2\uparrow \quad (17)$$

$$\text{(P)}-\underset{N}{\overset{H}{\underset{\diagdown}{N}\!\!=\!\!\diagdown}} + \begin{array}{c} O \\ \| \\ C-O-CH_2-CH_3 \\ O \\ C-O-CH_2-CH_3 \\ \| \\ O \end{array} \xrightarrow{pH\,4-7}$$

$$\text{(P)}-\underset{N}{\overset{N=\diagdown}{\underset{\diagdown}{N}}}-\overset{O}{\underset{\|}{C}}-OCH_2CH_3 + CH_3CH_2OH + CO_2\uparrow \quad (18)$$

Ethoxyformic anhydride is an inexpensive, commercially available liquid. It hydrolyzes slowly in water, giving 2 equivalents each of ethanol and carbon dioxide (or bicarbonate ion). At pH 4.0 it reacts principally with the amino and imidazole groups of proteins. Phenolic and aliphatic hydroxyl groups are not affected, with the exception of certain active serine residues as present in α-chymotrypsin. No reaction with sulfhydryl groups has been detected. The reaction of ethoxyformic anhydride with bovine serum albumin at pH 7 appears to involve tryptophan residues (Rosén and Fedorcsák, 1966).

Ethoxyformic anhydride is strongly bactericidal and has been used for cold sterilization of certain foods. It rapidly inactivates several enzymes, for example, ribonuclease and trypsin (Rosén and Fedorcsák, 1966). It has been used to acylate F- and G-actin (Mühlrad et al., 1969), ATP-creatine phosphotransferase, ATP-arginine phosphotransferase (Pradel and Kasseb, 1968), bovine pancreatic ribonuclease A, α-chymotrypsin, and swine pepsin (Melchior and Fahrney, 1970).

At pH 4.0, ethoxyformic anhydride reacts specifically with the single α-amino group of pepsin without affecting that enzyme's catalytic activity (Melchior and Fahrney, 1970). Under the same conditions, both amino and imidazole groups in pancreatic ribonuclease were modified. Three of the latter enzyme's four histidine residues could be rapidly acylated, causing a considerable loss of catalytic activity. The fourth histidine was inaccessible in the native protein (Melchior and Fahrney, 1970). Similar treatment with a low concentration of ethoxyformic anhydride resulted in the specific binding of a single ethoxyformyl group to the active-site serine of α-chymotrypsin. At higher reagent concentrations, amino groups were also modified. Both histidines were unreactive in the native protein (Melchior and Fahrney, 1970).

The reaction of ethoxyformic anhydride with imidazole groups of proteins is accompanied by an increase in absorbance at 230 to 240 mμ. This can be used to quantitate the extent of reaction using a molar extinction coefficient of $\varepsilon = 3.0 \times 10^3\ M^{-1}\ cm^{-1}$ at 230 mμ (Melchior and Fahrney, 1970) or $\varepsilon = 3.2 \times 10^3\ M^{-1}\ cm^{-1}$ at 240 mμ (Mühlrad et al., 1969). Its selective reaction with only accessible imidazole groups and the ease with which such reaction can be quantitated make ethoxyformic anhydride a convenient probe for determining the accessibility of histidine residues.

N-Ethoxyformylimidazole is nearly two orders of magnitude more stable than N-acetylimidazole in aqueous solutions. Its half-life at 25°C is about 2 hours at pH 2, 55 hours at pH 7, and 18 minutes at pH 10. Deacylation at neutral pH can be accomplished in a few minutes with 0.5 M hydroxylamine hydrochloride (Melchior and Fahrney, 1970).

5-1.9 MISCELLANEOUS ACYLATING AGENTS

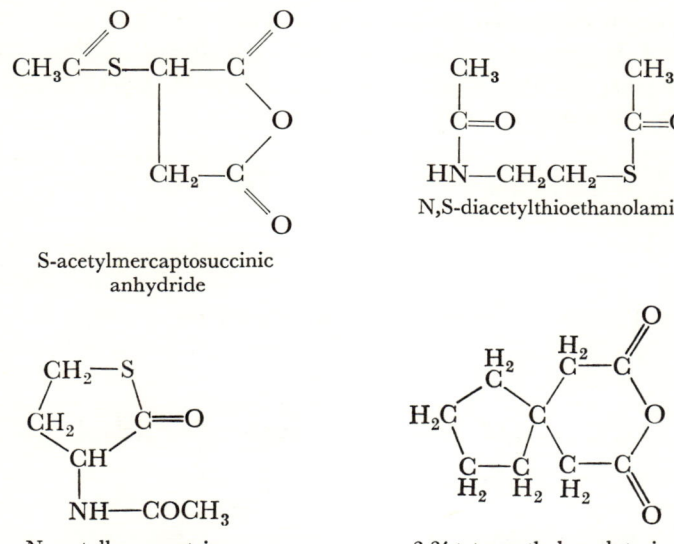

Other reagents that have been used to acylate proteins include N,S-diacetylthioethanolamine (Guldalian et al., 1965), 3,3'-tetramethyleneglutaric anhydride (Atassi, 1967), N-acetylhomocysteine thiolactone (Benesch and Benesch, 1956, Abadi and Wilcox, 1960), β-sulfopropionyl chloride (Terminiello et al., 1958); S-acetylmercaptosuccinic anhydride (Klotz and Heiney, 1962), and several p-nitrophenyl esters (Koltun et al., 1963; Mathew et al., 1967; Levy and Carpenter, 1967; O'Leary and Westheimer, 1968). For detailed information on these reagents, the original literature should be consulted.

5-1.10 QUANTITATION

The extent of amino-group acylation is frequently obtained by difference, by determining the amino groups remaining using ninhydrin (Moore and Stein, 1948), TNBS (Habeeb, 1966), nitrous acid (Oppenheimer et al., 1966), or by amino acid analysis after treatment with nitrous acid (Anfinsen et al., 1962) or dinitrofluorobenzene (Wofsy and Singer, 1963). The extent of polypeptidylation can be determined by amino acid analysis after acid hydrolysis. The number of ε-amino groups so blocked, and hence the average polypeptidyl chain length, can be determined from the lysine content remaining after reaction with nitrous acid (Anfinsen et al., 1962). The acylation of tyrosine is accompanied by a decrease in absorbance near 278 mμ which can be used to determine the extent of such reaction using $\Delta\varepsilon_{278} = 1160\ M^{-1}\ cm^{-1}$ (see Figure 5-2). O-Acyltyrosines react with hydroxylamine to give acylhydroxamates which can be quantitatively determined in the presence of $FeCl_3$ from their absorbance at 540 mμ (Balls and Wood, 1956). Total acyl groups introduced can be obtained using radioactively labeled reagent or by determination of the respective acids produced by hydrolysis, either titrametrically (Kabat and Mayer, 1961; Vithayathil and Richards, 1960) or by gas-liquid chromatography (Ward et al., 1966).

5-2 CYANATE

$$\text{(P)}-NH_2 + HNCO \xrightarrow{pH \geq 7} \text{(P)}-NH-\underset{O}{\overset{\parallel}{C}}-NH_2 \qquad (1)$$

$$\text{(P)}-S^- + HNCO + H_2O \underset{}{\overset{pH\ 6\ to\ 8}{\rightleftharpoons}} \text{(P)}-S-\underset{O}{\overset{\parallel}{C}}-NH_2 + OH^- \qquad (2)$$

$$\text{(P)}-\text{C}_6H_4-O^- + HNCO + H_2O \overset{pH > 5}{\rightleftharpoons}$$
$$\text{(P)}-\text{C}_6H_4-O-\underset{O}{\overset{\parallel}{C}}-NH_2 + OH^- \qquad (3)$$

$$\text{(P)}-C(=O)O^- + HNCO + H_2O \overset{pH \sim 5}{\rightleftharpoons} \text{(P)}-C(=O)-O-\underset{O}{\overset{\parallel}{C}}-NH_2 + OH^- \qquad (4)$$

$$\text{(P)}-\text{imidazole}-NH + HNCO \overset{pH \sim 8}{\rightleftharpoons} \text{(P)}-\text{imidazole}-N-\underset{O}{\overset{\parallel}{C}}-NH_2 \qquad (5)$$

Cyanate ion or cyanic acid (HNCO) reacts with amino, sulfhydryl, imidazole, tyrosyl, and carboxyl groups of proteins. Only the reaction with amino groups results in the formation of a stable product, however. Those with sulfhydryl, imidazole, tyrosyl, and carboxyl groups give relatively unstable adducts which decompose upon dilution or with a change of pH.

$$\text{(P)}-\overset{H}{\underset{H}{N:}}\overset{HO}{\underset{}{\overset{H}{\underset{}{\overset{}{C}}}}}\overset{H}{\underset{O}{}} \underset{k_{-1}}{\overset{k_1}{\rightleftarrows}} \text{(P)}-\overset{H}{\underset{H}{\overset{+}{N}}}-\overset{HO^-}{\underset{}{\overset{H}{\underset{O}{C}}}}\overset{H}{\underset{}{N-H}} \overset{k_2}{\longrightarrow} \text{(P)}-\overset{H}{\underset{H}{N}}-\overset{HO}{\underset{O}{\overset{H}{\underset{}{C}}}}\overset{H}{\underset{}{N-H}} \qquad (6)$$

The reaction of cyanate with amino groups appears to involve the unprotonated amine and electrically neutral cyanic acid (Johncock et al., 1958; Stark, 1965a). It is mechanistically analogous to the reaction of amines with alkylisocyanates (RNC=O). The reaction of lysine residues with cyanate converts them into homocitrulline. Homocitrulline is quite stable under ordinary conditions but is partially degraded to lysine during acid hydrolysis (i.e., ~24% in 6 N HCl at 110°C for 22 hours) (Stark et al., 1960).

Reaction rates of amines with cyanate are a function of the reaction medium pH and of the pK of the amine. A plot of second-order rate constants (k_2 = rate/[RNH$_2$][HNCO]) for the reaction of several amino acids with cyanate at pH 8.0 and 30°C versus their pK's gives a straight line. The slope of the line fits the empirical equation

$$\log k_2 (M^{-1}\ \text{min}^{-1}) = 7.94 - 0.71\ pK_a$$

From this equation, reaction rates of similar amines with cyanate can be predicted once their pK's are known (Stark, 1965a).

As a result of their lower pK values, α-amino groups react at neutral pH nearly 100 times more rapidly than do ε-amino groups (Stark, 1965a). As a consequence, α-amino groups of peptides and proteins can be selectively modified. Carbamylation of alanyllysine at pH 7 or lower, for example, takes place with 85 to 90% of the reaction occurring with α-amino groups (Stark, 1965a). The two α-amino groups of porcine insulin are approximately equal in susceptibility to reaction with cyanate at pH 6 in 8 M urea, whereas the single ε-amino group by comparison is unreactive (Cole, 1961). At pH 5.6 the α-amino group and six tyrosine residues of pepsin react with cyanate but the single ε-amino group does not (Rimon and Perlmann, 1968).

The reaction of cyanate with bovine pancreatic ribonuclease reduces its catalytic activity to less than 15% with the modification of seven of ten ε-amino groups (Stark et al., 1960). Carbamylation of nine of the ten ε-amino groups and four tyrosine residues of pepsinogen prevents its activation. With only three ε-amino groups blocked, complete activation can be obtained, but further modification parallels a loss in the potential for activation.

$$\text{P}-\text{S}-\overset{\overset{\displaystyle\text{HOH}\quad\text{H}}{\diagdown\;\diagup}}{\underset{\underset{\displaystyle\text{O}}{\parallel}}{\text{C}}}\quad\underset{k_{-1}}{\overset{k_1}{\rightleftharpoons}}\quad \text{P}-\text{S}-\text{C}\overset{\text{NH}_2}{\underset{\text{O}}{\diagdown}} + \text{OH}^- \qquad (7)$$

Cyanate reacts more rapidly with thiols than with amines, but the reaction is rapidly reversible (Stark et al., 1960; Stark, 1964). The rates of formation and

TABLE 5-1   Equilibrium and rate constants for the carbamylation of cysteine (from Stark, 1964)

pH	k_1	k_{-1}	K_{eq}
6	4.0	7.9×10^4	5.4×10^{-5}
8	2.8	7.3×10^4	3.8×10^{-5}

decomposition of S-carbamylcysteine are relatively constant between pH 6 and 8 (Table 5-1). Its half-life at 25°C varies in this pH range from approximately 11 to 6 minutes. At pH 5 and lower, it is relatively stable. Cyanate has been proposed to be useful for reversibly modifying protein sulfhydryl groups (Stark, 1964) but has not been employed for this purpose to any great extent.

Cyanate has been shown to react rapidly with papain at pH 6.0, effecting a loss of catalytic activity. Inactivation results from reaction of cyanate with the active-center sulfhydryl group and is 3000 times more rapid than its reaction with cysteine. Protection of the active site as indicated by reduced rates of inactivation is observed in the presence of substrates. The reaction is reversible, catalytic activity slowly returning upon dilution or removal of excess cyanate from the cyanate-inactivated enzyme (Sluyterman, 1967).

$$\text{P}-\!\!\!\bigcirc\!\!\!-\text{O}-\overset{\overset{\displaystyle\text{HOH}\quad\text{H}}{\diagdown\;\diagup}}{\underset{\underset{\displaystyle\text{O}}{\parallel}}{\text{C}}}\quad\underset{k_{-1}}{\overset{k_1}{\rightleftharpoons}}\quad\text{P}-\!\!\!\bigcirc\!\!\!-\text{O}-\text{C}\overset{\text{NH}_2}{\underset{\text{O}}{\diagdown}} + \text{OH}^- \qquad (8)$$

Phenolic hydroxyl groups readily react with cyanate (Equation 8). The reaction is reversible, similar to the reaction of cyanate with sulfhydryl groups. The phenolic anion and cyanic acid appear to be the reactive species (Smyth, 1965).

The rate of formation of O-carbamyltyrosine is nearly uniform between pH 5.0 and the pK of the phenolic group (Smyth, 1965). The half-life of formation for O-carbamyltyrosine at pH 7.0 and 37°C is approximately 2 hours in 0.4 M NaCNO. Once formed it is stable in acidic but decomposes rapidly in alkaline solutions. At pH 8.0 and 30°C, its half-life is approximately 13 minutes (Figure 5-6). Aliphatic hydroxyl groups are, in general, unreactive under even rather harsh conditions (Stark, 1965b). The serine proteases—chymotrypsin, trypsin, and subtilisin—are all rapidly inactivated by cyanate, presumably due to the reaction of cyanate with the active-site serines (Shaw et al., 1964).

At pH 8.8, four tyrosine residues in pepsinogen are carbamylated after treatment with 0.5 M cyanate for 24 to 27 hours at 30°C. During the same treatment, nine of the ten ε-amino groups are blocked, but the α-amino group is not affected (Rimon and Perlmann, 1968). Blockage of tyrosine residues is slowly reversed upon removal of the reagent, as indicated by recovery of the absorbance at 278 mμ lost during the blockage. Removal of O-carbamyl groups from tyrosine can be accomplished more rapidly and at a lower pH by treating with hydroxylamine (Rimon and Perlmann, 1968). At pH 5.6, treatment with 1 M hydroxylamine at 30°C for 180 minutes restored 80% of the lost catalytic activity of cyanate-treated pepsin, concomitant with the freeing of six O-carbamyltyrosine residues.

$$\underset{}{R-\overset{O}{\overset{\|}{C}}-O^-} + \underset{}{\overset{HOH\ \ H}{\overset{\diagdown\ /}{\underset{\|}{\underset{O}{N}}}}}C \underset{k_{-1}}{\overset{k_1}{\rightleftharpoons}} R-\overset{O}{\overset{\|}{C}}-O-C\overset{NH_2}{\underset{O}{\diagdown}} + OH^- \quad (9)$$

$$\text{P}-NH_2 + R-\overset{O}{\overset{\|}{C}}-O-C\overset{NH_2}{\underset{O}{\diagdown}} \longrightarrow \text{P}-NH-\overset{O}{\overset{\|}{C}}-R + HNCO + H_2O \quad (10)$$

Cyanate reacts with carboxyl groups to form mixed anhydrides of carbamic acid (Equation 9). Reaction takes place at low pH (i.e., near pH 5.3) and appears to be between the carboxyl anion and cyanic acid (Stark, 1965a). At low pH, cyanate appears to react with acetic acid to form carbamyl acetate. Like most acid anhydrides, these mixed carbonic acid anhydrides are highly reactive with many nucleophiles. Reaction with water results in regeneration of the carboxylic acid and cyanate. Amines may be acylated by such anhydrides (Stark, 1965a). Treatment of proteins or simple amines with cyanate in the presence of acetate or other carboxylic buffers can lead to the acylation of susceptible residues. Acylations of this type constitute the major evidence for anhydride formation (Stark, 1965a).

$$\underset{H}{\underset{N}{\boxed{}}}\!\!\!\!{-}N + HNCO + H_2O \underset{k_{-1}}{\overset{k_1}{\rightleftharpoons}} \underset{H}{\underset{N}{\boxed{}}}\!\!\!\!{-}\underset{+}{N}{-}\underset{\underset{O}{\parallel}}{C}{-}NH_2 + OH^- \qquad (11)$$

In aqueous solution, imidazolium cyanate is in equilibrium with carbamylimidazole (Equation 11). The equilibrium constant

$$K_{eq} = \frac{(ImCONH_2)}{(ImH^+)(NCO^-)}$$

at 25°C and pH 8 is approximately 0.3 mole^{-1} (Stark, 1965c). Unlike other acylimidazoles, carbamylimidazole is not an effective acylating (i.e., carbamylating) agent

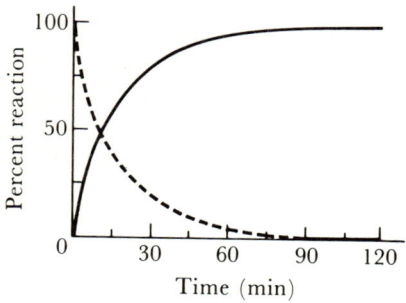

FIGURE 5-6 Rate of formation of tyrosine (solid curve) from O-carbamyltyrosine (dashed curve) at pH 8.0 and 30°C. (Adapted from Smyth, 1965.)

(Stark, 1965c). At neutral pH, imidazole does not alter the rate of carbamylation of amines by cyanate, nor does it catalyze decomposition of cyanate to carbon dioxide and ammonia. Decomposition of carbamylimidazole is catalyzed by acid but not base.

$$NH_2{-}\underset{\underset{O}{\parallel}}{C}{-}NH_2 + HCNO \overset{pH\ 6}{\rightleftharpoons} NH_2{-}\underset{\underset{O}{\parallel}}{C}{-}NH{-}\underset{\underset{O}{\parallel}}{C}{-}NH_2 \qquad (12)$$

Reaction of cyanate with urea or amides can take place at low pH to form biuret or, respectively, a substituted urea (Equation 12) (Smyth, 1965). At pH 6 and 30°C, urea in 0.4 M cyanate can be converted to biuret at a rate of 2 to 3% per hour. At lower pH's, more rapid conversion can be expected. Similar reaction with protein amide groups has not been demonstrated.

At equilibrium, the cyanate concentration in 8 M urea is approximately 0.02 M. Treatment of proteins with such can result in carbamylation of amino acid side chains, in addition to the physical disruption by urea. Freshly crystallized urea or acid-treated or deionized urea should be employed to avoid this problem (Stark et al., 1960). At low pH, cyanate is unstable and breaks down to carbon dioxide and ammonium ion (Lister, 1955; Stark et al., 1960). The presence of cyanate in urea may be determined by the colorimetric procedure of Werner (1923).

While ε-carbamylamino groups slowly revert to free amino groups in 6 N HCl at 110°C, α-carbamylamino acids cyclize to hydantoins (Equation 13). This sequence of reactions can be employed for the determination of NH_2-terminal amino acids (Stark and Smyth, 1963). As a method to determine NH_2-terminal residues the procedure is similar to the Edman sequential degradation procedure (Edman, 1956), but with a hydantoin rather than a phenylthiohydantoin being formed. It is not designed for sequential degradation but only for determination of the NH_2-terminal group.

$$\text{\tiny{WWW}}NH-\overset{O}{\underset{\|}{C}}-\overset{R}{\underset{|}{CH}}-NH_2 + HNCO \xrightarrow[\text{urea}]{pH\ 6}$$

$$\text{\tiny{WWW}}NH-\overset{O}{\underset{\|}{C}}-\overset{R}{\underset{|}{CH}}-NH-\underset{\underset{O}{\|}}{C}-NH_2 \xrightarrow[\text{1 hr, 100°C}]{6\ N\ HCl}$$

$$\text{\tiny{WWW}}NH_3^+ + \quad \underset{\underset{H}{N}}{\overset{R}{\underset{O}{\overset{CH-NH}{\underset{C}{\diagdown}}\diagup\underset{\|}{C}}}}\overset{}{\underset{O}{=}} \tag{13}$$

The hydantoin formed upon heating in 6 N HCl is separated from amino acids and peptides by ion exchange chromatography, hydrolyzed to the free amino acid, and the latter determined with an amino acid analyzer. Amino-terminal analysis of proteins by the cyanate procedure has recently been reviewed (Stark, 1967).

5-3 AMIDINATION

5-3.1 IMIDOESTERS

$$\boxed{P}-NH_2 + \underset{R'O}{\overset{H_2\overset{+}{N}}{\diagdown}}C-R \xrightarrow{pH>8.5} \boxed{P}-NH-\overset{\overset{+}{NH_2}}{\underset{\|}{C}}-R + R'OH \tag{1}$$

Imidoesters react in alkaline solution with amines to form imidoamides, so-called amidines (Equation 1). For protein modification, this reaction possesses many advantages. Reaction occurs at moderately alkaline pH's (i.e., between 7 and 10, depending upon the pK of the amine), in aqueous solvent, and at room temperature. Of the many reactive groups in proteins, only amino groups will react, and the products, like the amino groups they replace, are protonated at physiological pH. Both ε- and α-amino groups may be modified (Hunter and Ludwig, 1962).

$$\text{(reaction scheme)} \quad (2)$$

The reaction rate is strongly pH dependent (Hand and Jencks, 1962; Hunter and Ludwig, 1962). Kinetic data suggest the cationic form of the imidoester and the unprotonated amine to be the reactive species. The rate-determining step, depending upon the pH, is either addition of amine (free base) to the cationic imidoester (k_1) or breakdown of the tetrahedral orthoamide to amidine and alcohol (k_2) (see Equation 2). In moderately alkaline solution, the last step (k_2) is subject to general base catalysis. At higher pH, the overall second-order rate constants (corrected to include only unprotonated amine) increase with increasing nucleophilicity of the amine (Hunter and Ludwig, 1962). At pH 9.7, where both ε-aminocaproic acid and glycylglycine are largely unprotonated, the former reacts with methyl acetimidate approximately seven times faster than glycylglycine. At lower pH, the ε-amino groups are largely protonated, and α-amino groups react more rapidly than ε-amino groups (Hunter and Ludwig, 1962) (Figure 5-7).

Imidoesters are unstable and slowly hydrolyze in aqueous solution to the corresponding amide and alcohol. Between pH 7 and 11, the rate of hydrolysis is slow compared to the rate of their reaction with amines. Hydrolysis is not subject to general base catalysis in the moderately alkaline region and is not enhanced by the presence of protein.

With pK's above 11 [pK_a acetamidine = 12.5, pK_a benzamidine = 11.6 (Hunter and Ludwig, 1962)], amidino groups, like amino groups, are positively charged at neutral pH. For this reason, acetamidinated proteins have electrophoretic mobilities very much like the unmodified protein (Wofsy and Singer, 1963). Other physical properties, likewise, are usually found to differ relatively little from the

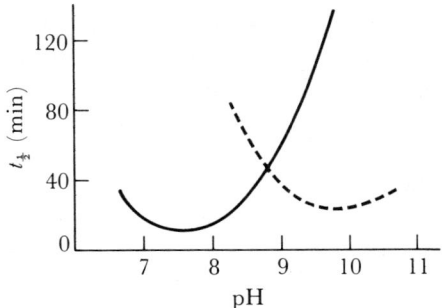

FIGURE 5-7   Half-times of reaction of glycylglycine (solid curve) and ε-aminocaproic acid (dashed curve) with methyl benzimidate at various pH's at 39°C. (Adapted from Hunter and Ludwig, 1962.)

native protein. Small but apparently significant changes have been observed, however, in the optical rotation, ultraviolet absorption, and electrophoresis patterns of very extensively amidinated bovine serum albumin (Wofsy and Singer, 1963).

methyl acetimidate
hydrochloride

methyl benzimidate
hydrochloride

Substitution of all the amino groups of the more stable protein bovine pancreatic ribonuclease by acetamidino groups has little effect on most of its physical and chemical properties. In spite of its resemblance to the unmodified protein, the product is catalytically inert. Nearly complete restoration of catalytic activity could be obtained by deacetamidination with pH 11.3 ammonium acetate (Reynolds, 1968).

Acetamidination of lima bean trypsin inhibitor, by maintaining the positive charge of the combining-site ε-ammonium group, reduced but did not abolish its trypsin-inhibitory activity. Similar treatment of several other lysine-type trypsin inhibitors eliminated trypsin-inhibitory activity (Haynes and Feeney, 1968).

Both acetamidino and benzamidino groups are quantitatively removed from α-amino groups under conditions normally employed for protein hydrolysis (6 N HCl, 110°C, 18 to 24 hours). ε-Amidino groups are only partially converted to ε-amino groups under the same conditions (Hunter and Ludwig, 1962). Quantitation of the extent of amidination cannot therefore be done simply by amino acid analysis of the acid-hydrolyzed protein. Quantitation can be accomplished, however, by treating the amidinated protein with dinitrofluorobenzene prior to hydrolysis and determining the amount of ε-dinitrophenyllysine present (Wofsy and Singer, 1963). Alternately, the amino groups remaining after modification can be determined using trinitrobenzenesulfonic acid (Haynes et al., 1967) or by formol titration (Hunter and Ludwig, 1962). Amidines give an orange-to-pink color upon treatment with nitroprusside reagent (Hunter and Ludwig, 1962).

Amidines are stable in neutral or acidic solution, but slowly hydrolyze at high pH. Removal of amidino groups and regeneration of the original amino groups can be carried out at pH 11.3 in concentrated ammonia–acetic acid (15:1) or at pH 9 using 0.6 to 1.2 M hydrazine (Ludwig and Byrne, 1962). Possible deleterious side reactions during such treatment have not been ruled out.

Trypsin does not catalyze the hydrolysis of ε-acetamidolysyl peptide bonds. Tryptic digestion of completely amidinated proteins, therefore, results in cleavage at only arginyl peptide bonds and hence in the formation of a proportionately smaller number of peptide products. Selective and limited cleavage of this sort permits easier isolation and purifications of the initial digestion products. Subsequent removal of the acetamidine groups and redigestion with trypsin greatly facilitates sequencing of the tryptic peptides.

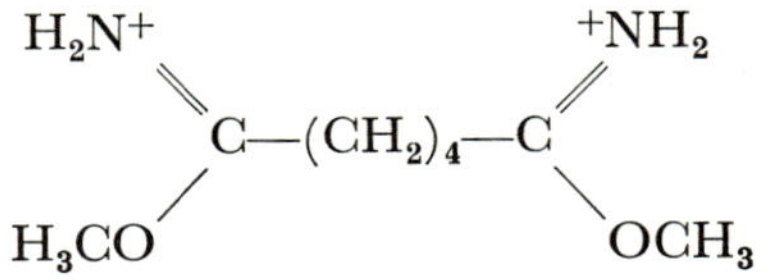

$$\underset{H_3CO}{H_2N^+}\!\!=\!\!C\!-\!(CH_2)_4\!-\!C\!\!=\!\!\underset{OCH_3}{^+NH_2}$$

dimethyl adipimidate

Dimethyl adipimidate, a bifunctional imidoester, has been used to form cross-linkages in several proteins (see Table 3-1). Its reaction with bovine pancreatic ribonuclease has been shown to give both inter- and intramolecular cross-linked products (Hartman and Wold, 1967). An intramolecularly cross-linked fraction has been shown to have catalytic activity 160% that of unmodified ribonuclease as assayed against cytidine 2′,3′-cyclic phosphate. Bifunctional imidoesters appear to be extremely useful as protein cross-linking agents and have been discussed in this regard by Hartman and Wold (1967) and Wold (1967).

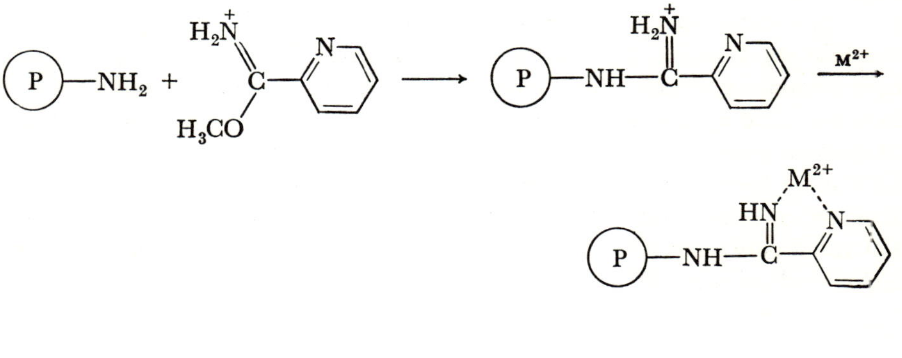

(3)

A possible approach to the preparation of heavy-atom-containing proteins has been described wherein methyl picolinimidate, after reaction with an amino group, is used as a covalently attached heavy-metal-chelating group (Benisek and Richards, 1968). The picolinamidine groups which are formed have been shown to strongly bind d-orbital transition-metal ions. If such derivatives can be induced to form crystals, they may be of use for X-ray crystallographic investigations.

5-3.2 Guanidination

$$\text{(P)}-NH_2 + \underset{RO}{\overset{H_2N^+}{\underset{\|}{C}}}-NH_2 \xrightarrow{pH > 9.5} \text{(P)}-NH-\overset{^+NH_2}{\underset{\|}{C}}-NH_2 + ROH \qquad (4)$$

Treatment of proteins with O-methylisourea or S-methylisothiourea allows one to selectively replace ε-amino groups of proteins with even more basic guanidino groups. Unfortunately, conditions necessary to carry out the reaction require a high pH, unsuitable for many proteins. Early workers (Schutte, 1943; Christensen, 1945; Roche and Mourgue, 1946; Roche et al., 1954) used S-methylisothiourea, but more recently O-methylisourea has been favored because of its greater reactivity (Hughes et al., 1949) and, perhaps in part due to strong odors associated with the former. No comprehensive study of the reaction mechanism has been published, but the pH dependence of the reaction indicates that at least one of the reacting species must be unprotonated (Hughes et al., 1949). At the required alkalinity, above pH 9.5, appreciable concentrations of protonated and unprotonated forms of both the reagent and ε-amino groups are present. It is unclear, therefore, which of these are the reactive pair, although a mechanism similar to that proposed for the reaction of amines with imidoesters (Equation 2) appears reasonable. The effect of pH upon the guanidination of ribonuclease with O-methylisourea is illustrated in Figure 5-8.

Guanidination of proteins proceeds in high yield but is applicable only to those which are stable in strongly alkaline solution. In several cases, near-complete modification of all ε-amino groups has been obtained (Chervenka and Wilcox, 1956; Hettinger and Harbury, 1965; Kassell and Chow, 1966). If stable under the conditions employed for the reaction, guanidination, like amidination, usually has little effect on the physical and chemical properties of proteins. Fully guanidinated tuna heart cytochrome c, for example, differs little from the unmodified enzyme. It maintains almost full activity in the succinate oxidase system, and its redox potential and visible spectrum are almost indistinguishable from those of unmodified cytochrome c.

Guanidination, in contrast to amidination, usually affects only ε-amino groups; α-amino groups are normally resistant. With insulin, however, O-methylisourea also reacts with the α-amino of the NH_2-terminal glycine residue (Evans and Saroff, 1957).

Treatment of β-lipoprotein with O-methylisourea has been shown to result in the modification of histidine residues (Margolis and Langdon, 1966). Histidine has not been found to react in other proteins and the reagent can usually be considered specific for ε-amino groups (Chervenka and Wilcox, 1956; Geshwind and Li, 1957).

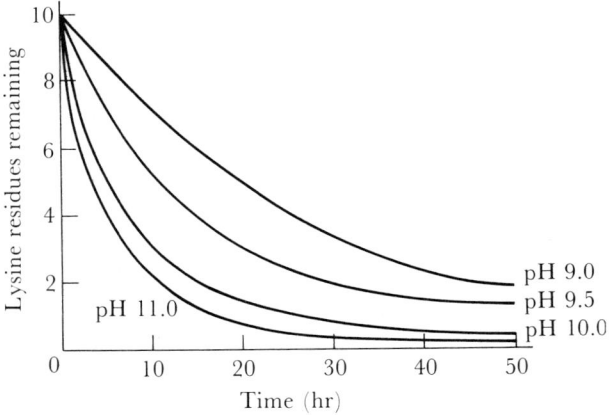

O-methylisourea   S-methylisothiourea   1-guanyl-3,5-dimethylpyrazole

Habeeb (1959, 1960) has investigated the reagent 1-guanyl-3,5-dimethylpyrazole nitrate for its ability to guanidinate protein amino groups and has found it

FIGURE 5-8 Reaction of amino groups with O-methylisourea in ribonuclease. Numbers on curves are pH values. (From Klee and Richards, 1957.)

to react at a lower pH than O-methylisourea (9.5 versus 10–11, see Figure 5-9). Like O-methylisourea, it appeared to modify only amino groups. However, both α- and ε-amino groups were modified (Habeeb, 1959; Fasold et al., 1961). This reagent has not been widely employed. It is available commercially.

Guanidino groups are strongly basic and, like amino groups, are protonated in the physiological pH range; charge relationships in guanidinated proteins are therefore very much like those in the native molecule. In some cases, it has thus been possible, upon guanidination, to distinguish with less ambiguity between simply the need to maintain a positive charge and the specific necessity of lysine residues for the expression of physiological activity (Geshwind and Li, 1957; Hettinger and Harbury, 1965; Kassell and Chow, 1966).

The product arising from guanidination of the ε-amino group of lysine is homoarginine. It is stable to acid hydrolysis and may be quantitated by amino acid analysis as a peak emerging from the amino acid analyzer shortly after arginine (Kassell and Chow, 1966; Shields et al., 1959). The extent of guanidination may also be determined by the Sakaguchi reaction (Chervenka and Wilcox, 1956; Sakaguchi, 1950), or subtractively by the determination of lysine (Fraenkel-Conrat et al., 1955; Moore and Stein, 1948; Van Slyke, 1929). Homoarginyl residues are not substrates for trypsin (Habeeb, 1960; Shields et al., 1959; Weil and Telka, 1957), but can be cleaved by papain (Shields et al., 1959).

5-3.3 RELATED REAGENTS

S-Methylglucosylisothiourea has been employed as a reagent to modify protein amino groups. The reagent is not readily available and has not been widely used. Its synthesis has been described by Maekawa and Ishimoto (1956). It appears to react with histidine in addition to amino groups (Maekawa and Ishimoto, 1961; Maekawa and Liener, 1960a). The glucosylguanidino group appears to be strongly basic. Reaction of S-methylglucosylisothiourea with trypsin takes place at pH 8 to 8.4 over a period of four days introducing six to seven glucosylguanidino groups per mole. The modified enzyme's properties were little different from the unmodified enzyme. Catalytic activity was not appreciably affected by this extent of modification (Maekawa and Liener, 1960b).

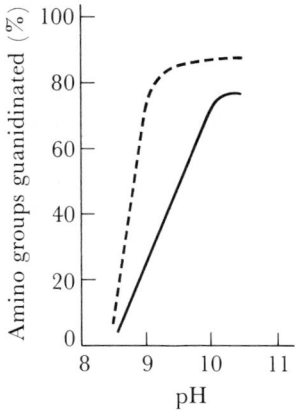

FIGURE 5-9 pH dependence for the reaction of O-methylisourea (solid curve) and 1-guanyl-3,5-dimethylpyrazole nitrate (dashed curve) with BSA. (From Habeeb, 1959.)

[Reaction scheme showing: P-NH₂ + S-methyl thiopseudourea sugar derivative → P-NH-C(=NH₂⁺)-NH-sugar + CH₃SH   (5)]

Nitroguanidino groups may also be introduced in place of protein amino groups by procedures analogous to those for guanidination. Saroff and Evans (1959) have used S-methyl-1-nitro-2-thiopseudourea for this purpose. They found it to react with protein amino groups, probably both α and ε, at pH 9.0 to give, after repeated treatments, upward to 70% modification. The extent of modification varied with the pH; higher pH's gave more extensive modification (see Figure 5-9). They reported that acid hydrolysis of nitroarginine led to a large number of products including arginine, ornithine, and argininic acid (Saroff and Evans, 1959).

S-methyl-1-nitro-2-thiopseudourea

1-nitroguanyl-3,5-dimethylpyrazole

Nitroguanidination of proteins has been studied by Habeeb (1964) using 1-nitroguanyl-3,5-dimethylpyrazole. At pH 9.5, 92% of the amino groups of BSA were modified without apparent effect upon other amino acids. The α-amino group was extensively substituted. Acid hydrolysis of nitroguanidinated bovine serum albumin resulted in a conversion of cystine to cysteic acid, and nitrohomoarginine to homoarginine.

Nitroguanidino groups absorb strongly between 251 and 267 mμ and, unlike the amino groups they replace, are electrically neutral at moderate pH. This is reflected by altered electrophoretic mobility and solubility properties of nitroguanidinated proteins.

5-4 SULFONYL HALIDES AND SULFONATES

Sulfonyl halides react with proteins in much the same way as acyl halides. They are less liable to nucleophilic attack and hence are more stable than are analogous carboxylates. Sulfonate esters and sulfonamides in the same way hydrolyze more slowly than the corresponding esters and amides of carboxylic acids. S-Sulfonylcysteine and imidazolesulfonylhistidine are stable in neutral solutions unlike the corresponding acyl compounds. Neither sulfonamides nor O-sulfonyltyrosine are greatly affected by hot acid under conditions that hydrolyze most peptide bonds.

$$\text{(P)}-NH_2 + ClSO_2-\text{[dansyl chloride]}-N(CH_3)_2 \xrightarrow{pH>7}$$

dansyl chloride

$$\text{(P)}-NH-SO_2-\text{[naphthalene]}-N(CH_3)_2 + HCl \quad (1)$$

dansyl conjugate

5-4.1 Sulfonyl Halides

The reactions of sulfonyl halides with proteins have not been systematically studied. Only sulfonyl chlorides and fluorides are commonly used, and of these the former are much more reactive. From its reaction with several amines, dansyl chloride (5-dimethylaminonaphthalenesulfonyl chloride) was originally thought to react only with the amino groups of bovine serum albumin (Weber, 1952). Subsequent work however, has shown that it also reacts with sulfhydryl, imidazole, and phenolic hydroxyl groups (Hartley and Massey, 1956). Its reaction with aliphatic hydroxyl groups is slow except in the special case of some serine proteases whose active-site serines are rapidly and sometimes almost specifically sulfonylated (Gold, 1965). Extinction coefficients of dansyl-protein conjugates (i.e., sulfonamides) are quite variable. A value of $3.4 \pm 0.4 \times 10^3$ M^{-1} cm^{-1} at 340 mμ appears to be the most common (Hartley and Massey, 1956; Chen, 1968).

The fluorescence of sulfonamides formed when dansyl chloride reacts with terminal amino groups of peptides and proteins enables one to detect them in very small amounts. They are stable in hot acid and permit dansyl chloride to be used for NH_2-terminal determinations as dinitrofluorobenzene is used. The intense fluorescence of the dansyl conjugates, however, gives nearly 100-fold greater sensitivity. As little as 0.1 to 0.01 nanomoles is sufficient for the identification of a single NH_2-terminal group (Gray and Hartley, 1963; Gray, 1967).

Covalent dansyl-protein conjugates have been used to study physical properties of proteins and other macromolecules whose natural fluorescence is weak or absent (Weber, 1952). Such conjugates can be used to detect and follow changes in conformation which lead to an increase or decrease in quenching and thereby reflect changes in the immediate molecular vicinity of the dansyl group (Frattali et al., 1965). By determining the depolarization of fluorescence of such derivatives, rotational relaxation times can be determined and used to estimate molecular weights and axial ratios (Weber, 1952; Johnson and Massey, 1957) (see Subsection 3-2.3).

The high reactivity of many sulfonyl halides with serine proteases is similar to that of diisopropylfluorophosphate with the same enzymes. Large hydrophobic sulfonyl halides are usually the most specific, but with acetylcholinesterase, smaller cationic derivatives work best (Kitz and Wilson, 1962; Alexander et al., 1963). Sulfonyl fluorides are normally much more specific than sulfonyl chlorides. In each case, inhibition is brought about by the sulfonylation of an active-site serine hydroxyl group. In this way, dansyl chloride rapidly inactivates α-chymotrypsin with little reaction occurring outside the active site (Gold, 1965). Phenylmethanesulfonyl fluoride is even more specific and appears to react exclusively with the active-site serine even when used in large excess (Gold and Fahrney, 1964).

$$\text{(serine protease)}-OH + FSO_2CH_2-C_6H_5 \longrightarrow$$

$$\text{(serine protease)}-OSO_2CH_2-C_6H_5 + HF \quad (2)$$

Inactivation of rennin by dansyl chloride results from specific reaction with one ε-amino group (Hill and Laing, 1967). Much more extensive modification of the enzyme with toluenesulfonyl chloride gives a similar inactivation.

5-4.2 Sulfonate Esters

$$X:\ \ \underset{H}{\overset{H}{C}}-OSO_2Ar \longrightarrow X-\underset{H}{\overset{H}{C}} + ArSO_3^- \quad (3)$$

$$\overset{OH^-}{\underset{H\ H}{\overset{H\ H}{C-C}}}_{OSO_2Ar} \longrightarrow \underset{H}{\overset{H}{C}}=\underset{H}{\overset{H}{C}} + ArSO_3^- + H_2O \quad (4)$$

The chemistry of sulfonate esters is quite different from that of carboxylate esters. Nucleophilic attack on a sulfonyl group is extremely difficult as compared to attack on a carboxyl group. With sulfonates it is the attached alkyl carbon which is most susceptible to nucleophilic attack (Equation 3). Cleavage of the carbon-oxygen bond of a weakly basic sulfonate ion, either via β-elimination (Equation 4) or simple S_N2 displacement (Equation 3), is much easier than are similar cleavages involving the release of a more basic carboxylate ion.

This lability of sulfonate esters has permitted the synthesis of many unusual enzyme derivatives.

Treating phenylmethanesulfonyl-chymotrypsin (i.e., PMS-chymotrypsin) with dilute alkali, for example, brings about, through β-elimination, the formation of a derivative in which the active-site serine residue has been converted to dehydro-alanine (Equation 5, **B**) (Strumeyer et al., 1963). This derivative, known as anhydro-chymotrypsin, has no detectable catalytic activity. Treatment of PMS-chymotrypsin with 2-mercaptoethylamine leads to the formation of another inactive derivative, in which the active-site serine residue is replaced by S-aminoethylcysteine (Equation 5, **C**) (Gold, 1965). Direct displacement of the sulfonate group in PMS-chymotrypsin by hydroxide ion, to regenerate active enzyme, has not been successful. Upon such treatment, elimination to give the anhydro derivative occurs instead (Equation 5, **B**). Brief treatment of PMS-chymotrypsin at pH 2, however, leads to the formation of an oxazoline derivative (Equation 5, **A**), which at pH 7 rapidly rearranges to give active enzyme (Gold and Fahrney, 1964). Treatment of PMS-chymotrypsin with hydrogen peroxide brings about the release of phenylmethanesulfonate ion and regeneration of proteolytic activity (Gibian et al., 1969). The enzyme obtained in this way is thought to have a hydroperoxy serine residue in place of the active-site serine of α-chymotrypsin.

$$\text{R SO}_2\text{—O—CH}_2\text{—subtilisin} + {}^-\text{S—C(=O)—CH}_3 \longrightarrow$$

$$\text{RSO}_3^- + \text{H—C(S—C(OH^-)(CH}_3\text{)(=O))—subtilisin} \longrightarrow \text{H—C(SH)—subtilisin} + \text{O=C(O^-)—CH}_3 \quad (6)$$

Treating PMS-subtilisin with thioacetate ion gives an inactive enzyme, wherein the sulfonate group is replaced by thioacetate (Equation 6). Thioacetyl-subtilisin rapidly deacylates at neutral pH to give a subtilisin analog, wherein the active-site serine residue has been converted to cysteine (Neet and Koshland, 1966; Polgar and Bender, 1967). Although strikingly like the native enzyme in most respects, this analog (with only a single atom replaced by a chemically similar atom) is catalytically inactive against typical subtilisin substrates. It retains activity, however, against substrates possessing a good leaving group like *p*-nitrophenylacetate. Kinetic analysis has shown the loss of catalytic activity to result from an inadequacy in the acylation step of the catalytic mechanism (Neet and Koshland, 1966; Polgar and Bender, 1967; Neet et al., 1968; Polgar and Bender, 1969). Extensive disulfide interchange reactions in alkaline solutions of thioacetate have prevented the application of this reaction sequence to several other enzymes. Subtilisin has no disulfide groups.

Phenylmethanesulfonyl fluoride at pH 7 reacts with and inactivates papain (Whitaker and Perez-Villasenor, 1968). Attempts to effect transformation of the enzyme's active-site thiol group to a hydroxyl group via this intermediate were unsuccessful. Phenylmethanesulfonyl-papain could be completely reactivated, however, by treatment with a thiol.

$$\text{papain}-S-SO_2-CH_2-\phenyl \xrightarrow{H^+ \; {}^-SR} \text{papain}-SH + R-S-SO_2-CH_2-\phenyl \quad (7)$$

Many proteins form noncovalent complexes with particular sulfonate ions. By examining the fluorescent properties of the reversibly bound probes, it is possible to gain some insight into the surface environment of a protein. Changes in the fluorescence of 2-toluidinylnaphthalene-6-sulfonate (TNS) bound to α-chymotrypsin have been used to detect changes in conformation which accompany denaturation, activation from zymogen, and interaction with various ligands (McClure and Edelman, 1967; Edelman and McClure, 1968). 1-Anilinonaphthalene-4-sulfonate (ANS) has been used in a similar fashion to study, for example, the binding properties of isolated immunoglobulin light chains (Yoo et al., 1967).

REFERENCES

Abadi, D. M., and P. E. Wilcox (1960): *J. Biol. Chem.*, **235**, 396.
Acher, R., J. Chauvet, R. Arnon, and M. Sela (1968): *Eur. J. Biochem.*, **3**, 476.
Alexander, J., I. B. Wilson, and R. Kitz (1963): *J. Biol. Chem.*, **238**, 741.
Anfinsen, C. B., M. Sela, and J. P. Cooke (1962): *J. Biol. Chem.*, **237**, 1825.
Atassi, M. Z. (1967): *Biochem. J.*, **102**, 488.
Balls, A. K., and H. N. Wood (1956): *J. Biol. Chem.*, **219**, 245.
Benesch, R., and R. E. Benesch (1956): *J. Am. Chem. Soc.*, **78**, 1597.
Benisek, W. F., and F. M. Richards (1968): *J. Biol. Chem.*, **243**, 4267.
Bethune, J. L., D. D. Ulmer, and B. L. Vallee (1964): *Biochemistry*, **3**, 1764.
Bezkorovainy, A., R. Zschocke, and D. Grohlich (1969): *Biochim. Biophys. Acta*, **181**, 295.
Braunitzer, V. G., K. Beyreuther, H. Fujiki, and B. Schrank (1968): *Hoppe-Seyl. Z. Physiol. Chem.*, **349**, 265.
Butler, P. J. G., J. I. Harris, B. S. Hartley, and R. Leberman (1967): *Biochem. J.*, **103**, 78P.
Butler, P. J. G., J. I. Harris, B. S. Hartley, and R. Leberman (1969): *Biochem. J.*, **112**, 679.
Chapman, A., T. Sanner, and A. Pihl (1969): *Eur. J. Biochim.*, **7**, 588.
Chen, R. F. (1968): *Anal. Biochem.*, **25**, 412.
Chervenka, C. H., and P. E. Wilcox (1956): *J. Biol. Chem.*, **222**, 635.
Chevallier, J., J. Yon, and J. Labouesse (1969): *Biochim. Biophys. Acta*, **181**, 73.
Christensen, H. N. (1945): *J. Biol. Chem.*, **160**, 75.
Cole, R. D. (1961): *J. Biol. Chem.*, **236**, 2670.
Cooke, J. P., C. B. Anfinsen, and M. Sela (1963): *J. Biol. Chem.*, **238**, 2034.
Davies, R. C., and A. Neuberger (1969): *Biochim. Biophys. Acta*, **178**, 306.
Dixon, G. H., and H. Neurath (1957): *J. Biol. Chem.*, **225**, 1049.

Dixon, H. B. F., and R. N. Perham (1968): *Biochem. J.*, **109**, 312.
Edelman, G. M., and W. O. McClure (1968): *Accounts Chem. Res.*, **1**, 65.
Edman, P. (1956): *Acta Chem. Scand.*, **10**, 761.
Evans, R. L., and H. A. Saroff (1957): *J. Biol. Chem.*, **228**, 295.
Fanger, M. W., and H. A. Harbury (1965): *Biochemistry*, **4**, 2541.
Fasold, H., F. Turba, and W. Wirsching (1961): *Biochem. Z.*, **335**, 86.
Fraenkel-Conrat, H. (1959): *Methods Enzymol.*, **4**, 247.
Fraenkel-Conrat, H., and M. Colloms (1967): *Biochemistry*, **6**, 2740.
Fraenkel-Conrat, H., J. I. Harris, and A. L. Levy (1955): *Methods of Biochemical Analysis*, edited by D. Glick, vol. 2, Academic Press, New York, p. 359.
Frattali, V., R. F. Steiner, and H. Edelhoch (1965): *J. Biol. Chem.*, **240**, 112.
Freedman, M. H., A. L. Grossberg, and D. Pressman (1968): *Biochemistry*, **7**, 1941.
Freedman, M. H., and M. Sela (1966): *J. Biol. Chem.*, **241**, 5225.
Freisheim, J. H., K. A. Walsh, and H. Neurath (1967): *Biochemistry*, **6**, 3010.
Fridovich, I., and F. H. Westheimer (1962): *J. Am. Chem. Soc.*, **84**, 3208.
Geshwind, I. I., and C. H. Li (1957): *Biochim. Biophys. Acta*, **25**, 171.
Gibian, M. J., D. L. Elliott, and W. R. Hardy (1969): *J. Am. Chem. Soc.*, **91**, 7528.
Glazer, A. N., A. Bar-Eli, and E. Katchalski (1962): *J. Biol. Chem.*, **237**, 1832.
Gold, A. M. (1965): *Biochemistry*, **4**, 897.
Gold, A. M., and D. Fahrney (1964): *Biochemistry*, **3**, 783.
Goldberger, R. F. (1967): *Methods Enzymol.*, **11**, 317.
Goldberger, R. F., and C. B. Anfinsen (1962): *Biochemistry*, **1**, 401.
Gounaris, A. D., and G. E. Perlmann (1967): *J. Biol. Chem.*, **242**, 2739.
Gray, W. R. (1967): *Methods Enzymol.*, **11**, 469.
Gray, W. R., and B. S. Hartley (1963): *Biochem. J.*, **89**, 59P.
Guldalian, J., W. B. Lawson, and R. K. Brown (1965): *J. Biol. Chem.*, **240**, PC2757.
Habeeb, A. F. S. A. (1959): *Biochim. Biophys. Acta*, **34**, 294.
Habeeb, A. F. S. A. (1960): *Can. J. Biochem. Physiol.*, **38**, 493.
Habeeb, A. F. S. A. (1964): *Biochim. Biophys. Acta*, **93**, 533.
Habeeb, A. F. S. A. (1966): *Anal. Biochem.*, **14**, 328.
Habeeb, A. F. S. A. (1967): *Arch. Biochem. Biophys.*, **121**, 652.
Habeeb, A. F. S. A., H. G. Cassidy, and S. J. Singer (1958): *Biochim. Biophys. Acta*, **29**, 587.
Hand, E. S., and W. P. Jencks (1962): *J. Am. Chem. Soc.*, **84**, 3505.
Hartley, B. S., and V. Massey (1956): *Biochim. Biophys. Acta*, **21**, 58.
Hartman, F. C., and F. Wold (1967): *Biochemistry*, **6**, 2439.
Haynes, R., and R. E. Feeney (1968): *Biochemistry*, **7**, 2879.
Haynes, R., D. T. Osuga, and R. E. Feeney (1967): *Biochemistry*, **6**, 541.
Hettinger, T. P., and H. A. Harbury (1965): *Biochemistry*, **4**, 2585.
Hill, R. D., and R. R. Laing (1967): *Biochim. Biophys. Acta*, **132**, 188.
Hollands, T. R., and J. S. Fruton (1968): *Biochemistry*, **7**, 2045.
Hughes, W. L., H. A. Saroff, and A. L. Carney (1949): *J. Am. Chem. Soc.*, **71**, 2476.
Hunter, M. J., and M. L. Ludwig (1962): *J. Am. Chem. Soc.*, **84**, 3491.
Jencks, W. P., and J. Carriuolo (1959): *J. Biol. Chem.*, **234**, 1272.
Johncock, P., G. Kohnstam, and D. Speight (1958): *J. Chem. Soc. (London)*, 2544.
Johnson, P., and V. Massey (1957): *Biochim. Biophys. Acta*, **23**, 544.
Kabat, E. A., and M. M. Mayer (1961): *Experimental Immunochemistry*, 2d ed., Thomas, Springfield, Ill.
Kassell, B., and R. Chow (1966): *Biochemistry*, **5**, 3449.
Katchalski, E., M. Sela, H. I. Silman, and A. Berger (1964): in *Proteins: Composition, Structure, and Function*, edited by H. Neurath, vol. 2, 2d ed., Academic Press, New York, p. 405.

Kitz, R., and I. B. Wilson (1962): *J. Biol. Chem.*, **237,** 3245.
Klee, W. A., and F. M. Richards (1957): *J. Biol. Chem.*, **229,** 489.
Klotz, I. M., and R. E. Heiney (1962): *Arch. Biochem. Biophys.*, **96,** 605.
Klotz, I. M., and S. Keresztes-Nagy (1962): *Nature*, **195,** 900.
Koltun, W. L., L. Ng, and F. R. N. Gurd (1963): *J. Biol. Chem.*, **238,** 1367.
Komatsu, S. K., and R. E. Feeney (1967): *Biochemistry*, **6,** 1136.
Labouesse, J., and M. Gervais (1967): *Eur. J. Biochem.*, **2,** 215.
Lee, M. J., A. J. Vithayathil, and F. F. Buck (1963): *Arch. Biochem. Biophys.*, **100,** 150.
Levy, D., and F. H. Carpenter (1967): *Biochemistry*, **6,** 3559.
Lindsay, D. G., and S. Shall (1969): *Biochem. J.*, **115,** 587.
Lister, M. W. (1955): *Can. J. Chem.*, **33,** 426.
Ludwig, M. L., and R. Byrne (1962): *J. Am. Chem. Soc.*, **84,** 4160.
Maekawa, K., and K. Ishimoto (1956): *Bull. Chem. Soc. (Japan)*, **77,** 999.
Maekawa, K., and K. Ishimoto (1961): *Bull. Chem. Soc. (Japan)*, **34,** 1221.
Maekawa, K., and I. E. Liener (1960a): *Arch. Biochem. Biophys.*, **91,** 101.
Maekawa, K., and I. E. Liener (1960b): *Arch. Biochem. Biophys.*, **91,** 108.
Margolis, S., and R. G. Langdon (1966): *J. Biol. Chem.*, **241,** 477.
Marzotto, A. (1969): *Experientia*, **25,** 1016.
Marzotto, A., P. Pajetta, and E. Scoffone (1967): *Biochem. Biophys. Res. Commun.*, **26,** 517.
Marzotto, A., P. Pajetta, L. Galzigna, and E. Scoffone (1968): *Biochim. Biophys. Acta*, **154,** 450.
Mathew, E., B. P. Meriwether, and J. H. Park (1967): *J. Biol. Chem.*, **242,** 5024.
McClure, W. O., and G. M. Edelman (1967): *Biochemistry*, **6,** 567.
Meighen, E. A., and H. K. Schachman (1970): *Biochemistry*, **9,** 1163.
Meighen, E. A., V. Pigiet, and H. K. Schachman (1970): *Proc. Nat. Acad. Sci.*, **65,** 234.
Melchior, W. B., and D. Fahrney (1970): *Biochemistry*, **9,** 251.
Moore, S., and W. H. Stein (1948): *J. Biol. Chem.*, **176,** 367.
Mühlrad, A., G. Hegyi, and M. Horanyi (1969): *Biochim. Biophys. Acta*, **181,** 184.
Neet, K. E., and D. E. Koshland (1966): *Proc. Nat. Acad. Sci.*, **56,** 1606.
Neet, K. E., A. Nanci, and D. E. Koshland (1968): *J. Biol. Chem.*, **243,** 6392.
Nishikawa, A. H., R. Y. Morita, and R. R. Becker (1968): *Biochemistry*, **7,** 1506.
O'Leary, M. H., and F. H. Westheimer (1968): *Biochemistry*, **7,** 913.
Oppenheimer, H. L., B. Labouesse, and G. P. Hess (1966): *J. Biol. Chem.*, **241,** 2720.
Patchornik, A., and Y. Shalitin (1961): *Anal. Chem.*, **33,** 1887.
Perham, R. N., and G. M. T. Jones (1967): *Eur. J. Biochem.*, **2,** 84.
Polgar, L., and M. L. Bender (1967): *Biochemistry*, **6,** 610.
Polgar, L., and M. L. Bender (1969): *Biochemistry*, **8,** 136.
Pradel, L. A., and R. Kassab (1968): *Biochim. Biophys. Acta*, **167,** 317.
Pugh, E. L., and B. L. Horecker (1967): *Arch. Biochem. Biophys.*, **122,** 196.
Reynolds, J. H. (1968): *Biochemistry*, **7,** 3131.
Rimon, S., and G. E. Perlmann (1968): *J. Biol. Chem.*, **243,** 3566.
Riordan, J. F., and B. L. Vallee (1963): *Biochemistry*, **2,** 1460.
Riordan, J. F., and B. L. Vallee (1964): *Biochemistry*, **3,** 1768.
Riordan, J. F., W. E. C. Wacker, and B. L. Vallee (1965): *Biochemistry*, **4,** 1758.
Roche, J., and M. Mourgue (1946): *Compt. Rend. Acad. Sci. (Paris)*, **222,** 1142.
Roche, J., M. Mourgue, and R. Baret (1954): *Bull. Soc. Chim. Biol.*, **36,** 85.
Rosén, C. G., and I. Fedorcsák (1966): *Biochim. Biophys. Acta*, **130,** 401.
Sakaguchi, S. (1950): *J. Biochem. (Tokyo)*, **37,** 231.
Saroff, H. A., and R. L. Evans (1959): *Biochim. Biophys. Acta*, **36,** 511.
Schallenberg, E. E., and M. Calvin (1955): *J. Am. Chem. Soc.*, **77,** 2779.
Schmid, A., P. Christen, and F. Leuthardt (1966): *Helv. Chim. Acta*, **49,** 281.

Schutte, E. (1943): *Hoppe-Seyl. Z. Physiol. Chem.*, **279,** 59.
Sela, M., and R. Arnon (1967): *Methods Enzymol.*, **11,** 580.
Shaw, D. C., W. H. Stein, and S. Moore (1964): *J. Biol. Chem.*, **239,** PC671.
Shields, G. S., R. L. Hill, and E. L. Smith (1959): *J. Biol. Chem.*, **234,** 1747.
Sia, C. L., and B. L. Horecker (1968): *Biochem. Biophys. Res. Commun.*, **31,** 731.
Simpson, R. T., J. F. Riordan, and B. L. Vallee (1963): *Biochemistry*, **2,** 616.
Sluyterman, L. A. E. (1967): *Biochim. Biophys. Acta*, **139,** 439.
Smyth, D. G. (1965): *Acta Chim. Hung.*, **44,** 197.
Smyth, D. G. (1967): *J. Biol. Chem.*, **242,** 1592.
Stark, G. R. (1964): *J. Biol. Chem.*, **239,** 1411.
Stark, G. R. (1965a): *Biochemistry*, **4,** 1030.
Stark, G. R. (1965b): *Biochemistry*, **4,** 2363.
Stark, G. R. (1965c): *Biochemistry*, **4,** 588.
Stark, G. R. (1967): *Methods Enzymol.*, **11,** 125.
Stark, G. R., and D. G. Smyth (1963): *J. Biol. Chem.*, **238,** 214.
Stark, G. R., W. H. Stein, and S. Moore (1960): *J. Biol. Chem.*, **235,** 3177.
Strumeyer, D. H., W. N. White, and D. E. Koshland (1963): *Proc. Nat. Acad. Sci.*, **50,** 931.
Terminiello, L., M. Bier, and F. F. Nord (1958): *Arch. Biochem. Biophys.*, **73,** 171.
Uraki, Z., L. Terminiello, M. Bier, and F. F. Nord (1957): *Arch. Biochem. Biophys.*, **69,** 644.
Uyeda, K. (1969): *Biochemistry*, **8,** 2366.
Van Slyke, D. D. (1929): *J. Biol. Chem.*, **83,** 425.
Vithayathil, P. J., and F. M. Richards (1960): *J. Biol. Chem.*, **235,** 1029.
Ward, D. N., J. Coffey, D. B. Ray, and W. M. Lamkin (1966): *Anal. Biochem.*, **14,** 243.
Warren, S., B. Zerner, and F. H. Westheimer (1966): *Biochemistry*, **5,** 817.
Weber, G. (1952): *Biochem. J.*, **51,** 155.
Weil, L., and M. Telka (1957): *Arch. Biochem. Biophys.*, **71,** 473.
Werner, E. A. (1923): *J. Chem. Soc. (London)*, 2577.
Whitaker, J. R., and J. Perez-Villasenor (1968): *Arch. Biochem. Biophys.*, **124,** 70.
Wofsy, L., and S. J. Singer (1963): *Biochemistry*, **2,** 104.
Wold, F. (1967): *Methods Enzymol.*, **11,** 617.
Yamasaki, N., K. Hayashi, and M. Funatsu (1968): *Agr. Biol. Chem. (Tokyo)*, **32,** 64.
Yoo, T. J., O. A. Roholt, and D. Pressman (1967): *Science*, **157,** 707.

6
alkylating and similar reagents

6-1 HALOACETATES

$$\text{P}-S^- + ICH_2COO^- \xrightarrow{pH \geq 7} \text{P}-S-CH_2COO^- + I^- \qquad (1)$$

$$\text{P}-\underset{\underset{H}{N}}{\overset{N}{\diagup\!\!\!\diagdown}} + ICH_2COO^- \xrightarrow{pH > 5.5} \text{P}-\underset{N}{\overset{N-CH_2COO^-}{\diagup\!\!\!\diagdown}} + I^- + H^+ \qquad (2)$$

$$\text{P}-S{-}CH_3 + ICH_2COO^- \xrightarrow{pH\ 2-8.5} \text{P}-S^+(CH_2COO^-)(CH_3) + I^- \qquad (3)$$

$$\text{P}-NH_2 + ICH_2COO^- \xrightarrow{pH > 8.5} \text{P}-NH-CH_2COO^- + I^- + H^+ \qquad (4)$$

Thunberg (1911) suggested that the toxicity of certain haloacids was a result of their reaction with tissue sulfhydryl groups. Iodoacetic, bromoacetic, and chloroacetic acids are the most widely used protein alkylating agents. Under various conditions, they have been shown to react with sulfhydryl, imidazole, thioether, and amino groups of proteins. With rather harsh treatment, phenolic side chains may also react (Korman and Clarke, 1956). The reaction of a carboxyl group of ribonuclease T1 with iodoacetate has been attributed to the unusual reactivity of that particular group (Takahashi et al., 1967). A review of haloacetates and their reactions with proteins has been published by Gurd (1967).

Nucleophilic substitution reactions of alkyl halides are among the most studied reactions in organic chemistry. Such reactions occur by two distinct mechanistic pathways. An α-carbonyl group, like that of haloacetates, enhances reactivity by direct displacement or what is known as the bimolecular mechanism. The reactivity of these compounds depends on the halide (i.e., I > Br > Cl ≫ F) such that iodo- and bromoacetates are highly useful protein reagents, chloroacetate is less reactive, and fluoroacetate is quite unreactive and of little value for modifying proteins. Haloacetamides are often slightly more reactive with proteins than are the corresponding acids and are used more or less interchangeably with them. The resulting different ionic character of the products formed may or may not be advantageous insofar as such modifications affect the properties of a particular protein. In some instances, the negative charge of haloacetates appears to enhance reactivity with a particular group by directing it to a positively charged center. Such an effect has, for example, been observed with pancreatic ribonuclease and will be discussed later. In most cases, and for most of the following discussion, the two kinds of reagent give equivalent results.

Rates of carboxymethylation may be determined from the rate of halide ion release (Stark and Stein, 1964) or from the release of protons (Gundlach et al., 1959). Halide ions are formed in each of the reactions shown above, and hydrogen ions in all but one.* Reaction with water is an important side reaction which decreases the effective amount of reagent and contributes a background release of both halide and hydrogen ions.

Of the several nucleophilic groups commonly found in proteins which are known to react with haloacetates, sulfhydryl groups are intrinsically the most reactive. Reactivity increases with pH since the anion is the reactive species. To prevent unwanted reactions with amino groups, however, unnecessarily high pH's should be avoided. Most often, a pH near neutrality under conditions compatible with physiological activity is best as a first try to determine the effects of haloacetate modification of sulfhydryl groups.

Under denaturing conditions, reactivities of a protein's sulfhydryl groups are similar to those of low-molecular-weight sulfhydryl compounds. In the absence of such conditions, considerable variations in reactivity are observed, reflecting the varied environments of the reacting groups. Neighboring groups can either enhance or suppress the reactivity of a group. Greatly enhanced reactivity is often observed, for example, for groups in or near catalytic centers. Such a sulfhydryl group in streptococcal proteinase, for example, is 50 to 100 times more reactive than that of glutathione (Gerwin, 1967). A similarly activated sulfhydryl group has been observed in papain (Sluyterman, 1968). Reactivities of sulfhydryl groups of proteins can sometimes be either enhanced or suppressed by substrates, allosteric effectors, and other specifically interacting substances. For example, the active-center sulfhydryl

---

* No release of hydrogen ions is obtained upon reaction with methionine side chains. At the pH's normally employed, at least a fractional release of hydrogen ions is obtained with each of the other groups (Equations 1, 2, and 4).

groups in both ficin and papain, in the presence of certain substrates, experience a several-fold stimulation of their reactivity with haloacetates (Whitaker, 1969). Conversely, the coenzyme nicotinamide adenine dinucleotide protects the two reactive sulfhydryl groups of horse liver alcohol dehydrogenase from iodoacetate and, in so doing, prevents loss of catalytic activity (Li and Vallee, 1965).

S-Carboxymethylcysteine is stable under conventional acid-hydrolysis conditions, so long as oxygen is carefully excluded, and can be quantitated by amino acid analysis; it elutes very rapidly, usually before aspartic acid. Carboxymethylation with iodoacetate or carboxyamidation with iodoacetamide, following reaction with excess thiol reducing agent (see Section 8-1), is frequently used to convert all cysteine into stable derivatives for various analytical purposes. Such reactions occur rapidly under mild conditions (i.e., $\sim$pH 8). To prevent reaction with other nucleophilic groups, only the stoichiometric amount, or a slight excess, of alkylating agent should be used. It is often convenient to perform the reaction without removing the reducing agent, in which case sufficient reagent for the total thiol should be employed.

Compared to their reaction with sulfhydryl groups, haloacetates react with imidazole groups of proteins very slowly. At pH 5.5 and slightly above, where imidazole groups are at least partially unprotonated, their modification is sufficiently fast to be the principal reaction in those proteins not having sulfhydryl groups. In several enzymes, histidine residues have been shown to have unusually high reactivities, considerably more reactive than even the imidazole group of histidine itself. In pancreatic ribonuclease, for example, two active-center imidazole groups are several orders of magnitude more reactive with bromoacetate than is histidine under similar conditions (Heinrikson et al., 1965). Optimum reaction with this enzyme occurs near pH 5.5, below the optimum pH for reaction with histidine, but where the anion affinity of the enzyme is greatest. High reactivity in this case has been attributed to the affinity of this site for the haloacetate anion.

With both swine heart fumarase (Bradshaw et al., 1969) and human carbonic anhydrase B (Bradbury, 1969), resemblance of the haloacetate to the enzyme's substrate (X—COO^-) appears to account for the observed high reactivity of histidine. The reaction of iodoacetamide with trypsin is enhanced in the presence of methylguanidine (Inagami, 1965; Inagami and Hatano, 1969). Its reaction with a histidine residue appears to require binding of methylguanidine in the substrate recognition site of the enzyme, such that an overall structure resembling that of the affinity-labeling reagent TLCK is achieved (see Chapter 2).

Reaction with imidazole groups can proceed in two steps to a 1,3-dicarboxymethyl derivative (Equation 5). It and each of the two monocarboxymethyl derivatives are stable compounds which can be isolated after acid hydrolysis and quantitated by conventional amino acid analysis (Gundlach et al., 1959; Stark and Stein, 1964). 1-Carboxymethylhistidine, upon amino acid analysis, elutes near proline, while 3-carboxymethylhistidine elutes nearer cystine.

Haloacetates react with methionyl residues of proteins over a wide pH range (Gundlach et al., 1959). At low pH, where other reactive groups are protonated

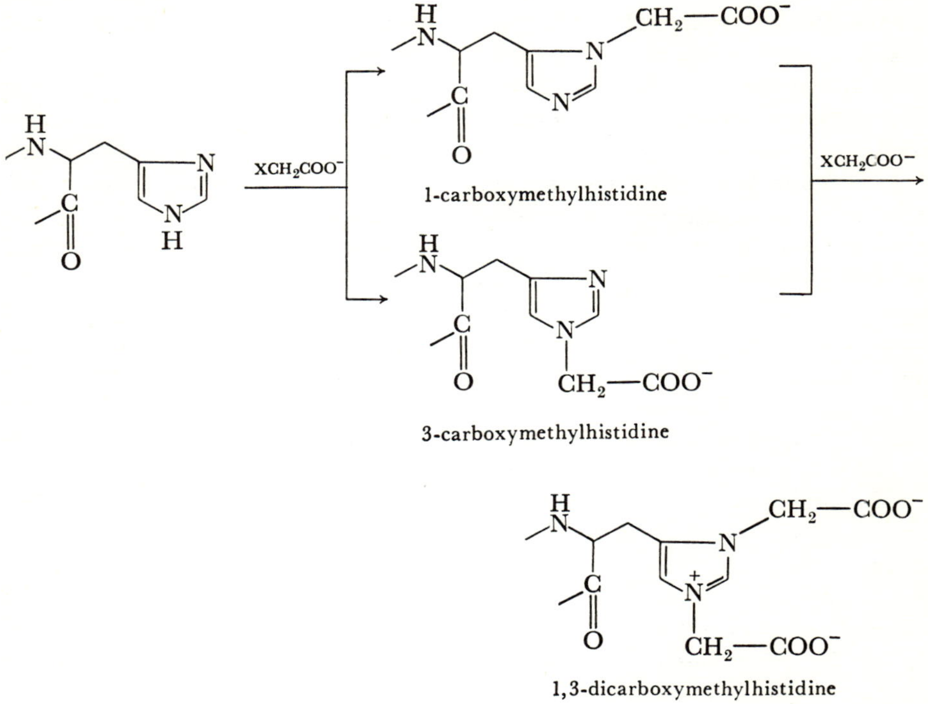

$$[\text{structures shown: starting histidine} \xrightarrow{XCH_2COO^-} \text{1-carboxymethylhistidine and 3-carboxymethylhistidine} \xrightarrow{XCH_2COO^-} \text{1,3-dicarboxymethylhistidine}] \quad (5)$$

(i.e., pH 2 to 3), haloacetate treatment may be relatively specific for methionine. Methionine is often situated in the hydrophobic interior of proteins, and some disruption of the tertiary structure is often a prerequisite for rapid reaction. In such cases, urea, guanidine hydrochloride, and other common protein denaturants are necessary to increase the rate of reaction. In some cases, a low pH will render them accessible. The product of the reaction, an S-carboxymethylmethionylsulfonium cation, is destroyed during acid hydrolysis, but its three decomposition products S-carboxymethylhomocysteine, homoserine lactone, and methionine are all measurable by amino acid analysis (Equation 6) (Gundlach et al., 1959). Treatment of the carboxymethylated protein with performic acid prior to its hydrolysis can be used to distinguish between that methionine which is regenerated and that remaining after carboxymethylation (Vithayathil and Richards, 1960). S-Carboxymethylmethioninesulfonium is not affected by this treatment; the methionine content of the modified protein is therefore obtained from the amount of methionine sulfone detected by amino acid analysis, and the observed methionine is attributed to the breakdown of its sulfonium derivative.

$$\underset{\substack{\text{S-carboxymethylmethionine-}\\\text{sulfonium}}}{\begin{array}{c}CH_3\diagdown\;\;\diagup CH_2COO^-\\ S_+\\ |\\ CH_2\\ |\\ CH_2\\ |\\ CHNH-\\ |\\ C=O\\ /\end{array}}\xrightarrow[H_2O]{H^+,\,\Delta}\begin{array}{c}CH_3\\ |\\ S\\ |\\ CH_2\\ |\\ CH_2\\ |\\ CHNH_3^+\\ |\\ COO^-\\ +\\ HOCH_2COO^-\end{array}+\begin{array}{c}CH_2COO^-\\ |\\ S\\ |\\ CH_2\\ |\\ CH_2\\ |\\ CHNH_3^+\\ |\\ COO^-\\ +\\ CH_3OH\end{array}+\begin{array}{c}CH_2\!-\!CH_2\\ |\quad\quad\;\;|\\ O\quad\;\; NH_3^+\\ \diagdown\diagup\\ C\\ \|\\ O\\ +\\ CH_3SCH_2COO^-\end{array}\quad(6)$$

Iodoacetate has been used to modify methionyl residues of pancreatic ribonuclease (Gundlach et al., 1959; Neumann et al., 1962; Goren and Barnard, 1970), trypsinogen and trypsin (Holeysovsky and Lazdunski, 1968), horse heart cytochrome c (Tsai and Williams, 1965), and isocitrate dehydrogenase (Colman, 1969). Its reaction and that of bromoacetate with pancreatic ribonuclease have been studied as a function of pH. At low pH, unfolding of the protein permits all four methionine groups to react irreversibly, destroying its catalytic activity (Gundlach et al., 1959; Neumann et al., 1962). Between pH 3 and 7, a single methionine reacts without decreasing catalytic activity. Below pH 6, iodoacetate reacts specifically with one methionine residue of isocitrate dehydrogenase to give a derivative having its dehydrogenase activity lowered more than its oxalosuccinate decarboxylase activity (Colman, 1968, 1969). The modified enzyme is still capable of binding substrate as strongly as the unmodified enzyme.

Amino groups of proteins react with haloacetates at high pH's, where they are unprotonated, but still at a rate less than 1/100 as fast as sulfhydryl groups. The reaction can proceed in two steps to dialkylated derivatives (Equation 7). Both mono- and dialkyl derivatives are stable during, and can be isolated and quantitated after, acid hydrolysis. Reaction with amino groups is most often observed as a side reaction accompanying extensive modification of sulfhydryl groups. Prevention of this side reaction depends upon using as low pH as is compatible with the desired reaction.

$$\boxed{P}\!-\!NH_2 + ICH_2COO^- \longrightarrow \boxed{P}\!-\!NHCH_2COO^- \xrightarrow{ICH_2COO^-}$$
$$+ HI$$

$$\boxed{P}\!-\!N\!\!\begin{array}{c}\diagup CH_2COO^-\\ \diagdown CH_2COO^-\end{array} + HI \quad (7)$$

At pH 8.5 and above, a single ε-amino group becomes the most reactive group with haloacetates in pancreatic ribonuclease (Heinrikson, 1966). Its reaction eliminates catalytic activity. The reaction of haloacetates with amino groups of most proteins is too slow and nonspecific to be of great value.

Other Alkyl Halides. In addition to haloacetates and haloacetamides, many other alkyl halides are extremely useful for protein-modification studies. They vary in complexity from methyl iodide (Link and Stark, 1968) to compounds like 2-bromoacetamido-4-nitrophenol (Burr and Koshland, 1964; Kirtley and Koshland, 1967) and tosyllysinechloromethyl ketone (TLCK) (Schoellmann and Shaw, 1962; Shaw, 1967). Many possess either the same reactive structure

$$-\underset{\underset{O}{\|}}{C}-CH_2-X$$

as haloacetates or an analogous reactive structure. Because such groups, like haloacetates, possess the ability to react with many protein groups, they have been incorporated into many special-purpose reagents, such as those for affinity labeling (see Table 2-1), for introducing reporter groups (see Subsection 3-2.3), and for other purposes. 2-Hydroxy-5-nitrobenzyl bromide has been shown to possess special reactivity with tryptophan residues, a characteristic which sets it apart from other alkyl halides; it is discussed in a separate section (see Section 6-6). The use of benzyl halides, including 2-hydroxy-5-nitrobenzyl bromide and 2-methoxy-5-nitrobenzyl bromide, as reagents for protein modification has been reviewed by Horton and Koshland (1967). The use of 2-bromoacetamido-4-nitrophenol for modification of proteins has been reviewed by Kirtley and Koshland (1967).

6-2 MALEIMIDES

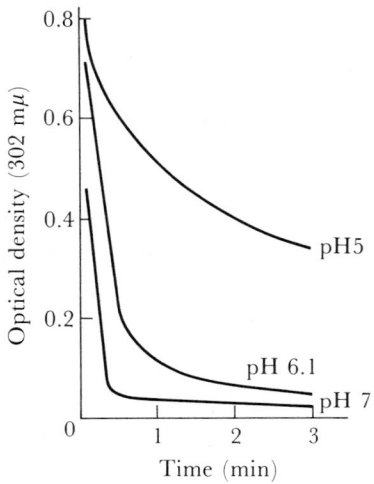

FIGURE 6-1 Rate of reaction of NEM with glutathione (GSH), both $10^{-3}\,M$ in $0.1\,M$ sodium acetate, pH 5.0; sodium phosphate, pH 6.1 and pH 7.0. (Adapted from Gregory, 1955.)

N-Ethylmaleimide (NEM) is a widely used sulfhydryl reagent. It has been used to determine the effect of sulfhydryl-group modification and to determine the number of sulfhydryl groups present in a wide variety of proteins. Its reaction with amino groups is also now widely recognized.

NEM was shown by Friedman et al. (1949) to be a powerful antimitotic agent, whose activity correlated with the ability to react with sulfhydryl groups. Maleic acid, maleimide, and citraconimide all had lower antimitotic activity and lower reactivity toward sulfhydryl groups. Succinimide, the fully saturated analog of maleimide, was without antimitotic activity. The reaction of maleimides with sulfhydryl groups involves addition of mercaptide ion to the olefinic double bond, similar to the Michael reaction. Reaction with amino groups occurs more slowly by the same general mechanism. Products of both reactions have been isolated and identified (Equations 1 and 2) (Marrian, 1949; Smyth et al., 1960).

NEM has a broad absorption band around 300 mμ ($\varepsilon = 620\,M^{-1}\,cm^{-1}$), which can be used to follow the NEM concentration during reactions and as the basis for determining sulfhydryl contents. Quantitative procedures have been used to detect as little as 0.1 mM thiol and are applicable to tissue extracts as well as to purified protein solutions (Gregory, 1955; Alexander, 1958; Roberts and Rouser, 1958). Benesch et al. (1956) observed a reddish color when the products from reaction of NEM with thiols were treated with alkali; they used this as the basis for a rather sensitive test for such compounds on paper chromatograms. The specificity of this test, however, has not been clearly defined. The same test has been shown to give a dark red color when it is applied to imidazole (Smyth et al., 1960).

Since mercaptide ion is the most reactive species, the reaction experiences a marked enhancement with increasing pH. NEM is a highly specific reagent for protein sulfhydryl groups below pH 7. At pH 7, its reaction rate with simple thiols is approximately 1000-fold greater than that with corresponding simple amino compounds. At higher pH's, reaction with amino groups becomes increasingly more significant. For proteins having no sulfhydryl groups, NEM can be used as a relatively specific reagent for amino-group modification (Table 6-1).

TABLE 6-1 Lysine, histidine, ethylamine, and ε-N-(2-succinyl)lysine content of ribonuclease A after treatment with 0.1 M NEM at the indicated pH's for 4 hours at 25°C, found after acid hydrolysis (from Brewer and Riehm, 1967)

	Residues (*moles/mole protein*)			
	Control	pH 8	pH 7	pH 6
Lysine	9.7 (10)[a]	2.2	4.6	7.1
Histidine	3.7 (4)[a]	3.4	3.2	3.8
Ethylamine	—	6.2	3.8	2.5
ε-N-(2-succinyl)lysine	—	6.4	4.0	2.3

[a] Theoretical values.

The secondary amino group of proline reacts with NEM more rapidly than do primary amino groups. The imidazole group of histidine under relatively vigorous conditions reacts very slowly with NEM (Sharpless and Flavin, 1966; Brewer and Riehm, 1967). Alkylation of imidazole groups may occur at either heteroatom, resulting therefore in the formation of two monoalkylated products. Hydrolysis of these presumably yields corresponding 1- or 3-(2-succinyl)histidines (Brewer and Riehm, 1967). These reactions are not of great significance under the conditions employed for sulfhydryl determinations.

Other nitrogen-substituted maleimides behave similarly to NEM; N-(4-dimethylamino-3,5-dinitrophenyl)maleimide (Tuppy, 1959; Tsunoda and Yasunobu, 1966), N-(4-hydroxy-1-naphthyl)maleimide (Seligman, 1954), and N-2,4-dinitroanilinomaleimide (Clark-Waler and Robinson, 1961) have all been used because they form colored adducts. N,N'-(1,3-diphenylene)-bis-maleimide has been employed as a cross-linking agent (Moore and Ward, 1956).

N-(4-hydroxy-1-naphthyl)-
maleimide

N-(4-dimethylamino-3,5-
dinitrophenyl)maleimide

N-2,4-dinitroanilinomaleimide

N,N'-(1,3-diphenylene)-bis-maleimide

Long-chain alkylmaleimides are sometimes more effective reagents than NEM for reaction with sulfhydryl groups in apolar environments. The inactivation of D-amino acid oxidase by N-octylmaleimide and several other long-chain alkylmaleimides, for example, is manyfold more rapid than with hydrophilic NEM (Fonda and Anderson, 1969). A similar effect of chain length on the inactivation rate of yeast alcohol dehydrogenase has been described (Figure 6-2) (Heitz et al., 1968). This effect of alkyl chain length was not observed, however, for the inactivation of horse liver alcohol dehydrogenase and several other enzymes.

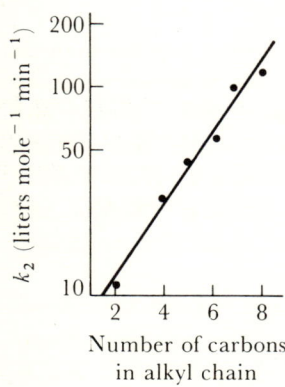

FIGURE 6-2 Effect of chain length on inactivation of yeast alcohol dehydrogenase by N-alkylmaleimides. (From Heitz et al., 1968.)

In mildly alkaline solution, NEM slowly hydrolyzes to N-ethylmaleamic acid (Gregory, 1955). This base-catalyzed reaction has an apparent rate of 3.2×10^{-4} min^{-1} at pH 7.0 in 0.1 M sodium pyrophosphate buffer at 20°C (Heitz et al., 1968). At higher pH's, hydrolysis becomes appreciable. The extremely slow hydrolysis at pH 7 would not be expected to interfere with the reaction of NEM with most proteins. The inactivation of yeast alcohol dehydrogenase by NEM, for example, is more than 100-fold faster at pH 7 than the rate of hydrolysis (Heitz et al., 1968). No other reactions involving its carbonyl carbons are known to occur under mild conditions. NEM is not an effective acylating agent.

$$\text{maleimide-NC}_2\text{H}_5 \xrightarrow[\text{OH}^-]{\text{H}_2\text{O}} \text{N-ethylmaleamic acid} \quad (3)$$

Hydrolysis of NEM-treated proteins converts the modified cysteine and lysine residues to S-(2-succinyl)cysteine and ε,N-(2-succinyl)lysine, respectively. Both can be detected and quantitated by amino acid analysis. The former is rapidly eluted and usually precedes aspartic acid, and the latter comes in the vicinity of proline. Ethylamine formed in the hydrolysis can also be used for quantitation. It elutes in the region between lysine and arginine upon amino acid analysis and has a ninhydrin color at 570 mμ, approximately 0.40 that of leucine (Brewer and Riehm, 1967).

$$\text{RS-adduct-NC}_2\text{H}_5 \xrightarrow[\text{H}_2\text{O}]{\Delta,\ \text{HCl}} \text{RS-succinate} + \text{C}_2\text{H}_5\text{NH}_2 \quad (4)$$

Quantitative analysis for the extent of N-ethylmaleylation can be obtained by difference, by determining the unreacted groups remaining. This can be done by amino acid analysis after hydrolysis, or by examining the intact protein using one of the several spectrophotometric procedures for sulfhydryl or amino groups (see Table 4-1).

6-3 ACRYLONITRILE

$$\text{(P)}-\text{S}^- + \text{CH}_2\!\!=\!\!\text{CHCN} + \text{H}^+ \underset{}{\overset{\text{pH} > 6.5}{\rightleftharpoons}} \text{(P)}-\text{S}-\text{CH}_2\text{CH}_2\text{CN} \quad (1)$$

$$\text{(P)}-\text{NH}_2 + \text{CH}_2\!\!=\!\!\text{CHCN} \xrightarrow{\text{pH} > 8} \text{(P)}-\text{NH}-\text{CH}_2\text{CH}_2\text{CN} \quad (2)$$

Acrylonitrile reacts with both amino and sulfhydryl groups of proteins. The products of the reaction are ε-N-cyanoethyllysine, ε-N,N-dicyanoethyllysine, and S-cyanoethylcysteine. α-N-Cyanoethylamino acids are produced from NH_2-terminal amino groups. Acrylonitrile can be used for the determination of NH_2-terminal amino acids (Fletcher, 1966). It has also been shown to react with histidine imidazole groups (McKinney et al., 1951) and those of histidine residues in insulin (Bosshard et al., 1969).

$$\text{(P)}-\overset{H}{\underset{H}{N}}: \curvearrowright CH_2=CH-C\overset{O\uparrow}{\underset{X}{\diagdown}} \longrightarrow$$

$$\text{(P)}-\underset{H}{N}-CH_2-CH=C\overset{OH}{\underset{X}{\diagdown}} \rightleftharpoons \text{(P)}-\underset{H}{N}-CH_2-CH_2-C\overset{O}{\underset{X}{\diagdown}} \quad (3)$$

Acrylonitrile and certain other activated vinyl compounds react rapidly with both amino and sulfhydryl groups of proteins by a mechanism analogous to the well-known Michael reaction (Equation 3). Reaction rates increase with increasing pH (Figure 6-3). The intrinsic rate of reaction with anionic sulfur is approximately 300-fold faster than with unprotonated amines. Reaction with amino groups is enhanced in dimethyl sulfoxide and other low-dielectric media (Friedman, 1967). The reaction of β-lactoglobulin with acrylonitrile was limited solely to modification of sulfhydryl groups at pH 8 using a 2:1 mole ratio of reagent to sulfhydryl groups (Weil and Seibles, 1961). Even greater specificity for sulfhydryl groups can be obtained at pH 7. For specific alkylation of sulfhydryl groups in proteins reduced with β-mercaptoethanol and in the presence of excess β-mercaptoethanol, pH 7 and a 1:1 ratio of vinyl compound to total sulfhydryl content appears optimum (Cavins and Friedman, 1968). Under these conditions, the reaction is completed in 5 minutes with methyl acrylate, in 20 minutes with acrylonitrile, and in 100 minutes with acrylamide. Under the specified conditions, which include 6 M urea, protein sulfhydryl groups are modified more rapidly than is β-mercaptoethanol.

Acrylonitrile has been widely used to block sulfhydryl groups in proteins in order to convert these groups into stable derivatives for subsequent analytical procedures, but has only in a few cases been used as a selective reagent to determine the effects of substitution on biological activities. Its reaction with pancreatic ribonuclease at pH 9.2, for example, abolishes that enzyme's catalytic activity through reaction with amino groups. The enzyme so inactivated largely retains physical-chemical properties like the unmodified enzyme (Riehm and Scheraga, 1966).

Second-order rate constants for the reaction of acrylonitrile and other α,β-unsaturated compounds with different amino and sulfhydryl compounds vary

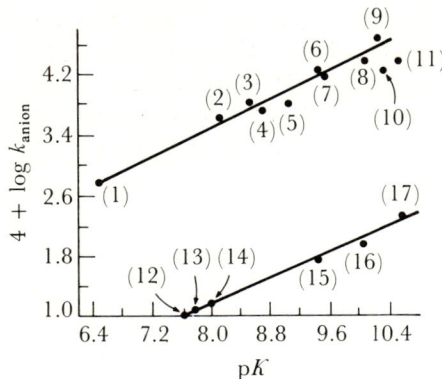

FIGURE 6-3 Plot of log second-order anion rate constants versus pK for the reaction of mercaptide groups [(1)–(11)] and amino groups [(12)–(17)] with acrylonitrile at 30°C. Upper plot, mercaptide groups attached to primary carbon atoms: (1) $^-$SCH$_2$CH(NH$_3^+$)COOC$_2$H$_5$, (2) $^-$SCH$_2$CH(NH$_3^+$)COO$^-$, (3) $^-$S—glutathione—NH$_3^+$, (4) $^-$SCH$_2$CH$_2$CH(NH$_3^+$)COO$^-$, (5) $^-$SCH$_2$CH(NH$_2$)COOC$_2$H$_5$, (6) $^-$S—CH$_2$CH(NHCOCH$_3$)COO$^-$, (7) $^-$S—glutathione—NH$_2$, (8) $^-$SCH$_2$CH$_2$COO$^-$, (9) $^-$SCH$_2$COO$^-$, (10) $^-$SCH$_2$CH(NH$_2$)COO$^-$, (11) $^-$SCH$_2$CH$_2$CH(NH$_2$)COO$^-$. Lower plot, amino groups attached to primary carbon atoms: (12) tetraglycine, (13) triglycine, (14) diglycine, (15) glycine, (16) β-alanine, (17) ε-aminocaproic acid. (From Friedman et al., 1965.)

TABLE 6-2 Second-order rate constants for the reaction of NH$_2$CH$_2$CO$_2^-$ with α,β-unsaturated compounds at pH 8.75 and 30° ($\mu = 1.2$) (from Friedman and Wall, 1966)

Compound CH$_2$=CHX	$K_2 \times 10^5$ (liters/mole sec)
—CONH$_2$	0.63
—C$_5$H$_4$N (pyridyl)	4.2
—C≡N	5.0
—CO$_2$CH$_3$	18.2
—SO$_2$CH$_3$	30.6
—COCH$_3$	400

considerably due to both steric and inductive effects but have been interpreted in terms of linear free energy parameters (Friedman and Wall, 1964; Friedman and Wall, 1966). The range of reactivities of several different α,β-unsaturated compounds are indicated in Table 6-2 by their second-order rate constants for reaction with glycine.

Cyanoethylation of amines lowers their pK by 2 to 2.5 pH units for each added cyanoethyl group. Upon hydrolysis, acrylonitrile-modified proteins yield S- and/or N-carboxyethylamino acids, resulting, respectively, from the modified cysteinyl and/or lysyl residues. A true picture of the amount of dicyanoethyllysine may not be obtained, however, due to its partial decarboxyethylation during acid hydrolysis. Alkaline hydrolysis is reported to effect a partial reversion to lysine (Cavins and Friedman, 1967). S-Carboxyethylcysteine, N,N-dicarboxyethyllysine, and N-carboxyethyllysine can be detected and quantitated by amino acid analysis. S-Carboxyethylcysteine elutes near glutamic acid by standard analysis procedures, while N,N-di- and N-carboxyethyllysine elute near lysine.

$$\begin{array}{c} CH_2\!-\!SCH_2CH_2C\!\equiv\!N \\ | \\ CHNH_3^+ \\ | \\ COOH \end{array} \xrightarrow[\Delta]{HCl/H_2O} \begin{array}{c} CH_2\!-\!SCH_2CH_2COOH \\ | \\ CHNH_3^+ \\ | \\ COOH \end{array} + NH_4^+ \qquad (4)$$

6-4 ETHYLENIMINE

$$\text{(P)}\!-\!SH + \begin{array}{c} H_2C \\ | \\ H_2C \end{array}\!\!\!\!>\!NH \xrightarrow{pH\ 8.6} \text{(P)}\!-\!S\!-\!CH_2CH_2NH_2 \qquad (1)$$

Ethylenimine reacts with the sulfhydryl groups of proteins converting cysteine residues into S-(2-aminoethyl)cysteine (Raftery and Cole, 1963). No other groups are known to react under the mild, slightly alkaline conditions normally used. At acidic pH, methionine reacts slowly to give S-(2-aminoethyl)methioninesulfonium ion (Schroeder et al., 1967). Aminoethylcysteine closely resembles lysine but with a sulfur atom (—S—) taking the place of a methylene group (—CH_2—). Aminoethylcysteine residues in proteins are susceptible to tryptic hydrolysis although at a slower rate than lysine (Lindley, 1956; Plapp et al., 1967; Wang and Carpenter, 1968). The reaction is commonly used to introduce additional trypsin-susceptible bonds into proteins, usually to facilitate degradation in sequence studies. Carboxyl-terminal aminoethylcysteine residues are susceptible to cleavage by carboxypeptidase

B (Tietze et al., 1957). A colorimetric procedure for its detection has been described (Rothfus, 1969).

$$\text{CH}_2\text{—Br} \atop \text{CH}_2\text{—NH}_2 \xrightarrow{\text{OH}^-} \underset{\text{H}}{\overset{}{\text{CH}_2\text{—CH}_2 \atop \diagdown\text{N}\diagup}} \xrightarrow{\text{P—SH}} \text{P—S—CH}_2\text{CH}_2\text{NH}_2 \quad (2)$$

Aminoethylation to introduce additional trypsin-susceptible bonds is usually done after complete reduction of all protein disulfide bonds. This is usually accomplished in 8 M urea or another denaturing solvent through the action of a large excess of β-mercaptoethanol or other thiol (see Subsection 8-1.1). Ethylenimine is then added directly to the reducing-reaction solution much in excess of its total thiol content. A reducing atmosphere is desirable throughout these procedures. The extent of reaction can be determined by amino acid analysis. Aminoethylcysteine is stable to conditions for acid hydrolysis of proteins and emerges from the standard cation-exchange columns near lysine and histidine. Its color value after reaction with ninhydrin has been reported to be 2–5% lower than that of lysine (Cole, 1967).

2-Bromoethylamine is slowly converted to ethylenimine in slightly alkaline solutions (Equation 2). Modification of sulfhydryl groups of proteins with 2-bromoethylamine is not as rapid as with ethylenimine, and some losses of methionine and histidine have been found to occur (Cole, 1967).

6-5 ARYL HALIDES

$$\text{P—NH}_2 + \text{X—C}_6\text{H}_4\text{—NO}_2 \longrightarrow \text{P—NH—C}_6\text{H}_4\text{—NO}_2 + \text{X}^- + \text{H}^+ \quad (1)$$

$$\text{P—S}^- + \text{X—C}_6\text{H}_4\text{—NO}_2 \longrightarrow \text{P—S—C}_6\text{H}_4\text{—NO}_2 + \text{X}^- \quad (2)$$

$$\text{P—C}_6\text{H}_4\text{—O}^- + \text{X—C}_6\text{H}_4\text{—NO}_2 \longrightarrow \text{P—C}_6\text{H}_4\text{—O—C}_6\text{H}_4\text{—NO}_2 + \text{X}^- \quad (3)$$

$$\text{P—imidazole(NH)} + \text{X—C}_6\text{H}_4\text{—NO}_2 \longrightarrow \text{P—imidazole(N—C}_6\text{H}_4\text{—NO}_2) + \text{X}^- + \text{H}^+ \quad (4)$$

Dinitrofluorobenzene (DNFB) was used by Sanger to identify the NH_2-terminal amino acids of insulin (Sanger, 1945). Its principal use has been for the determination of NH_2-terminal residues of polypeptides and proteins. Its reaction with amino groups yields bright-yellow-colored dinitrophenyl (DNP) amino compounds. Reaction is usually carried out at mildly alkaline pH. In addition to amino groups, DNFB also reacts with sulfhydryl and imidazole groups and with the phenolic hydroxyl of tyrosine.

Kinetic studies of the reaction of DNFB with nucleophiles indicate a simple bimolecular mechanism similar to bimolecular nucleophilic displacement (S_N2) at aliphatic carbons (Equation 5). The order of reactivity in terms of the group being displaced is $F > Cl \approx Br > SO_3H$. In protic solvents, the reaction of DNFB with some amines appears to be base catalyzed (Bunnett and Randall, 1958), while its reaction with others, including apparently most amino acids, is not (Bunnett and Hermann, 1970). Reaction requires the unprotonated nucleophile and hence is accelerated at high pH. The presence of 80% dimethyl sulfoxide accelerates the reaction rate of DNFB with glycine more than 100-fold as compared to the rate in water (Bunnett and Hermann, 1970). Trinitro compounds of this type are more reactive than the corresponding dinitro compounds.

$$NO_2\text{-}C_6H_3(NO_2)\text{-}X + :Nuc \rightleftharpoons [\text{intermediate}] \longrightarrow NO_2\text{-}C_6H_3(NO_2)\text{-}Nuc + X: \tag{5}$$

Reactions with α-amino groups are more rapid than with ε-amino groups at pH 8.5 and below, a reflection of the α-amino group's lower pK. In some proteins, environmental factors alter this intrinsic difference in reactivity. DNFB has been shown, for example, to react preferentially with one ε-amino group in bovine pancreatic ribonuclease effecting the loss of that enzyme's catalytic activity (Hirs, 1962; Hirs et al., 1965). Reaction with the terminal α-amino group occurs more slowly. The product of the former reaction, dinitrophenyllysine-41-ribonuclease, has been extensively characterized and the modified group appears to be in or immediately adjacent to the enzyme's positively charged active center. An abnormally low pK of 8.8 has been estimated to explain the great reactivity of this ε-amino group. Conformation changes and decreased affinity for substratelike anions appear to account for the enzyme's lost catalytic activity (Ettinger and Hirs, 1968). The dinitrophenyl group has been shown by X-ray-crystallographic studies of the crystallized protein to reside in a fold outside the active center (Richards et al., 1969).

Although usually considered an amino-group reagent DNFB also reacts readily with other common amino acid side chains in proteins. For example, at pH 7.5, DNFB preferentially reacts with one sulfhydryl group of rabbit liver fructose-1,6-diphosphatase and in so doing increases that enzyme's catalytic activity more than twofold when Mn^{++} is used as the activating cation (Pontremoli et al., 1965). This reaction is blocked in the presence of substrate, fructose-1,6-diphosphate. Changes in catalytic activity brought about by the reaction resemble those observed with the unmodified enzyme at high pH and may mimic the effects of an unknown physiological allosteric activator. At higher pH, DNFB reacts more extensively with the enzyme, modifying sulfhydryl and amino groups.

Treatment of rabbit muscle glyceraldehyde-3-phosphate dehydrogenase with small amounts of DNFB brings about the specific substitution of a sulfhydryl group in the coenzyme binding site (Shaltiel and Soria, 1969). The group appears to be in the coenzyme binding site since its reaction can be blocked by the coenzyme nicotinamide adenine dinucleotide. Treatment of the inactive S-dinitrophenyl enzyme at pH 8.0 with excess β-mercaptoethanol brings about displacement of the dinitrophenyl group (Equation 6) and restoration of the catalytic activity. Displacement of dinitrophenyl-substituted imidazole and phenolic groups, but not amino groups, can be accomplished under similar conditions (Shaltiel, 1967).

$$\text{(P)}-S-\underset{NO_2}{\overset{NO_2}{\bigcirc}}-NO_2 + R-SH \underset{}{\overset{pH\ 8}{\rightleftharpoons}} \text{(P)}-SH + R-S-\underset{NO_2}{\overset{NO_2}{\bigcirc}}-NO_2$$

(6)

Dinitrophenylamines absorb maximally near 365 mμ and have an extinction coefficient of approximately 1.6×10^4 M^{-1} cm^{-1}. They are light sensitive and should be protected from direct illumination (Pollara and Von Korff, 1960). Other dinitrophenyl derivatives usually have absorption maxima at shorter wavelengths. Extinction coefficients have not been well characterized. S-DNP groups have absorption maxima near 330 mμ (Mahowald, 1965; Shaltiel, 1967) and the DNP derivatives of imidazole and tyrosine side chains usually absorb between 330 and 360 mμ (Hirs, 1967; Shaltiel, 1967). The strong electron-withdrawing dinitrophenyl group greatly lowers the pK of substituted amines. DNP amino groups are not protonated at neutral or even moderately acidic pH's, and proteins so modified have altered electrostatic properties and electrophoretic mobilities.

Dinitrobenzenesulfonic acid reacts with amino groups more slowly than does DNFB (Eisen et al., 1953). Longer reaction times or more alkaline solutions are usually required to achieve similar degrees of modification. Dinitrochlorobenzene is somewhat less reactive than DNFB but very similar in its side-chain specificity (Bunnett and Zahler, 1953). Its inactivation of *Candida utilis* transaldolase, for example, comes about from reaction with two lysine side chains, whereas under similar conditions rabbit muscle aldolase is inactivated by reaction with cysteine side chains (Kowal et al., 1965). Trinitrochlorobenzene is similar but, being more reactive than

DNFB, reacts more readily with phenolic and imidazole groups. It also appears to react with guanidino groups.

$$\text{NO}_2\text{-C}_6\text{H}_2(\text{NO}_2)_2\text{-SO}_3\text{H} + \text{NH}_2\text{R} \xrightarrow{\text{pH}>7} \text{NO}_2\text{-C}_6\text{H}_2(\text{NO}_2)_2\text{-NHR} + \text{SO}_3\text{H}^- + \text{H}^+ \quad (7)$$

Trinitrobenzenesulfonic acid (TNBS) has been used by Okuyama and Satake (1960) and by Satake et al. (1960) as a colorimetric reagent for the determination of primary amines, amino acids, and peptides. They showed it to react only very weakly or not at all with the amino nitrogen of proline, with the imidazole nitrogens of histidine, or with the hydroxyl groups of tyrosine, threonine, and serine. No reaction with the guanidino group of arginine was detected. TNBS has been shown to react with the ribose and guanine moieties of nucleic acids (Azegami and Iwai, 1964). It has been shown not to react with secondary amines (Means and Feeney, 1968a). Habeeb (1966) has suggested that it reacts with both urea and guanidine.

Kotaki et al. (1964) have shown TNBS to be less reactive with sulfhydryl groups than with amino groups. They showed S-trinitrophenyl (S-TNP) derivatives to be labile at physiological pH and showed sulfhydryl groups to result from their breakdown. The sulfhydryl group of N-acetylcysteine has been shown to react with TNBS at pH 7.4 faster than do most amines (Freedman and Radda, 1968). The resultant S-TNP group was unstable at alkaline pH (Equation 8). In high concentrations of β-mercaptoethanol, we have observed the reduction of TNBS to give either 2- or 4-aminodinitrobenzenesulfonic acid (Equation 9). This new compound has a strong absorption at 480 mμ. The reactivity of TNBS with sulfhydryl groups should not be dismissed without further study.

The reaction of TNBS with amines is strongly pH-dependent, since only the unprotonated amino groups react (Goldfarb, 1966a; Freedman and Radda, 1968). Similar pH dependencies have been found for its reaction with human serum albumin (Goldfarb, 1966b; Goldfarb, 1970) and several other proteins (Freedman and Radda, 1968). The intrinsic reactivities of amino acids increase with increasing nucleophilicity of their amino groups. Logarithms of second-order rate constants, taking into consideration only the unprotonated amine, increase linearly with the increase in pK (see Figure 2-3).

The spectra of ε-N-TNP-lysine and α-TNP amino acids are similar (Satake et al., 1960). Both have absorption maxima at approximately 345 mμ with a broad shoulder to 420 mμ. Sulfite ion, formed in the reaction of TNBS with amines, complexes with trinitrophenylamines with an association constant greater than 10^4 at neutral pH. The presence of sulfite thus affects the absorption spectrum of trinitrophenylamino groups, increasing it at 420 mμ and decreasing it at 345 mμ.

$$\text{P}\!-\!\text{S}^- + {}^-\text{SO}_3\!-\!\underset{\text{NO}_2}{\overset{\text{NO}_2}{\bigcirc}}\!-\!\text{NO}_2 \longrightarrow \text{P}\!-\!\text{S}\!-\!\underset{\text{NO}_2}{\overset{\text{NO}_2}{\bigcirc}}\!-\!\text{NO}_2 + \text{SO}_3^= \xrightarrow{\text{OH}^-}$$

$$\text{P}\!-\!\text{S}^- + \text{HO}\!-\!\underset{\text{NO}_2}{\overset{\text{NO}_2}{\bigcirc}}\!-\!\text{NO}_2 \qquad (8)$$

$${}^-\text{SO}_3\!-\!\underset{\text{NO}_2}{\overset{\text{NO}_2}{\bigcirc}}\!-\!\text{NO}_2 + 4\,\text{RSH} \longrightarrow {}^-\text{SO}_3\!-\!\underset{\text{NO}_2}{\overset{\text{NH}_2}{\bigcirc}}\!-\!\text{NO}_2 + 2\,\text{RSSR} + \text{H}_2\text{O}$$

(9)

In the presence of sulfite ion, absorption varies with pH, indicating sulfite ($SO_3^=$) rather than bisulfite (HSO_3^-) to be the complexing species. The molar extinction coefficient of ε-TNP-α-acetyllysine in the presence of 1 molar equivalent of sulfite at pH 8.5 is $1.19 \times 10^4\ M^{-1}\ cm^{-1}$ at 345 mμ and increases to a value of $1.42 \times 10^4\ M^{-1}\ cm^{-1}$ at pH 4.5. Over the same pH range, the 420-mμ absorption decreases from $1.37 \times 10^4\ M^{-1}\ cm^{-1}$ to $6.46 \times 10^3\ M^{-1}\ cm^{-1}$ (Goldfarb, 1966a). In the absence of sulfite, the extinction coefficient of ε-TNP-α-acetyllysine is $1.45 \times 10^4\ M^{-1}\ cm^{-1}$ at 345 mμ. That of α-TNP-glycine is $1.55 \times 10^4\ M^{-1}\ cm^{-1}$ (Satake et al., 1960). Extinction coefficients of trinitrophenylglycine peptides at 345 mμ decrease with increasing numbers of residues approaching $\sim 1.0 \times 10^4\ M^{-1}\ cm^{-1}$. A sensitive procedure for the quantitative determination of amino groups in proteins using TNBS has been described by Habeeb (1966).

Spectral properties of trinitrophenylamino groups are such as to make the detection of these groups very easy. TNBS has thus afforded a useful means for determining reactivities of amino groups in proteins. It has been possible, for example, to evaluate reactivities of amino groups in many proteins in terms of several reactive classes. From changes in properties accompanying such modification, it has been possible in some cases to identify groups in such a class relative to their functional importance. Thus, the rapid loss of trypsin-inhibitory activity of turkey ovomucoid upon treatment with TNBS appears to result from reaction with a single fast-reacting amino group (see Figure 2-3) (Haynes et al., 1967). The reaction of

TNBS with amino groups of human serum albumin (Goldfarb, 1966b, 1970), bovine chymotrypsinogen, pancreatic ribonuclease, insulin, glutamic dehydrogenase (Freedman and Radda, 1968), and many other proteins have been described in terms of two or three exponential-rate curves, based on their different reactivities.

TNP amino acids are more labile to hot acid than are analogous DNP amino acids (Okuyama and Satake, 1960). They can be cleaved by ammonia to give picramide and the free amino acid in near-quantitative yield (Satake et al., 1961). TNP-amines are more photosensitive than the corresponding DNP-amines, photolysis of the former apparently yielding both picramide and picric acid among the decomposition products (Okuyama and Satake, 1960).

The reaction of TNBS with proteins proceeds under mild conditions, converting positively charged amino groups into neutral TNP-amino groups. The mild side reaction with sulfhydryl groups is not a serious drawback, due to spontaneous lysis of the TNP-sulfur bond at physiological pH. Like dinitrophenylation, trinitrophenylation of amino groups is observable spectrophotometrically and is therefore easily quantitated. Because there appear to be fewer side reactions with other protein groups and with components of the solvent, TNBS is considerably superior to DNFB for the quantitative determination of amino groups in proteins. Acid hydrolysis followed by amino acid analysis affords an additional means for quantitation of both di- and trinitrophenylated proteins, since none of the modified groups revert to the original group during the hydrolysis. Quantitation is obtained by difference as compared to the unmodified protein.

Several reactive nitrofluoroaromatic compounds have been used as cross-linking reagents for proteins. The compounds p,p'-difluoro-m,m'-dinitrodiphenyl sulfone, and 1,5-difluoro-2,4-dinitrobenzene, with brief mention of their use as cross-linking reagents, are discussed in Subsection 3-1.4.

6-6 2-HYDROXY-5-NITROBENZYL BROMIDE

(1)

2-Hydroxy-5-nitrobenzyl bromide (HNBB) reacts in neutral and acidic solutions only with the tryptophan residues of α-chymotrypsin (Koshland et al., 1964). Under the same conditions, it reacts with cysteine less than one-fifth as rapidly as with tryptophan. It reacts specifically with tryptophan in those proteins such as α-chymotrypsin which contains no —SH groups. In alkaline solutions, it can also react with tyrosine (Horton and Koshland, 1965).

$$(2)$$

The great reactivity of HNBB appears due to resonance stabilization of its incipient carbonium ion (Equation 2). Only tryptophan and (to a lesser extent) cysteine are able to effectively compete with water for this reactive intermediate. The half-life for hydrolysis of HNBB in aqueous solutions is less than 1 minute.

The reaction of HNBB with N-acetyltryptophan gives three products, one of which is disubstituted (Barman and Koshland, 1967; Spande et al., 1968). Its reaction with porcine pepsin appears to give more than one derivative with each tryptophan (Dopheide and Jones, 1968). Three products of the reaction with N-acetyltryptophan methyl ester have been identified (Spande et al., 1968) (Equation 3), and the structure of the main product has been independently confirmed (Chan and Schellenberg, 1968).

$$(3)$$

$$R = H \text{ or } -CH_2\text{-}C_6H_3(OH)(NO_2)$$

Reaction with HNBB can be used to quantitatively determine tryptophan in proteins (Barman and Koshland, 1967). The procedure is simple, involving measurement of the absorbance at 410 mμ after treatment with HNBB and removal of hydrolyzed reagent. An extinction coefficient of 18,000 M^{-1} cm^{-1} has been employed at pH values above 10. The possible formation of more than one product necessitates careful control of the reaction conditions. Cysteine residues have not been found to interfere in these determinations.

The accessibility of tryptophan residues in α-chymotrypsin to reaction with HNBB has been examined as a function of pH (Oza and Martin, 1967). At pH 2, all eight tryptophans are reactive, whereas at pH 4 only one is available. This one does not appear to be essential for catalytic activity. In porcine pepsin, only two of the four tryptophans are reactive at pH 3.5, and modification of these has only a slight effect on catalytic activity (Dopheide and Jones, 1968). Disruption by reduction and alkylation (Chapter 3 and Section 8-1) or alkali treatment, respectively, render the third and fourth tryptophans reactive. Modification of one tryptophan with HNBB has little effect on the properties of apomyoglobin whereas modification of the second residue is accompanied by drastic changes in its physical properties (Atassi and Caruso, 1968).

HNBB is useful for quantitatively determining tryptophan in proteins, for determining their relative accessibilities under various conditions, and for studying the effects of such modification of tryptophan on the properties of proteins. It may also be used as an environmental probe, its spectrum being sensitive to changes in local environment, especially in pH (Koshland et al., 1964; Horton and Koshland, 1965).

The related compound 2-methoxy-5-nitrobenzyl bromide is considerably less reactive, similar to unsubstituted benzyl bromide. It reacts with tryptophan, cysteine, and methionine in neutral or acidic solutions, although at a much slower rate than HNBB. Its spectral properties are very sensitive to changes in solvent polarity but not to pH. It has been used as a pH-insensitive environmental probe to follow and assess conformational changes of proteins (see Subsection 3-2.3) (Horton et al., 1965).

6-7 FORMALDEHYDE AND SOME OTHER CARBONYL COMPOUNDS

6-7.1 FORMALDEHYDE

In neutral or alkaline solutions, the amino groups of proteins can react in a readily reversible manner with 2 moles of formaldehyde (Equation 1) giving, with the addition of each, an apparent lowering in amino-group pK of 2 to 3 pH units. The resulting displacement of the acid-base equilibrium by formaldehyde is the basis of the well-known formol titration of amino groups (French and Edsall, 1945; Kallen and Jencks, 1966).

$$\text{(P)}-NH_2 + \begin{matrix} H \\ \end{matrix}\!\!\!\!\!>\!\!C\!=\!O \rightleftharpoons$$

$$\text{(P)}-NHCH_2OH \underset{H}{\overset{H\,>\!C=O}{\rightleftharpoons}} \text{(P)}-N\!\!\!<\!\!\!\begin{matrix} CH_2OH \\ CH_2OH \end{matrix} \quad (1)$$

Formaldehyde is available from commercial sources in 37–40% aqueous solutions, containing small amounts of methanol and known as *formalin*. In such solutions, it exists primarily as a number of low-molecular-weight polymers of the type $H(OCH_2)_nOH$. Formaldehyde may also be obtained as a stable solid known as *paraformaldehyde*, composed of high-molecular-weight polymers of the same type. Heating of paraformaldehyde can be used to generate pure gaseous formaldehyde.

Both sources of polymeric formaldehyde revert to monomer in dilute aqueous solutions. Under such conditions, formaldehyde is in very rapid equilibrium with a hydrated form (Equation 2), and more than 99.9% is in the form of the hydrate.

$$\begin{matrix} H \\ \end{matrix}\!\!\!\!\!>\!\!C\!=\!O + H_2O \rightleftharpoons HO-\underset{H}{\overset{H}{C}}-OH \quad (2)$$

Compounds I through V (see below) have been shown to result from the reaction of formaldehyde with tyrosine, tryptophan, histidine, asparagine, and cysteine, respectively (French and Edsall, 1945). Presumably, their formation is preceded by the formation of very electrophilic immonium cations (Equation 3),

$$\underset{H}{\overset{H}{\stackrel{+}{N}}}\!\!\!\!<\!\!\!\begin{matrix}CH_2OH \\ \end{matrix} \underset{+H_2O}{\overset{-H_2O}{\rightleftharpoons}} \overset{H}{\underset{+N}{}}\!\!\!\!=\!\!CH_2 \longleftrightarrow \overset{H}{\underset{N}{}}\!\!\!\!-\!\!\overset{+}{C}H_2 \quad (3)$$

which react with the adjacent amino acid side chains. Similar reactions have been postulated to occur in proteins linking the ε-amino groups of lysine residues via methylene bridges to neighboring side chains (Fraenkel-Conrat et al., 1947; Fraenkel-Conrat and Olcott, 1948a, 1948b; Fraenkel-Conrat and Mecham, 1949). The postulated reactions are mechanistically similar to the widely studied Mannich reaction (Fernandez and Fowler, 1964; Alexander and Underhill, 1949). One such product, the methylene-bridged lysine-tyrosine compound (VI), has been isolated from acid hydrolysates of formalin-treated tetanus and diphtheria toxins (Blass et al., 1965).

I II III IV V VI

The reaction of formaldehyde with thiols (Equation 4) is analogous to, and very much more rapid than, its reaction with amines (Lewin, 1956; Barnett and Jencks, 1967). Histidine side chains undergo a similar reaction (Equation 5) (Levy, 1935; Kallen and Jencks, 1966), which is thought to be responsible for the decrease in catalytic activity of α-chymotrypsin in the presence of formaldehyde (Martin and Marini, 1967). The formation of such adducts is accompanied by an increase in absorbance near 230 mμ (Saidel and Carino, 1966; Martin and Marini, 1967). The reaction of formaldehyde with peptide bonds brings about an increase in absorbance near 220 mμ (Saidel et al., 1965) and is presumably similar to its reaction with urea (Smythe, 1953). Arginine residues are also thought to react with formaldehyde, but the nature of the reaction is not known (Fraenkel-Conrat and Olcott, 1948a).

$$\text{P}-\text{SH} + \text{CH}_2\text{O} \rightleftharpoons \text{P}-\text{SCH}_2\text{OH} \qquad (4)$$

$$\text{P}-\text{Im} + \text{CH}_2\text{O} \rightleftharpoons \text{P}-\text{Im}-\text{CH}_2\text{OH} \qquad (5)$$

Although the reactions of formaldehyde have been used industrially for many years, little is really known about the many underlying reactions. Its potential reaction with so many different amino acid side chains and the difficulties surrounding the detection of such have largely precluded its use as a specific protein reagent. Under carefully controlled conditions, it can be safely used as the carbonyl component for the reductive alkylation of proteins (Means and Feeney, 1968) (see Section 6-8).

6-7.2 Other Aldehydes and Ketones

The reactions of other aldehydes and ketones with proteins have not been widely studied but are generally considered to be much less extensive than those of formaldehyde. The isolation of ε-N-isopropyllysine from proteins exposed to low concentrations of acetone and reducing agents, however, affords clear evidence that the reaction of acetone with primary amino groups is much more extensive than generally believed (Means and Feeney, 1968). The same can presumably be said for a large number of aldehydes and ketones. Oxytocin reacts with acetone to give what is thought to be a 2,2-dimethyl-4-imidazolidone derivative (Ferrier et al., 1965; Yamashiro et al., 1965; Yamashiro and du Vigneaud, 1968). This is a specific reaction of oxytocin and certain closely related peptides with acetone, but it illustrates the kind of unusual reactions that may occur under certain circumstances.

Salicylaldehyde reacts readily with amino groups of proteins at moderately alkaline pH giving salicylidene adducts (Equation 6) resembling those formed from amino groups and pyridoxal phosphate (Section 6-9) (Williams and Jacobs, 1966, 1968). The reaction is easily reversed by dilution or by removing excess reagent. Upon incubation with salicylaldehyde at pH 9.6, horse heart cytochrome c, for example, has been shown to lose its ability to reduce DPN in a rat heart cytochrome c reductase system. If high ratios of salicylaldehyde are used, all amino groups can be modified. Dialysis to remove the reagent gives a 20% return of DPN reductase activity and, if followed by brief exposure to either pH 3 or 11, nearly complete recovery occurs.

$$\text{P}-NH_2 + \text{(HO-C}_6H_4\text{-CHO)} \rightleftharpoons \text{P}-N=CH-\text{C}_6H_4\text{-OH} \quad (6)$$

2,4,6-Trinitrobenzaldehyde reacts at neutral pH with bovine serum albumin to give an unstable colored complex with absorption maxima at 440 and 520 mμ. Amino groups of the protein are necessary for the reaction (Whitehouse and Skidmore, 1966).

Benzoquinone reacts with amino groups of proteins to give disubstituted quinones. Optimum reaction is obtained at pH 8.0. These substituted quinones can be reduced by ascorbate and reoxidized by oxygen to produce hydrogen peroxide. Under appropriate conditions the rate of ascorbate oxidation can be used to determine the amount of 1,4-addition product. Cytochrome c at pH 7.4 reacts with benzoquinone to give a derivative containing 2 to 4 moles of benzoquinone. The product was oxidizable by cytochrome oxidase but at a much reduced rate as compared to unmodified cytochrome c. In the presence of phenazine methosulfate, benzoquinone-treated cytochrome c was reduced but required greater amounts of NADH than untreated cytochrome c (Morrison et al., 1969).

$$\text{(P)}-NH_2 + CH_3O-\text{[tropone]}-NO_2 \xrightarrow{pH\ 8.5} \text{(P)}-NH-\text{[tropone]}-NO_2 + CH_3OH \tag{7}$$

Amino groups of proteins react with 2-methoxy-5-nitrotropone at room temperature in slightly alkaline solutions (pH 7.0 to 8.5) to give an adduct absorbing maximally at 420 mμ ($\varepsilon = 2.07 \times 10^4\ M^{-1}\ cm^{-1}$) (Tamaoki et al., 1967) (see Equation 7). No other amino acid side chains are believed to react under these mild conditions. With taka-amylase A, 9 out of 19 amino groups were readily modified, whereas the remainder were relatively resistant. Such modification brought about a progressive decrease in amylase activity but slightly increased the maltosidase activity. When the modified enzyme was incubated for several hours at 38°C in 1 to 2 M hydrazine at pH 8.5 to 9.0, the 420-mμ absorption band was replaced by a new maximum at 354 mμ, due to release of the blocking group upon reaction with 2 moles of hydrazine. This treatment substantially restored the enzyme's amylase activity.

2-Methoxy-5-nitrotropone is a commercially available, stable crystalline compound. It decomposes slowly in neutral aqueous solutions and quite rapidly in solutions more alkaline than pH 8.5. Although it has not been widely used, it appears promising as an easily reversible amino-group reagent.

Certain 1,2- and 1,3-dicarbonyl compounds have been used to modify arginine residues of proteins (see Section 10-1). Glutaraldehyde, a 1,5-dialdehyde, has been used as a bifunctional reagent to form cross-links between various side chains in proteins. It has been thus used to insolubilize crystals of carboxypeptidase A, in which case only a slight loss of catalytic activity was observed (Quiocho and Richards, 1966). The chemical bases of the reactions remain obscure, but they appear to involve the side chains of lysine, cysteine, histidine, and tyrosine residues (Habeeb and Hiramoto, 1968).

6-8 REDUCTIVE ALKYLATION

$$\boxed{P}-NH_2 + RHCO \underset{+H_2O}{\overset{-H_2O}{\rightleftharpoons}} \boxed{P}-N=CHR \xrightarrow{[H]} \boxed{P}-NH-CH_2R \quad (1)$$

Aliphatic aldehydes and ketones react very rapidly and in a highly reversible manner with amino groups of proteins (see Section 6-7). The adducts so formed can be reduced by mild reducing agents to stable alkylamino groups (Equation 1) (Means and Feeney, 1968a). Under the mild, slightly alkaline conditions required for extensive alkylation of protein amino groups, other side-chain groups do not give stable derivatives. Both α- and ε-amino groups react. The procedure is applicable to a wide range of proteins.

Extensive modification is easily obtainable under mild conditions with sodium borohydride as the reductant. The required concentration of sodium borohydride is low so as not to cleave disulfide bonds (Section 8-1). The reaction is strongly pH-dependent, best results being obtained near pH 9. At pH 7, similar modification can be obtained, but with rather inefficient utilization of the reagents. Low temperatures (0°–4°C) favor more extensive reaction.

Using formaldehyde as the carbonyl component, the reaction proceeds rapidly giving ε-N,N-dimethyllysine residues as the principal products. Monomethyllysine is formed initially, but its conversion to dimethyllysine is very fast. It predominates only at very early or incomplete stages of reaction. With other aldehydes and ketones, reaction with a second carbonyl molecule is greatly retarded such that predominantly monoalkyllysine residues are obtained. When acetone is the carbonyl component, for example, only the monoalkyl derivative, ε-N-isopropyllysine, is obtained. Mixed ε-N-methylalkyllysines, such as ε-N-methyl-N-isopropyllysine, can be prepared by treating first with the higher carbonyl compound and then, in a second step, with formaldehyde.

$$\boxed{P}-NH_2 + (CH_3)_2CO + \tfrac{1}{4} NaBH_4 \xrightarrow{\substack{0°C, \\ pH\ 9}}$$

$$\boxed{P}-NHCH(CH_3)_2 + \tfrac{1}{4} NaH_2BO_3 + \tfrac{1}{4} H_2O \quad (2)$$

$$\boxed{P}-NH_2 + 2\,H_2CO + \tfrac{1}{2} NaBH_4 \xrightarrow{\substack{0°C, \\ pH\ 9}} \boxed{P}-N(CH_3)_2 + \tfrac{1}{2} NaH_2BO_3 + \tfrac{1}{2} H_2O \quad (3)$$

Alkylation of amino groups in this way does not greatly affect their basicity. The pK values of dimethylamino groups (i.e., tertiary amines) are about 0.4 to 0.6 pH

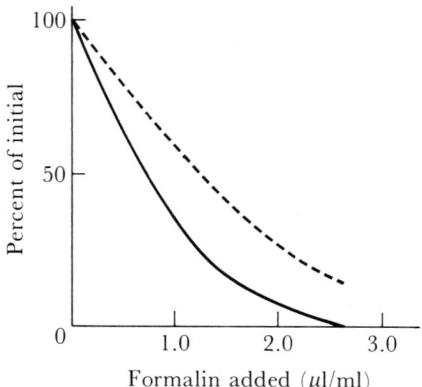

FIGURE 6-4 Reductive methylation of pancreatic ribonuclease; solid line indicates enzymatic activity, dashed line shows amino groups. A solution of pancreatic ribonuclease (5 mg/ml) was treated with small increments of 37% formalin in pH 9.0, 0.2 M borate buffer, and a small excess of sodium borohydride. (Adapted from Means and Feeney, 1968.)

units below those of primary amino groups. Monoalkylamino groups (i.e., secondary amines) usually have pK's from 0.1 to 0.7 pH units higher than primary amino groups. These differences result in slight alterations in the titration curves and electrophoretic mobility of modified proteins at alkaline pH. Because of the relatively small change in basicity and the relatively small space occupied by the methyl groups, ε-N,N-dimethyllysine residues are more similar to unmodified lysines than most other derivatives of lysine.

When bovine pancreatic ribonuclease is reductively methylated, catalytic activity is lost with little change in the protein's physical properties (Means and Feeney, 1968b). Its ability to bind 3'-cytidylic acid is impaired slightly and its high reactivity with iodoacetate is increased. The loss of catalytic activity appears to result from minor alterations of its active site. A similar loss of catalytic activity occurs with reductive isopropylation (i.e., using acetone as the carbonyl component) and is accompanied by slightly greater impairment of the ability to bind 3'-cytidylic acid. A somewhat greater enhancement of its reactivity with iodoacetate is also obtained. In this instance, small changes in optical rotation show it to have experienced some change in tertiary structure.

Turkey ovomucoid, in which a lysine residue is essential for trypsin-inhibitory activity, loses its antitrypsin activity upon reductive methylation or reductive isopropylation. Both derivatives retain their ability to inhibit α-chymotrypsin, showing the two inhibitory sites to be independent (Means and Feeney, 1968a).

When reductively methylated bovine chymotrypsinogen A is activated by trypsin, chymotryptic activity is obtained against benzoyltyrosine ethyl ester equivalent to that obtained by activation of the unmodified zymogen. The terminal amino group formed upon activation is resistant to subsequent alkylation (Means and Feeney, 1970).

Simple mono- and dialkyllysines are stable under conditions for acid hydrolysis of proteins and are easily quantitated by slight modifications of currently used amino-acid-analysis systems. The optical absorption factors at 570 mμ for mono- and dimethyllysine after reaction with ninhydrin are 0.881 and 0.822, respectively, as compared to a value of 1.00 for lysine (Means and Feeney, 1968a). Mono- and dialkyllysine residues do not react with trinitrobenzenesulfonic acid (TNBS) (Habeeb, 1966), and their sum can be obtained colorimetrically from the decrease in TNBS-reactive amino groups (Means and Feeney, 1968a). Borohydride concentrations can be easily determined by a simple colorimetric procedure which also uses TNBS (Means and Feeney, 1968a) (Sections 4-1, 6-5). Trypsin does not attack internal ε-N,N-dimethyllysine residues (Gorecki and Shalitin, 1967; Lin et al., 1969a) and carboxypeptidase B does not cleave carboxy-terminal dimethyllysine (Lin et al., 1969b).

6-9 PYRIDOXAL PHOSPHATE

Reduction with sodium borohydride has been used to stabilize linkages between pyridoxal phosphate (PLP) and many enzymes in which it occurs. The procedure has been used to label the PLP binding sites of these enzymes (Schirich and Mason, 1963; Fischer et al., 1958; Hughes et al., 1962; Wilson and Kornberg, 1963; Anderson and Chang, 1965; Phillips and Wood, 1965; Wilson and Crawford, 1965; Matsuo and Greenberg, 1959). As employed for rabbit muscle glycogen phosphorylase (Fischer et al., 1958), urea, detergents, or extremes of pH (pH < 4.5 or > 9.5) are required to induce the enzyme-PLP complex into a reactive state. This process, presumably, involves its transformation from an unreactive aldamine form, I or II ($\lambda_{max} = 330$ mμ), into a Schiff-base form, III ($\lambda_{max} = 415$ mμ), which is then reduced (Equation 2). Borohydride-reduced glycogen phosphorylases a and b have absorption maxima at 325 mμ and retain most of their catalytic activity. The effects of various conditions upon the reduction of glycogen phosphorylase have recently been described (Strausbauch et al., 1967). Many PLP-containing enzymes are found primarily in the form of a Schiff-base-type complex and are easily reduced by sodium borohydride at neutral pH without deforming agents (Anderson and Song, 1967; Dempsey and Snell, 1963; Klein and Sagers, 1967).

$$\text{(2)}$$

Many proteins which do not normally contain PLP nevertheless can, under appropriate conditions, form specific complexes with it, in some cases, perhaps reflecting an enzyme–allosteric effector relationship. These complexes may be either aldamines, Schiff bases, or mixed complexes. Bovine serum albumin, for example, binds 2 moles of PLP very tightly, one as a Schiff base and one as an aldamine. At higher PLP concentrations, additional PLP is bound (Dempsey and Christensen, 1962). The high affinity binding appears dependent upon specific interactions between the anionic phosphate ester moiety of PLP and certain cationic groups of the protein. The same can be said for most protein-PLP complexes.

Reduction of pig kidney fructose-1,6-diphosphatase in the presence of PLP gives an enzyme which is still active but no longer sensitive to allosteric AMP inhibition or to high substrate inhibition (Marcus and Hubert, 1968). Similar treatments inactivate *Candida utilis* phosphogluconate dehydrogenase (Rippa et al., 1967), rabbit muscle aldolase (Shapiro et al., 1968), bovine pancreatic ribonuclease (Means and Feeney, 1970), and bovine liver glutamic dehydrogenase (Anderson et al., 1966). PLP alone reversibly inhibits the catalytic activity of each of these enzymes. PLP lowers the oxygen affinity of hemoglobin, as do a number of other phosphate esters. Reduction with sodium borohydride has been used to determine the site of its interactions with hemoglobin (Benesch et al., 1969). In each of these cases, binding appears to be facilitated by, and the site largely determined by, the phosphate moiety.

Equimolar PLP and pancreatic ribonuclease in neutral solution react slowly, giving a complex which, after a few minutes, has a composite absorption maximum at a wavelength between that of PLP (~ 388 mμ) and its Schiff-base derivatives (415 to 425 mμ) under the same conditions. Addition of sodium borohydride decolorizes the yellow solution, shifting its absorption maximum to shorter wavelength and stabilizing the linkage between enzyme and PLP. After purification, the pyridoxamine ribonuclease ($\lambda_{max} = 324$ mμ, assumed $\varepsilon = 0.97 \times 10^4\ M^{-1}\ cm^{-1}$) can be oxidized with performic acid and digested with trypsin to give one major fluorescent peptide, the position of which in the linear sequence of ribonuclease has been determined. The PLP label is at lysine residue 7, a position known from other chemical and X-ray-diffraction studies to be in or immediately adjacent to the active site.

PLP labeling with sodium borohydride appears to have general utility as a means to specifically label lysine residues in or near phosphate binding sites of proteins. These labels are readily detected by their characteristic absorption and fluorescent properties (Klein and Sagers, 1967).

REFERENCES

Alexander, E. R., and E. J. Underhill (1949): *J. Am. Chem. Soc.*, **71**, 4014.
Alexander, N. M. (1958): *Anal. Chem.*, **30**, 1292.
Anderson, B. M., C. D. Anderson, and J. E. Churchich (1966): *Biochemistry*, **5**, 2893.
Anderson, J. A., and H. F. W. Chang (1965): *Arch. Biochem. Biophys.*, **110**, 346.
Anderson, J. A., and P. S. Song (1967): *Arch. Biochem. Biophys.*, **122**, 224.

Atassi, M. Z., and D. R. Caruso (1968): *Biochemistry*, **7,** 699.
Azegami, M., and K. Iwai (1964): *J. Biochem. (Tokyo)*, **55,** 346.
Barman, T. E., and D. E. Koshland (1967): *J. Biol. Chem.*, **242,** 5771.
Barnett, R., and W. P. Jencks (1967): *J. Am. Chem. Soc.*, **89,** 5963.
Benesch, R., R. E. Benesch, M. Gutcho, and L. Laufer (1956): *Science*, **123,** 981.
Benesch, R., R. E. Benesch, and C. I. Yu (1969): *Fed. Proc.*, **28,** 604.
Blass, J., B. Bizzini, and M. Raynaud (1965): *Compt. Rend. Acad. Sci. (Paris)*, **261,** 1448.
Bosshard, H. R., K. H. Jorgensen, and R. E. Humbel (1969): *Eur. J. Biochem.*, **9,** 353.
Bradbury, S. L. (1969): *J. Biol. Chem.*, **244,** 2002.
Bradshaw, R. A., G. W. Robinson, G. M. Hass, and R. L. Hill (1969): *J. Biol. Chem.*, **244,** 1755.
Brewer, C. F., and J. P. Riehm (1967): *Anal. Biochem.*, **18,** 248.
Bunnett, J. F., and D. H. Hermann (1970): *Biochemistry*, **9,** 816.
Bunnett, J. F., and J. J. Randall (1958): *J. Am. Chem. Soc.*, **80,** 6020.
Bunnett, J. F., and R. E. Zahler (1951): *Chem. Rev.*, **49,** 273.
Burr, M., and D. E. Koshland (1964): *Proc. Nat. Acad. Sci.*, **52,** 1017.
Cavins, J. F., and M. Friedman (1967): *Biochemistry*, **6,** 3766.
Cavins, J. F., and M. Friedman (1968): *J. Biol. Chem.*, **243,** 3357.
Chan, T. L., and K. A. Schellenberg (1968): *J. Biol. Chem.*, **243,** 6284.
Clark-Waler, G. D., and H. C. Robinson (1961): *J. Chem. Soc. (London)*, 2810.
Cole, R. D. (1967): *Methods Enzymol.*, **11,** 315.
Colman, R. F. (1968): *J. Biol. Chem.*, **243,** 2454.
Colman, R. F. (1969): *Biochim. Biophys. Acta*, **191,** 469.
Dempsey, W. B., and H. N. Christensen (1962): *J. Biol. Chem.*, **237,** 1113.
Dempsey, W. B., and E. E. Snell (1963): *Biochemistry*, **2,** 1414.
Dopheide, T. A. A., and W. M. Jones (1968): *J. Biol. Chem.*, **243,** 3906.
Eisen, H. N., S. Belman, and M. E. Carsten (1953): *J. Am. Chem. Soc.*, **75,** 4583.
Ettinger, M. J., and C. H. W. Hirs (1968): *Biochemistry*, **7,** 3374.
Fernandez, J. E., and J. S. Fowler (1964): *J. Org. Chem.*, **29,** 402.
Ferrier, B. M., D. Jarvis, and V. du Vigneaud (1965): *J. Biol. Chem.*, **240,** 4264.
Fischer, E. H., A. B. Kent, E. R. Snyder, and E. G. Krebs (1958): *J. Am. Chem. Soc.*, **80,** 2906.
Fletcher, J. C. (1966): *Biochem. J.*, **98,** 34C.
Fonda, M. L., and B. M. Anderson (1969): *J. Biol. Chem.*, **244,** 666.
Fraenkel-Conrat, H., B. A. Brandon, and H. S. Olcott (1947): *J. Biol. Chem.*, **168,** 99.
Fraenkel-Conrat, H., and D. K. Mecham (1949): *J. Biol. Chem.*, **177,** 477.
Fraenkel-Conrat, H., and H. S. Olcott (1948a): *J. Am. Chem. Soc.*, **70,** 2673.
Fraenkel-Conrat, H., and H. S. Olcott (1948b): *J. Biol. Chem.*, **174,** 827.
Freedman, R. B., and G. K. Radda (1968): *Biochem. J.*, **108,** 383.
French, D., and J. T. Edsall (1945): *Adv. Protein Chem.*, **2,** 277.
Friedman, E., D. H. Marrian, and I. Simon-Reuss (1949): *Brit. J. Pharmacol.*, **4,** 105.
Friedman, M. (1967): *J. Am. Chem. Soc.*, **89,** 4709.
Friedman, M., J. F. Cavins, and J. S. Wall (1965): *J. Am. Chem. Soc.*, **87,** 3672.
Friedman, M., and J. S. Wall (1964): *J. Am. Chem. Soc.*, **86,** 3735.
Friedman, M., and J. S. Wall (1966): *J. Org. Chem.*, **31,** 2888.
Gerwin, B. I. (1967): *J. Biol. Chem.*, **242,** 451.
Goldfarb, A. R. (1966a): *Biochemistry*, **5,** 2570.
Goldfarb, A. R. (1966b): *Biochemistry*, **5,** 2574.
Goldfarb, A. R. (1970): *Biochim. Biophys. Acta*, **200,** 1.
Gorecki, M., and Y. Shalitin (1967): *Biochem. Biophys. Res. Commun.*, **29,** 189.

Goren, H. J., and E. A. Barnard (1970): *Biochemistry*, **9**, 959.
Gregory, J. D. (1955): *J. Am. Chem. Soc.*, **77**, 3922.
Gundlach, H. G., W. H. Stein, and S. Moore (1959): *J. Biol. Chem.*, **234**, 1754.
Gurd, F. R. N. (1967): *Methods Enzymol.*, **11**, 532.
Habeeb, A. F. S. A. (1966): *Anal. Biochem.*, **14**, 328.
Habeeb, A. F. S. A., and R. Hiramoto (1968): *Arch. Biochem. Biophys.*, **126**, 16.
Haynes, R., D. T. Osuga, and R. E. Feeney (1967): *Biochemistry*, **6**, 541.
Heinrikson, R. L. (1966): *J. Biol. Chem.*, **241**, 1393.
Heinrikson, R. L., W. H. Stein, A. M. Crestfield, and S. Moore (1965): *J. Biol. Chem.*, **240**, 2921.
Heitz, J. R., C. D. Anderson, and B. M. Anderson (1968): *Arch. Biochem. Biophys.*, **127**, 627.
Hirs, C. H. W. (1962): *Brookhaven Symp. Biol.*, **15**, 154.
Hirs, C. H. W. (1967): *Methods Enzymol.*, **11**, 27.
Hirs, C. H. W., M. Halmann, and J. H. Kycia (1965): *Arch. Biochem. Biophys.*, **111**, 209.
Holeysovsky, V., and M. Lazdunski (1968): *Biochim. Biophys. Acta*, **154**, 457.
Horton, H. R., H. Kelley, and D. E. Koshland (1965): *J. Biol. Chem.*, **240**, 722.
Horton, H. R., and D. E. Koshland (1965): *J. Am. Chem. Soc.*, **87**, 1126.
Horton, H. R., and D. E. Koshland (1967): *Methods Enzymol.*, **11**, 556.
Hughes, R. C., W. T. Jenkins, and E. H. Fischer (1962): *Proc. Nat. Acad. Sci.*, **48**, 1615.
Inagami, T. (1965): *J. Biol. Chem.*, **240**, PC3453.
Inagami, T., and H. Hatano (1969): *J. Biol. Chem.*, **244**, 1176.
Kallen, R. G., and W. P. Jencks (1966): *J. Biol. Chem.*, **241**, 5864.
Kirtley, M. E., and D. E. Koshland (1967): *Methods Enzymol.*, **11**, 866.
Klein, S. M., and R. D. Sagers (1967): *J. Biol. Chem.*, **242**, 301.
Korman, S., and H. T. Clarke (1956): *J. Biol. Chem.*, **221**, 113.
Koshland, D. E., Y. D. Karkhanis, and H. G. Latham (1964): *J. Am. Chem. Soc.*, **86**, 1448.
Kotaki, A., M. Harada, and K. Yagi (1964): *J. Biochem. (Tokyo)*, **55**, 553.
Kowal, J., T. Cremona, and B. L. Horecker (1965): *J. Biol. Chem.*, **240**, 2485.
Levy, M. (1935): *J. Biol. Chem.*, **109**, 365.
Lewin, S. (1956): *Biochem. J.*, **64**, 30P.
Li, T. K., and B. L. Vallee (1965): *Biochemistry*, **4**, 1195.
Lin, Y., G. E. Means, and R. E. Feeney (1969a): *J. Biol. Chem.*, **244**, 789.
Lin, Y., G. E. Means, and R. E. Feeney (1969b): *Anal. Biochem.*, **32**, 436.
Lindley, H. (1956): *Nature*, **178**, 647.
Link, T. P., and G. R. Stark (1968): *J. Biol. Chem.*, **243**, 1082.
Mahowald, T. A. (1965): *Biochemistry*, **4**, 732.
Marcus, F., and E. Hubert (1968): *J. Biol. Chem.*, **243**, 4923.
Marrian, D. H. (1949): *J. Chem. Soc. (London)*, 1515.
Martin, C. J., and M. A. Marini (1967): *J. Biol. Chem.*, **242**, 5736.
Matsuo, Y., and D. M. Greenberg (1959): *J. Biol. Chem.*, **233**, 507.
McKinney, L. L., E. H. Uhing, E. A. Setzkorn, and J. C. Cowan (1951): *J. Am. Chem. Soc.*, **73**, 1641.
Means, G. E., and R. E. Feeney (1968a): *Biochemistry*, **7**, 2192.
Means, G. E., and R. E. Feeney (1968b): *156th Meeting of the American Chemical Society, Biological Chemistry Division*, Abstract #165.
Means, G. E., and R. E. Feeney (1970): unpublished data.
Moore, J. E., and W. H. Ward (1956): *J. Am. Chem. Soc.*, **78**, 2414.
Morrison, M., W. Steele, and D. J. Danner (1969): *Arch. Biochem. Biophys.*, **134**, 515.

Neumann, N. P., S. Moore, and W. H. Stein (1962): *Biochemistry*, **1**, 68.
Okuyama, T., and K. Satake (1960): *J. Biochem. (Tokyo)*, **47**, 454.
Oza, N. B., and C. J. Martin (1967): *Biochem. Biophys. Res. Commun.*, **26**, 7.
Phillips, A. T., and W. A. Wood (1965): *J. Biol. Chem.*, **240**, 4703.
Plapp, B. V., M. A. Raftery, and R. D. Cole (1967): *J. Biol. Chem.*, **242**, 265.
Pollara, B., and R. W. Von Korff (1960): *Biochim. Biophys. Acta*, **39**, 364.
Pontremoli, S., B. Luppis, W. A. Wood, S. Traniello, and B. L. Horecker (1965): *J. Biol. Chem.*, **240**, 3464.
Quiocho, F. A., and F. M. Richards (1966): *Biochemistry*, **5**, 4062.
Raftery, M. A., and R. D. Cole (1963): *Biochem. Biophys. Res. Commun.*, **10**, 467.
Richards, F. M., N. Allewell, D. Tsernoglou, K. Hardman, and H. W. Wyckoff (1969): *158th Meeting of the American Chemical Society, Biological Chemistry Division*, Abstract # 65.
Riehm, J. P., and H. A. Scheraga (1966): *Biochemistry*, **5**, 93.
Rippa, M., L. Spanio, and S. Pontremoli (1967): *Arch. Biochem. Biophys.*, **118**, 48.
Roberts, E., and G. Rouser (1958): *Anal. Chem.*, **30**, 1291.
Rothfus, J. A. (1969): *Anal. Biochem.*, **30**, 279.
Saidel, L. J., and R. L. Carino (1966): *Fed. Proc.*, **25**, 796.
Saidel, L. J., J. S. Satzman, and W. H. Elfring (1965); *Nature*, **207**, 169.
Sanger, F. (1945): *Biochem. J.*, **39**, 507.
Satake, K., T. Okuyama, M. Ohashi, and T. Shinoda (1960): *J. Biochem. (Tokyo)*, **47**, 654.
Satake, K., M. Tanaka, and H. Shino (1961): *J. Biochem. (Tokyo)*, **50**, 6.
Schirich, L. G., and M. Mason (1963): *J. Biol. Chem.*, **238**, 1032.
Schoellmann, G., and E. Shaw (1962): *Biochem. Biophys. Res. Commun.*, **7**, 36.
Schroeder, W. A., J. R. Shelton, and B. Robberson (1967): *Biochim. Biophys. Acta*, **147**, 590.
Seligman, A. M. (1954): in *Glutathione*, edited by S. P. Colowick, Academic Press, New York.
Shaltiel, S. (1967): *Biochem. Biophys. Res. Commun.*, **29**, 178.
Shaltiel, S., and M. Soria (1969): *Biochemistry*, **8**, 4411.
Shapiro, S., M. Enser, E. Pugh, and B. L. Horecker (1968): *Arch. Biochem. Biophys.*, **128**, 554.
Sharpless, N. E., and M. Flavin (1966): *Biochemistry*, **5**, 2963.
Shaw, E. (1967): *Methods Enzymol.*, **11**, 677.
Sluyterman, L. A. E. (1968): *Biochim. Biophys. Acta*, **151**, 178.
Smyth, D. G., A. Nagamatsu, and J. S. Fruton (1960): *J. Am. Chem. Soc.*, **82**, 4600.
Smythe, L. E. (1953): *J. Am. Chem. Soc.*, **75**, 574.
Spande, T. F., N. Wilchek, and B. Witkop (1968): *J. Am. Chem. Soc.*, **90**, 3256.
Stark, G. R., and W. H. Stein (1964): *J. Biol. Chem.*, **239**, 3755.
Strausbauch, P. H., A. B. Kent, J. L. Hedrick, and E. H. Fischer (1967): *Methods Enzymol.*, **11**, 671.
Takahashi, K., W. H. Stein, and S. Moore (1967): *J. Biol. Chem.*, **242**, 4682.
Tamaoki, H., Y. Murase, S. Minato, and K. Nakanishi (1967): *J. Biochem. (Tokyo)*, **62**, 7.
Thunberg, T. (1911): *Skand. Arch. Physiol.*, **25**, 343.
Tietze, F., J. A. Gladner and J. E. Folk (1957): *Biochim. Biophys. Acta*, **26**, 659.
Tsai, H. J., and G. R. Williams (1965): *Can. J. Biochem.*, **43**, 1409.
Tsunoda, J. N., and K. T. Yasunobu (1966): *J. Biol. Chem.*, **241**, 4610.
Tuppy, H. (1959): "The Role of Sulfur in Cytochrome c," in *Sulfur in Proteins*, edited by R. Benesch, Academic Press, New York.
Vithayathil, P. J., and F. M. Richards (1960): *J. Biol. Chem.*, **235**, 2343.
Wang, S. S., and F. H. Carpenter (1968): *J. Biol. Chem.*, **243**, 3702.
Weil, L., and T. S. Seibles (1961): *Arch. Biochem. Biophys.*, **95**, 470.
Whitaker, J. R. (1969): *Biochemistry*, **8**, 459.

Whitehouse, M. W., and I. F. Skidmore (1966): *Biochem. J.*, **100,** 52P.
Williams, J. N., and R. M. Jacobs (1966): *Biochem. Biophys. Res. Commun.*, **22,** 695.
Williams, J. N., and R. M. Jacobs (1968): *Biochim. Biophys. Acta*, **154,** 323.
Wilson, D. A., and I. P. Crawford (1965): *J. Biol. Chem.*, **240,** 4801.
Wilson, E. M., and H. L. Kornberg (1963): *Biochem. J.*, **88,** 578.
Yamashiro, D., H. L. Aanning, and V. du Vigneaud (1965): *Proc. Nat. Acad. Sci.*, **54,** 166.
Yamashiro, D., and V. du Vigneaud (1968): *J. Am. Chem. Soc.*, **90,** 487.

7
ester- and amide-forming reagents

7-1 ESTERIFICATION

7-1.1 METHANOL/HCl

$$\text{P}-\text{C}(=O)\text{OH} + CH_3OH \xrightarrow{0.02 \text{ to } 0.1 M \text{ HCl}} \text{P}-\text{C}(=O)\text{OCH}_3 + H_2O \quad (1)$$

Carboxyl groups of proteins can be converted to methyl esters in methanol containing small amounts of hydrochloric acid (0.02–0.1 M). Under such conditions most proteins experience considerable change in conformation but rather few chemical changes, other than the esterification. Only two minor side reactions, methanolysis of amide groups

$$-\text{C}(=O)\text{NH}_2 \xrightarrow{CH_3OH, HCl} -\text{C}(=O)\text{OCH}_3 + NH_3$$

and the so-called N → O acyl shift (see Section 4-2) can be detected under typical conditions. Conformational changes accompanying the reaction preclude its use with many proteins. It is well suited, however, for relatively stable proteins and has, for example, given valuable information concerning the roles of carboxyl groups in the structure and function of egg white lysozyme (Fraenkel-Conrat and Olcott, 1945; Frieden, 1956), chymotrypsinogen (Doscher and Wilcox, 1961), pancreatic ribonuclease (Broomfield et al., 1965), and bovine serum albumin (Ram and Maurer, 1959).

The reaction is usually performed by dissolving or, if it is insoluble, by suspending the protein in anhydrous methanol to which a small amount of concentrated hydrochloric acid or hydrogen chloride has been added. The amount of water contained in the reaction solution is of extreme importance to the course of reaction and should be kept to a minimum. Even low amounts decrease the extent of esterification. Rates of esterification are subject to considerable variation with changes in temperature and in acid concentration. To minimize unwanted changes in protein conformation and in the two aforementioned side reactions, the mildest conditions commensurate with the desired modification should be used. This normally means a temperature between 0° and 25°C and an acid concentration from 0.02 to 0.10 M. The time usually required to achieve extensive modification under such conditions varies from one to several days. Because some protein carboxyl groups are esterified much more readily than others, extensive modification is not always necessary. Esterification of a few carboxyl groups can sometimes give as much information as modification of all. When eight of the eleven carboxyl groups of pancreatic ribonuclease were esterified, for example, a catalytically inactive but water-soluble derivative was obtained having an apparent conformation similar to the unmodified enzyme. When using higher temperatures or acid concentrations, however, only an insoluble completely esterified derivative with an altered conformation could be obtained (Broomfield et al., 1965).

Protein methyl esters slowly hydrolyze in aqueous solutions at pH's greater than 7 (Ram and Maurer, 1959). In working up such derivatives from their reaction medium, it is therefore usually necessary to maintain a low pH during dialysis and throughout any chromatography steps. Esterified proteins have few or no anionic groups, and therefore have high isoelectric points; if dialyzed against weakly buffered solutions, they experience a rise in pH. It is therefore necessary, in order to avoid hydrolysis of the esterified groups, to dialyze against low-pH buffers or, more commonly, against dilute hydrochloric acid (i.e., 0.001 M).

$$\text{(P)}-\text{C}\begin{matrix}\text{O}\\\text{OCH}_3\end{matrix} + \text{OH}^- \xrightarrow{\text{pH}>10} \text{(P)}-\text{C}\begin{matrix}\text{O}\\\text{O}^-\end{matrix} + \text{CH}_3\text{OH} \quad (2)$$

Hydrolysis can be troublesome during purification and characterization of esterified proteins but can ultimately be of considerable value in assessing effects of the modification. In weak to moderately alkaline solutions, hydrolysis is sufficiently rapid to be useful for bringing about reversion to the unmodified state. By so doing, it may be possible to distinguish those effects of the modification procedure due specifically to the substitution of carboxyl groups. A pH of 10 or slightly higher is usually employed for such reversals. When the water-soluble, but catalytically inactive, methyl ester derivative of pancreatic ribonuclease is treated at pH 10.4

Ester- and Amide-Forming Reagents 141

and 24° for 26 hours, most of its initial catalytic activity is recovered (Broomfield et al., 1965).

7-1.2 DIAZOACETATES

$$\text{P}-\text{C}(=O)\text{OH} + \text{N}_2\text{CH}-\text{C}(=O)\text{NHR} \xrightarrow{\text{pH} \sim 5} \text{P}-\text{C}(=O)\text{OCH}_2-\text{C}(=O)\text{NHR} + \text{N}_2\uparrow \quad (3)$$

Diazomethane is a highly reactive compound widely used to esterify carboxylic acids. It was used by Herriott (1947) to esterify the carboxyl groups of pepsin, but such treatment also resulted in the modification of other groups. Diazoacetate esters and amides are more stable than diazomethane and react more specifically with carboxyl groups of proteins. They are easily prepared and can be stored as stable crystalline compounds. Their reaction with protein carboxyl groups takes place rapidly under mild conditions. Alkylation of sulfhydryl groups is an important side reaction.

Diazoacetates react with water and with many simple inorganic anions. In aqueous solutions, their rapid reaction with water limits modification to the more accessible or reactive carboxyl groups of proteins and makes the use of excess reagent necessary. With a typical protein, modification of only 20–30% of the carboxyl groups is usually possible. Sulfate, nitrate, and especially chloride ions promote decomposition of diazoacetates (Fraenkel, 1907). The reaction of ethyl diazoacetate with chloride ion presumably gives ethyl chloroacetate. Perchloric acid and its salts do not react with diazoacetates and can be used to adjust the pH or ionic strength of reaction solutions.

$$\text{N}_2\text{CHC}(=O)\text{NHR} + \text{H}_2\text{O} \longrightarrow \text{HOCH}_2\text{C}(=O)\text{NHR} + \text{N}_2\uparrow \quad (4)$$

$$\text{N}_2\text{CHC}(=O)\text{NHR} + \text{H}^+ + \text{Cl}^- \longrightarrow \text{Cl}-\text{CH}_2\text{C}(=O)\text{NHR} + \text{N}_2\uparrow \quad (5)$$

Diazoacetates react optimally with protein carboxyl groups near pH 5. With pancreatic ribonuclease and diazoacetylglycinamide, optimum reaction was at pH 4.5 (Riehm and Scheraga, 1965). The un-ionized carboxyl groups appear to be the reactive species (Doscher and Wilcox, 1961). At lower pH's, hydrolysis of the reagent becomes appreciable and limits the extent of modification. Carboxyl groups of proteins vary greatly in their reactivity with diazoacetates. Single carboxyl groups in both pancreatic ribonuclease (Riehm and Scheraga, 1965) and chymotrypsinogen A (Doscher and Wilcox, 1961), for example, have been shown to be manyfold more reactive than all others in those proteins.

Diazoacetyl-D,L-norleucine methyl ester has been shown to react with and inactivate pepsin (Rajagopalan et al., 1966; Lundblad and Stein, 1969). Cupric (Cu^{++}) and silver ions (Ag^+) have been shown to greatly facilitate the inactivation. Other diazoacetates, for example, diazoacetylglycine methyl ester and methyl diazoacetate, similarly inactivate pepsin showing the specificity not to reside with the reagent side chain. The reactive species has been postulated to be a metal ion (Cu^{++} or Ag^+) complexed carbene. It is thought to react with a protonated carboxyl group of the enzyme with the specificity resulting from an adjacent ionized carboxyl group, which serves to orient the positively charged complex. Other diazo compounds inactivate pepsin by reacting with carboxyl groups at or near its active site. These include both the D and L isomers of diazoacetylphenylalanine (Delpierre and Fruton, 1966), 1-diazo-4-phenyl-2-butanone (Fry et al., 1968), and diphenyldiazomethane (Delpierre and Fruton, 1965). Unlike diazoacetylnorleucine, reaction with these reagents is partly determined by their bulky side chains.

Methyl diazoacetate can be prepared from glycine methyl ester by treating with nitrous acid (Equation 6). Diazoacetamide, which is more soluble in water and for this reason usually preferred, is prepared by treatment of methyl diazoacetate with ammonia (Equation 7). It has a weak yellow color in aqueous solution ($\varepsilon = 18.4$ M^{-1} cm^{-1} at 375 mμ) and can be quantitated from this absorption or at its 250-mμ maximum ($\varepsilon = 1.78 \times 10^4$ M^{-1} cm^{-1}) (Doscher and Wilcox, 1961).

$$NH_2-CH_2-C(=O)OCH_3 + HNO_2 \xrightarrow{0°-4°C} N_2CH-C(=O)OCH_3 + 2 H_2O \quad (6)$$

$$N_2CH-C(=O)OCH_3 + NH_3 \longrightarrow N_2CH-C(=O)NH_2 + CH_3OH \quad (7)$$

Glycolate esters resulting from reaction of carboxyl groups with diazoacetates are readily saponified in alkaline solution (Equation 8). Incubation of a monoesterified ribonuclease derivative for 12 hours at pH 9.5, for example, resulted in regeneration of the unesterified enzyme (Riehm and Scheraga, 1965). Glycolesters and glycolamide residues may be determined after saponification at pH 10.2 and treatment with hot concentrated sulfuric acid. The latter treatment converts the released glycolamide molecules to carbon monoxide and formaldehyde. The formaldehyde is then determined using chromotropic acid (Doscher and Wilcox, 1961). The extent of esterification with diazoacetylglycine amide can be determined from the increase in glycine content by amino acid analysis (Riehm and Scheraga, 1965).

This procedure is particularly useful with proteins having naturally low glycine contents.

$$\text{P}-\text{C}(=\text{O})-\text{OCH}_2-\text{C}(=\text{O})\text{NH}_2 + \text{OH}^- \xrightarrow{\text{pH}>9.5} \text{P}-\text{C}(=\text{O})-\text{O}^- + \text{HOCH}_2-\text{C}(=\text{O})\text{NH}_2 \quad (8)$$

$$\text{P}-\text{C}(=\text{O})-\text{O}^- + \text{CH}_3\text{CH}_2-\overset{+}{\text{O}}(\text{CH}_2\text{CH}_3)_2\, \text{BF}_4^- \xrightarrow{\text{pH 4-5}}$$

$$\text{P}-\text{C}(=\text{O})-\text{O}-\text{CH}_2\text{CH}_3 + \text{O}(\text{CH}_2\text{CH}_3)_2 + \text{BF}_4^- \quad (9)$$

Triethyloxonium fluoroborate has been used by Parsons et al. (1969) to esterify the carboxyl groups of egg white lysozyme (Equation 9). When carried out at pH 4.5, the reaction with 1.0 M oxonium reagent gave a derivative having 5.5 esterified carboxyl groups and no other modified amino acid residues. At lower levels of reagent, a monoesterified derivative could be obtained with negligible catalytic activity. A different monoesterified derivative was formed at pH 4.0 which retained appreciable catalytic activity. This derivative was unusually labile, readily reverting at pH 7 to unmodified lysozyme. This reagent has not yet been used with proteins other than lysozyme but appears suitable as a general reagent for the preferential esterification of reactive protein carboxyl groups.

The use of methanol/HCl and of diazoacetates for the esterification of proteins has been reviewed by Wilcox (1967).

7-2 CARBODIIMIDE-PROMOTED AMIDE FORMATION

$$\text{P}-\text{C}(=O)-O^- + NH_2-R \xrightarrow[R'-N=C=N-R']{pH \sim 5} \text{P}-\text{C}(=O)-NHR \quad (1)$$

Carboxyl groups of proteins can be converted into amides by a two-step reaction with a water-soluble carbodiimide (WSC) and an amine (Equation 1). Under mild conditions, only the more accessible or reactive carboxyl groups react, while in the presence of denaturants, nearly quantitative substitution can be obtained. In the absence of an added amine, the formation of linkages between protein carboxyl and amino groups has been observed (Sheehan and Hlavka, 1957; Goodfriend et al., 1964). Unknown carboxylate derivatives, probably N-acylureas (Equation 4), have also been reported (Franzblau et al., 1963; Riehm and Scheraga, 1966).

$$\text{P}-\text{C}(=O)-O^- + \overset{R}{\underset{H}{N}}=C=\overset{+}{N}-R' + H^+ \longrightarrow \text{P}-\text{C}(=O)-O-C(=\overset{+}{N}HR)(NHR') \quad (2)$$

$$\text{P}-\text{C}(=O)-O-C(=\overset{+}{N}HR)(NHR') + HX \longrightarrow \text{P}-\text{C}(=O)-X + O=C(NHR)(NHR') + H^+ \quad (3)$$

$$\text{P}-\text{C}(=O)-O-C(=\overset{+}{N}HR)(NHR') \longrightarrow \text{P}-\text{C}(=O)-NR'-C(=O)-NHR + H^+ \quad (4)$$

Carbodiimide-catalyzed amide formation appears to involve the formation of an intermediate O-acylisourea (Equation 2). In aqueous solution, these highly reactive intermediates either slowly hydrolyze, condense with amines to yield corresponding amides, or condense with other nucleophiles to give a large number of different carboxylate derivatives (Equation 3) (Khorana, 1953). The WSC-promoted reaction between acetylglycine and glycine methyl ester has a broad flat optimum between pH 5 and 7 (Horinishi et al., 1968). However, with proteins the reaction appears to be best at the lower end of this range, between pH 4.5 and 5.0.

Ester- and Amide-Forming Reagents 145

Reactions of carbodiimides have recently been reviewed (Kurzer and Douraghi-Zadeh, 1967). Carbodiimides react not only with carboxylic acids but also with water, alcohols, amines, and many other compounds with an active hydrogen.

$$\text{P}-\text{C}_6\text{H}_4-\text{OH} + \underset{\underset{\text{R}'}{|}}{\underset{\text{N}}{\overset{\text{R}}{\overset{|}{\text{C}}}}}\overset{\text{N}}{\underset{\text{N}}{\|}} \longrightarrow \text{P}-\text{C}_6\text{H}_4-\text{O}-\text{C}\underset{\text{NHR}'}{\overset{\text{NR}}{\diagup}} \qquad (5)$$

$$\text{P}-\text{SH} + \underset{\underset{\text{R}'}{|}}{\underset{\text{N}}{\overset{\text{R}}{\overset{|}{\text{C}}}}}\overset{\text{N}}{\underset{\text{N}}{\|}} \longrightarrow \text{P}-\text{S}-\text{C}\underset{\text{NHR}'}{\overset{\text{NR}}{\diagup}} \qquad (6)$$

The reaction of WSC with phenolic groups of tyrosine residues is thought to give O-arylisoureas of the type shown in Equation 5 (Carraway and Koshland, 1968). These adducts are stable at neutral pH and are even moderately stable during acid hydrolysis of the protein, but they can be decomposed by treatment with hydroxylamine. Sulfhydryl groups of proteins react similarly but the resulting product appears to be more stable and is not decomposed upon treatment with β-mercaptoethanol or ammonia (Carraway and Triplett, 1970). These two reactions appear to be possible for all proteins containing sulfhydryl groups or tyrosine phenolic groups and should be considered when interpreting the results of WSC-promoted reactions with such proteins. Another reaction which may be important for proteins having active serine residues has been observed with α-chymotrypsin. Evidence has been presented to show that the WSC 1-cyclohexyl-3-(2-morpholinyl-4-ethyl)carbodiimide metho-p-toluenesulfonate (CMC) reacts with the active-center serine of this enzyme (Banks et al., 1969). Loss of catalytic activity accompanies the reaction, but treatment with hydroxylamine restores this activity. The kinetics of inactivation are consistent with the formation of a reversible enzyme-WSC absorption complex prior to the inactivation step.

The conversion of protein carboxyl groups into amides in near-quantitative yield provides a convenient means for determining numbers of carboxyl groups in proteins (Hoare and Koshland, 1967). The procedure involves treatment of the protein with excess WSC and an amine in the presence of a high concentration of urea or guanidine hydrochloride. It gives a measure of the number of carboxyl groups based upon the amount of the amine component bound to the protein. This amount can be determined after removing the unbound amine by hydrolysis and amino acid analysis. This is probably the simplest method to distinguish between the carboxyl and carboxamidyl groups in proteins and should have wide general utility.

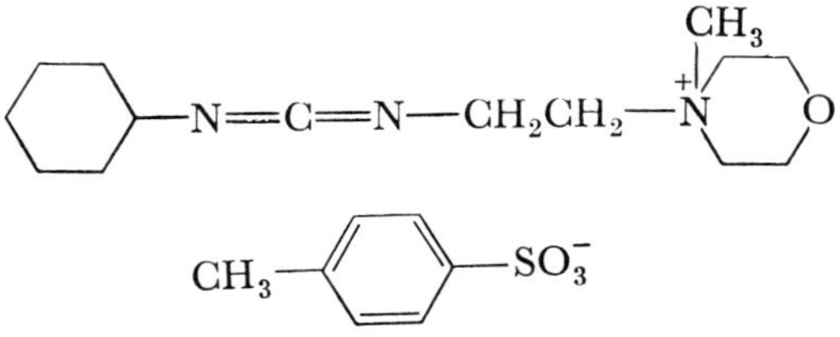

$$\text{cyclohexyl}-N=C=N-CH_2CH_2-\overset{+}{N}(CH_3)(\text{morpholinyl})\quad CH_3-C_6H_4-SO_3^-$$

1-cyclohexyl-3-(2-morpholinyl-4-ethyl)carbodiimide
metho-p-toluenesulfonate (CMC)

$$CH_3CH_2-N=C=N-CH_2CH_2CH_2N(CH_3)_2$$

1-ethyl-3-(3-dimethylaminopropyl)carbodiimide (EDC)

Several WSC's have been used for modification of proteins. CMC and EDC, the structures of which are shown above, are now commercially available and are the most widely used. 1-Benzyl-3-(3-dimethylaminopropyl)carbodiimide (BDC) is also frequently used. Its synthesis has been described (Hoare and Koshland, 1966). All three reagents react similarly although the smaller reagents might be expected to have greater access to partially buried carboxyl groups. This expectation appeared correct in a recent study with tobacco mosaic virus protein, wherein up to three additional carboxyl groups could be modified with EDC, which did not react with CMC.

Just as the WSC can be varied to suit different experimental situations, so can the amine component be varied. Depending upon the amine employed, the character of the product can be varied considerably. Its ionic character, for example, can be like that of the carboxyl group replaced (anionic), or it may be neutral or even cationic as desired (Equations 7, 8, and 9).

Because WSC-promoted amide formation is probably the mildest method for modifying protein carboxyl groups, it appears to be the method of choice for determining the relationship of carboxyl groups to the structure and function of proteins at this time. Modification of only 2.5 carboxyl groups of trypsinogen under very mild conditions, for example, eliminated the calcium-ion requirement for its activation to trypsin (Radhakrishnan et al., 1969). Trypsin isolated after the activation was unmodified. The modified groups appeared to be in those parts of the peptide chain cleaved and then released in the activation process. Under similar conditions trypsin can be extensively modified with the loss of its catalytic activity (Eyl and Inagami, 1970). Competitive inhibitors of trypsin protect several carboxyl groups, and, when the enzyme is modified in the presence of such inhibitors, partial activity is retained.

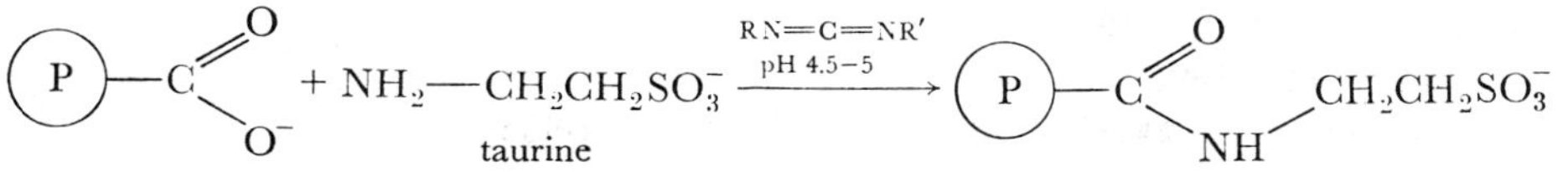

Ester- and Amide-Forming Reagents

$$\text{P}-\text{C}(=O)\text{O}^- + \text{NH}_2-\text{CH}_2\text{CH}_2\text{SO}_3^- \xrightarrow[\text{pH 4.5-5}]{\text{RN=C=NR'}} \text{P}-\text{C}(=O)\text{NH}-\text{CH}_2\text{CH}_2\text{SO}_3^-$$

taurine

(7)

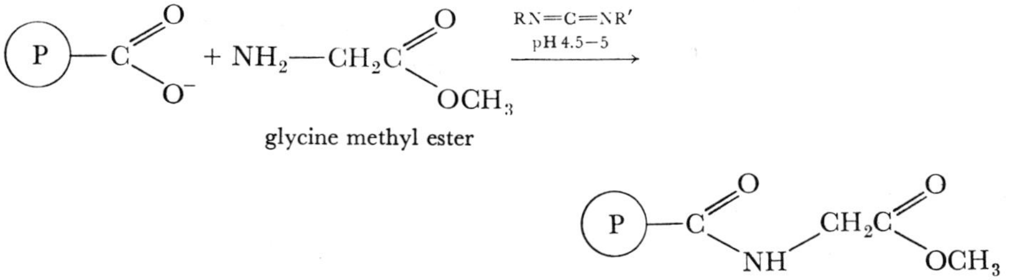

$$\text{P}-\text{C}(=O)\text{O}^- + \text{NH}_2-\text{CH}_2\text{C}(=O)\text{OCH}_3 \xrightarrow[\text{pH 4.5-5}]{\text{RN=C=NR'}}$$

glycine methyl ester

$$\text{P}-\text{C}(=O)\text{NH}-\text{CH}_2\text{C}(=O)\text{OCH}_3 \quad (8)$$

$$\text{P}-\text{C}(=O)\text{O}^- + \text{NH}_2-\text{CH}_2\text{CH}_2\overset{+}{\text{NH}}_3 \xrightarrow[\text{pH 4.5-5}]{\text{RN=C=NR'}}$$

ethylenediamine

$$\text{P}-\text{C}(=O)\text{NH}-\text{CH}_2\text{CH}_2\overset{+}{\text{NH}}_3 \quad (9)$$

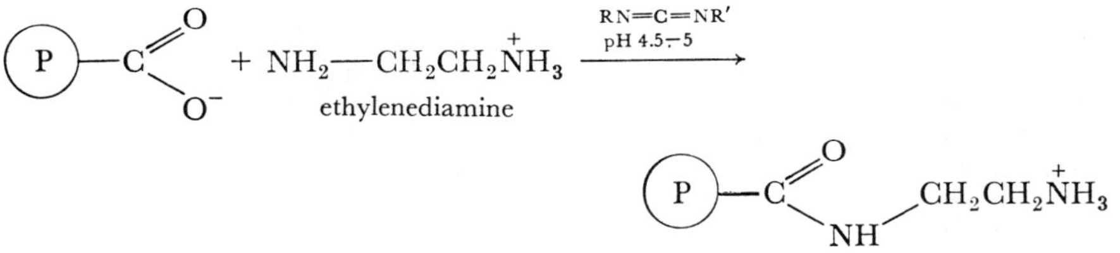

Complete substitution of the carboxyl groups of pancreatic ribonuclease with glycine residues was accomplished by a two-step sequence wherein first the enzyme carboxyl groups were coupled with glycine-N-phthalimidomethyl ester, followed by removal of the phthalimidomethyl group in 0.5 M piperidine (Wilchek et al., 1967). The gross ionic charge of the enzyme was retained, and the individual anionic loci were more or less uniformly displaced a few angstroms from the protein. Unlike other carboxyl-substituted ribonuclease derivatives, whose charge relationships have been altered, this derivative retained its catalytic activity.

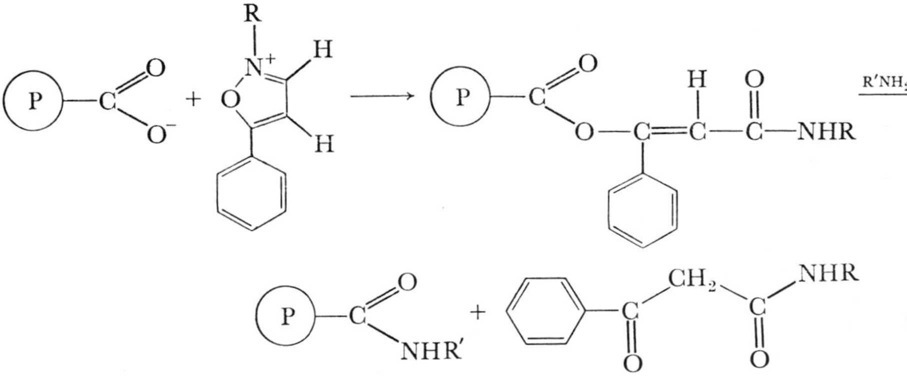

$$\text{P}-\text{C}(=O)\text{O}^- + \text{[N-alkyl-5-phenylisooxazolium]} \longrightarrow \text{P}-\text{C}(=O)-\text{O}-\text{C}(\text{Ph})=\text{CH}-\text{C}(=O)-\text{NHR} \xrightarrow{\text{R'NH}_2}$$

$$\text{P}-\text{C}(=O)\text{NHR'} + \text{Ph}-\text{C}(=O)-\text{CH}_2-\text{C}(=O)-\text{NHR} \quad (10)$$

Woodward et al. (1961, 1966) have shown N-alkyl-5-phenylisooxazolium salts to be useful for activating carboxylic acids for peptide formation. Marfey et al.

(1965) used such a component to promote intermolecular cross-linking of polypeptides. A thorough study of the action of isooxazolium salts on proteins in aqueous solutions has been reported only recently (Bodlaender et al., 1969; Feinstein et al., 1969). These compounds were shown to react rapidly under mild conditions with only carboxyl groups at pH values below 4.75. The enol esters formed in the initial step of the reaction (Equation 10) are sufficiently stable to allow isolation of the protein and subsequent reaction with a nucleophile. Displacement of the enol ester and substitution with many nucleophiles was shown possible.

REFERENCES

Banks, T. E., B. K. Blossey, and J. A. Shafer (1969): *J. Biol. Chem.*, **244**, 6323.
Bodlaender, P., G. Feinstein, and E. Shaw (1969): *Biochemistry*, **8**, 4941.
Broomfield, C. A., J. P. Riehm, and H. A. Scheraga (1965): *Biochemistry*, **4**, 751.
Carraway, K. L., and D. E. Koshland (1968): *Biochim. Biophys. Acta*, **160**, 272.
Carraway, K. L., and R. B. Triplett (1970): *Biochim. Biophys. Acta*, **200**, 564.
Delpierre, G. R., and J. S. Fruton (1965): *Proc. Nat. Acad. Sci.*, **54**, 1161.
Delpierre, G. R., and J. S. Fruton (1966): *Proc. Nat. Acad. Sci.*, **56**, 1817.
Doscher, M. S., and P. E. Wilcox (1961): *J. Biol. Chem.*, **236**, 1328.
Eyl, A., and T. Inagami (1970): *Biochem. Biophys. Res. Commun.*, **38**, 149.
Feinstein, G., P. Bodlaender, and E. Shaw (1969): *Biochemistry*, **8**, 4949.
Fraenkel, W. (1907): *Z. Physik. Chem.*, **60**, 202.
Fraenkel-Conrat, H., and H. S. Olcott (1945): *J. Biol. Chem.*, **161**, 259.
Franzblau, C., P. M. Gallop, and S. Seifen (1963): *Biopolymers*, **1**, 79.
Frieden, E. H. (1956): *J. Am. Chem. Soc.*, **78**, 961.
Fry, K. T., O. K. Kim, J. Spona, and G. A. Hamilton (1968): *Biochem. Biophys. Res. Commun.*, **30**, 489.
Goodfriend, T. L., L. Levine, and G. D. Fasman (1964): *Science*, **144**, 1344.
Herriott, R. M. (1947): *Adv. Protein Chem.*, **3**, 169.
Hoare, D. G., and D. E. Koshland (1966): *J. Am. Chem. Soc.*, **88**, 2057.
Hoare, D. G., and D. E. Koshland (1967): *J. Biol. Chem.*, **242**, 2447.
Horinishi, H., K. Nakaya, A. Tani, and K. Shibata (1968): *J. Biochem. (Tokyo)*, **63**, 41.
Khorana, H. G. (1953): *Chem. Rev.*, **53**, 145.
Kurzer, F., and K. Douraghi-Zadeh (1967): *Chem. Rev.*, **67**, 107.
Lundblad, R. L., and W. H. Stein (1969): *J. Biol. Chem.*, **244**, 154.
Marfey, P. S., T. J. Gill, and H. W. Kunz (1965): *Biopolymers*, **3**, 27.
Parsons, S. M., L. Jao, F. W. Dahlquist, C. L. Borders, T. Groff, J. Racs, and M. A. Raftery (1969): *Biochemistry*, **8**, 700.
Radhakrishnan, T. M., K. A. Walsh, and H. Neurath (1969): *Biochemistry*, **8**, 4020.
Rajagopalan, T. G., W. H. Stein, and S. Moore (1966): *J. Biol. Chem.*, **241**, 4295.
Ram, J. S., and P. H. Maurer (1959): *Arch. Biochem. Biophys.*, **85**, 512.
Riehm, J. P., and H. A. Scheraga (1965): *Biochemistry*, **4**, 772.
Riehm, J. P., and H. A. Scheraga (1966): *Biochemistry*, **5**, 99.
Sheehan, J. C., and J. J. Hlavka (1957): *J. Am. Chem. Soc.*, **79**, 4528.
Wilchek, M., A. Frensdorff, and M. Sela (1967): *Biochemistry*, **6**, 247.
Wilcox, P. E. (1967): *Methods Enzymol.*, **11**, 605.
Woodward, R. B., R. A. Olofson, and H. Mayer (1961): *J. Am. Chem. Soc.*, **83**, 1010.
Woodward, R. B., R. A. Olofson, and H. Mayer (1966): *Tetrahedron Suppl.*, **8**, 321.

8

reducing and oxidizing reagents

8-1 REDUCING AGENTS

$$\underset{S}{\overset{S}{P}}\!\!\diagdown\!\!\!\text{-SR} \rightleftharpoons \underset{S-SR}{\overset{S^-}{P}} \overset{RS^-}{\rightleftharpoons} \underset{S^-}{\overset{S^-}{P}} + R-S-S-R \qquad (1)$$

Disulfide bonds of proteins are reduced to sulfhydryl groups by relatively mild reducing agents. Among the most commonly used are low-molecular-weight simple thiols as typified by β-mercaptoethanol and thioglycolate. The reaction is illustrated in Equation 1. With simple monothiols, the equilibrium constants for interchange are near unity, and a large excess of thiol is required to force the reactions to completion. With dithiols, which upon oxidation form stable cyclic disulfides (for example, dithiothreitol and dithioerythritol), this equilibrium is displaced further to the right, and lower concentrations will usually suffice (Cleland, 1964). Irrespective of the reductant, urea or another denaturing agent is usually necessary to insure complete reduction.

$$\begin{array}{c} \text{CH}_2\text{SH} \\ | \\ \text{CHOH} \quad \text{CH}_2\text{SH} \\ \text{CHOH} \end{array} \rightleftharpoons \begin{array}{c} \text{CH}_2\text{S} \\ | \quad \diagdown \\ \text{CHOH} \quad \text{S} \\ \text{CHOH} \quad \text{CH}_2 \end{array} + 2\,[H] \qquad (2)$$

dithioerythritol (reduced)   dithioerythritol (oxidized)

Although many reducing agents are able to reduce disulfide bonds, thiols appear intrinsically the most specific. The reaction proceeds by an interchange mechanism involving two sequential nucleophilic attacks at sulfur by thiol anions, with mixed disulfides being formed as intermediates (Equation 1) (Fava et al., 1957; Eldjarn and Pihl, 1957). Rates of interchange are proportional to the concentration of thiol anion and are therefore strongly pH-dependent. Reoxidation is likewise prevented or severely retarded at low pH. Separation from excess thiol, to prevent reoxidation, can thus be accomplished at low pH (i.e., <4). In the absence of deforming agents, only more labile or exposed disulfide bonds are reduced. In lysozyme, prolactin, insulin, and in human growth hormone, complete reduction occurs at pH 8.1 with dithiothreitol in the absence of deforming agents. Under the same conditions, only 14 of the 18 disulfide bonds of bovine serum albumin can be reduced, whereas in pancreatic ribonuclease none are susceptible (Bewley et al., 1968; Bewley and Li, 1969). One of three disulfide bonds in bovine pancreatic trypsin inhibitor is subject to reduction under similar conditions (Liu and Meienhofer, 1968). All but one of the disulfide bonds in papain resist reduction by thiols even in 8 M urea, conditions which effect complete reduction of pancreatic ribonuclease (Shapira and Arnon, 1969). Reduction of disulfide bonds was not possible with the enzyme in its native conformation, which it appears to retain even in 8 M urea.

Alkylation of newly formed sulfhydryl groups prevents reoxidation and refolding of the protein subsequent to removal of reducing agents. Following the reduction of lysozyme with β-mercaptoethanol, for example, reoxidation was blocked by alkylating the newly formed sulfhydryl groups with iodoacetamide (see Section 6-1) (Fraenkel-Conrat et al., 1951). A similar procedure was used by Weil and Seibles (1961), but acrylonitrile (see Section 6-3) instead of iodoacetamide was used to alkylate the sulfhydryl groups. Both alkylation procedures are widely used and give approximately equivalent results. Both give derivatives which can be quantitated by amino acid analysis after acid hydrolysis. Acrylonitrile is sometimes preferred because it does not react with methionine and histidine residues, but the lower solubility of the resulting S-cyanoethyl proteins is sometimes a more important consideration.

Sodium borohydride ($NaBH_4$) has not been widely used for complete reduction of proteins although it appears to be well suited for this. Several reports of its ability to reductively cleave peptide bonds have no doubt contributed to its infrequent use (Crestfield et al., 1960; Seon, 1967). It has been shown to be rather useful for the limited selective reduction of more labile or accessible disulfide bonds. For example, it selectively reduces two of five disulfide bonds in both trypsin and trypsinogen, but under the same conditions fails to reduce any disulfide bonds in α-chymotrypsin or chymotrypsinogen A (Light and Sinha, 1967). Of the three S—S bonds in Kunitz bovine pancreatic trypsin inhibitor, one can be selectively reduced by sodium borohydride with no loss of trypsin-inhibitory activity (Kress and Laskowski, 1967). Further reduction of the two remaining disulfide bonds can be obtained in urea but is accompanied by complete loss of inhibitory activity.

An important advantage to the use of sodium borohydride derives from its instability at low pH. Acidification to below pH 7 results in its rapid decomposition to hydrogen gas and borate.

$$NaBH_4 + 3\ H_2O \xrightarrow{pH<7} NaH_2BO_3 + 4\ H_2 \uparrow$$

If air and other oxidizing agents are carefully excluded, or if the pH is kept below 4, reoxidation can be prevented and the newly formed —SH groups can be easily quantitated.

$$\begin{array}{c}\text{(P)}-S \\ | \\ \text{(P)}-S \leftarrow SPO_3^= \end{array} \rightleftharpoons \text{(P)}-S^- + \text{(P)}-S-SPO_3^= \qquad (3)$$

Sodium phosphorothioate (Na_3SPO_3) reacts reversibly with protein disulfide bonds as illustrated by Equation 3. The reaction is not a simple reduction, as

TABLE 8-1

Reducing agent	E_0 (volts)	Reference
Sodium phosphorothioate	~0.0	Neumann et al., 1965
Cysteine	−0.21	Cleland, 1964
Dithiothreitol	−0.33	Cleland, 1964
Sodium borohydride	−1.24	Stockmayer et al., 1955

attested by the high redox potential of phosphorothioate (Table 8-1); instead, it appears to result from simple nucleophilic displacement of thiol by phosphorylthioate (Neumann and Smith, 1967). The reaction is reversible and thiophosphoryl groups are easily displaced by treatment with β-mercaptoethanol. In the presence of oxygen, thiophosphorolysis proceeds until all available cysteine residues are substituted (Equation 4) (Neumann et al., 1964).

$$\tfrac{1}{2}O_2 + \text{(P)}-S-S-\text{(P)} + 2\ HPO_3S^= \rightleftharpoons 2\ \text{(P)}-S-SPO_3^= + H_2O \qquad (4)$$

Phosphorothioate in the presence of urea has been used to cleave all of the S—S bonds of lysozyme and ribonuclease (Neumann et al., 1964). Without urea, only two of the four S—S bonds of ribonuclease were cleaved, and the resulting product retained its enzymatic activity (Neumann et al., 1967a). Stoichiometric amounts of phosphorothioate are sufficient to activate ficin or papain (Neumann et al., 1967b). Phosphorothioate reacts with 2 moles of p-mercuribenzoate and with 1 mole of either iodoacetate or N-ethylmaleimide (Neumann and Smith, 1967).

Electrolytic reduction of proteins (Cecil and Weitzman, 1964; Markus, 1964) has not been widely used and will not be discussed. Metal hydrides other than $NaBH_4$ must be used in aprotic solvents and have also been shown to cleave susceptible peptide bonds (Chibnall and Rees, 1958). The reduction of methionine sulfoxide to methionine is discussed in Section 8-7.

8-2 SULFITE

$$\boxed{P}-SS-\boxed{P} + SO_3^= \rightleftharpoons \boxed{P}-S-SO_3^- + \boxed{P}-S^- \qquad (1)$$

Sulfite ions react with disulfides to form S-substituted thiosulfates, sometimes known as S-sulfonates, and a thiol (Equation 1). Sulfite ion ($SO_3^=$), not bisulfite (HSO_3^-), is the reactive species (Cecil and McPhee, 1955; McPhee, 1956). Above pH 9 where sulfite predominates, the reaction appears to be a simple bimolecular displacement. The equilibrium constant under these conditions is not highly favorable for cleavage ($K_{eq} < 10^{-1}$), and a high concentration of $SO_3^=$ is required to push the reaction to the right. Complex kinetics and a more favorable equilibrium are obtained near pH 7, where the thiol is protonated and yet much of the sulfite remains in the dianionic form.

$$HSO_3^- \xrightleftharpoons{pK \simeq 7} SO_3^= + H^+$$

The apparent equilibrium for the sulfitolysis of disulfide bonds is displaced to the right by anything that removes the thiol anion. Heavy-metal ions and oxidizing agents are commonly used for this purpose. If the reaction is carried out in the presence of cupric ion as an oxidant, all disulfide and thiol groups may be converted into S-sulfonates, accompanied by reduction of the cupric ion to cuprous ion (Equation 2). In this capacity, sulfite and cupric ion have been used both analytically, to determine the sum of thiol and disulfide groups (Kolthoff and Stricks, 1951; Swan, 1957), and preparatively (Dixon and Wardlaw, 1960; Richmond, 1966). o-Iodosobenzoate or dithionite can be used as oxidants in place of cupric ion (Bailey and Cole, 1959).

$$\boxed{P}-SH + Cu^{++} + SO_3^= \rightleftharpoons \boxed{P}-SSO_3^- + Cu^+ + H^+ \qquad (2)$$

The disulfide bonds of many proteins resist the action of sulfite. Bovine serum albumin, for example, is virtually unaffected by sulfite in the absence of other agents. Even in the presence of guanidine hydrochloride, 4 or 5 of its 17 S—S bonds fail to react. If a small amount of mercuric ion is added, however, all 17 bonds are cleaved readily (Kolthoff et al., 1958). Mercuric ion has a similar effect on many other seemingly resistant proteins although the mechanism of its action remains obscure (Cecil and Loening, 1960; Cecil and Wake, 1962).

The action of sulfite on proteins is complicated by an extreme dependence upon ionic environment in the vicinity of individual disulfide bonds. Anionic disulfides have been shown, for example, to have reaction rates very much slower than neutral disulfides. The cationic disulfide, cystamine, on the other hand, is many times more reactive than either (McPhee, 1956).

$$\overset{+}{N}H_3—CH_2CH_2—S—S—CH_2CH_2\overset{+}{N}H_3$$

cystamine

The course of the reaction is further governed very strictly by steric factors. Displacement of sulfide ion from a disulfide linkage appears to involve backside attack by sulfite at one or the other sulfur atom similar to S_N2 substitution at aliphatic carbons (Equation 3). With mixed disulfides, this results in a very strong preference for reaction at the least hindered sulfur atom (Van Rensburg and Swanepoel, 1967). The combination of steric and charge limitations presumably accounts for the great resistance of many proteins to the action of sulfite.

$$\underset{SO_3^=}{R\diagdown S—S\diagdown R'} \rightleftharpoons \underset{SO_3^-}{R\diagdown S} + {}^-S\diagdown R' \qquad (3)$$

If mediated by low-molecular-weight cationic thiols, atmospheric oxygen can be used as oxidant to convert sulfhydryl groups into S-sulfonates (Chan, 1968). The requirement for both oxygen and thiol indicates that the reaction proceeds via formation of disulfide bonds followed by sulfitolysis. With rabbit muscle aldolase, complete S-sulfonation was obtained within an hour in 8 M urea at pH's between 7.0

and 8.5. Little reaction was obtained at pH 9.5 or higher. Cysteine or β-mercaptoethylamine were required to catalyze the reaction and could not be replaced by β-mercaptoethanol or dithiothreitol. The specificity of the reaction and that of its reversal by β-mercaptoethanol were strikingly demonstrated by recovering full enzymatic activity.

$$\text{(P)}-SH + \begin{array}{c} \overset{+}{N}H_3 \\ | \\ CH_2 \\ | \\ CH_2 \\ | \\ HS \end{array} \xrightarrow{\frac{1}{2}O_2} \text{(P)}-S\underset{S-CH_2}{\overset{SO_3^=}{\diagdown}}\begin{array}{c} \overset{+}{N}H_3 \\ | \\ CH_2 \end{array} + H_2O \rightleftharpoons$$

$$\text{(P)}-S\diagup^{SO_3^-} + \begin{array}{c} \overset{+}{N}H_3 \\ | \\ CH_2 \\ | \\ CH_2 \\ | \\ HS \end{array} + OH^- \quad (4)$$

S-Sulfonates are unstable in strongly acidic and alkaline solutions decomposing either to disulfides or in some cases to a mixture of disulfide and dithionate (Rosnati, 1945). Heating S-sulfocysteine in 6 N HCl under conditions used for protein hydrolysis converts it to cystine. Treatment of S-sulfocysteine with reducing agents yields cysteine (—SH). Sulfite is easily displaced from S-sulfo linkage by cyanide, thiol, or sulfite anions. Exchange with the latter may be useful for introducing radioactively labeled sulfur into or out of S-sulfoproteins. The ease with which sulfite can be displaced from S-sulfo linkage by thiols can be a distinct advantage for regenerating the native polypeptide chain (Dixon and Wardlaw, 1960).

The cleavage of disulfide bonds by sulfite is a very useful method. It proceeds under mild conditions, and the resulting cleaved polypeptide chains are often more soluble than those produced by other methods. Unlike performic acid oxidation, sulfitolysis does not destroy tryptophan residues and does not therefore interfere in this respect with subsequent sequence determinations. The ease with which the S-sulfo linkage may be broken is an additional advantage for reversible modification. Its synthetic usefulness has been demonstrated by the complete synthesis of biologically active sheep insulin from fully synthetic S-sulfonated A and B chains (Kung et al., 1965) and later by the synthesis of an enzymatically active synthetic bovine pancreatic ribonuclease A from a synthetic S-sulfo polypeptide precursor (Gutte and Merrifield, 1969).

Further information on the reaction of sulfite with proteins has been described by Cecil (1963) and Cole (1967).

8-3 5,5'-DITHIOBIS(2-NITROBENZOIC ACID) (ELLMAN'S REAGENT)

$$\text{P}-\text{SH} + \text{NO}_2-\underset{^-\text{OOC}}{\bigcirc}-\text{S}-\text{S}-\underset{\text{COO}^-}{\bigcirc}-\text{NO}_2 \underset{\longleftarrow}{\overset{pH \gtrsim 6.8}{\longrightarrow}}$$

$$\text{P}-\text{S}-\text{S}-\underset{\text{COO}^-}{\bigcirc}-\text{NO}_2 + {}^-\text{S}-\underset{\text{COO}^-}{\bigcirc}-\text{NO}_2 \quad (1)$$
$$+ \text{H}^+$$

5,5'-Dithiobis(2-nitrobenzoic acid) (DTNB) reacts with free sulfhydryl groups of proteins, forming thionitrobenzoate protein and liberating for each —SH group 1 mole of thionitrobenzoate anion (Equation 1). This strongly colored anion can be determined from its absorption at 412 mμ ($\varepsilon = 1.36 \times 10^4\ M^{-1}\ \text{cm}^{-1}$ at pH 8). DTNB is thus widely used for quantitative determinations of —SH groups in proteins as well as in various biological fluids and tissues (Ellman, 1959; Butterworth et al., 1967).

DTNB can also be a rather mild reagent for the modification of —SH groups in proteins. Treatment of muscle phosphorylase b dimer with DTNB, for example, demonstrated the presence of two reactive classes of —SH groups. Two (or three) of the first class react rapidly with DTNB with little effect upon catalytic activity. The remainder are modified very slowly and parallel a decrease in catalytic activity (Damjanovich and Kleppe, 1966). The decrease in catalytic activity occurs concomitantly with a dissociation of the enzyme into subunits (Kastenschmidt et al., 1968) similar to that observed with other —SH reagents (Madsen and Cori, 1956; Chignell et al., 1968).

Phosphorylase b dimer $\xrightarrow{\text{DTNB}}$ Phosphorylase b monomer (2)

(active) (inactive)

Monomers containing 2 moles of thiophenylate were completely inactive but could be restored to nearly full activity and to an apparently "native" quaternary structure by treatment with dithiothreitol (Kastenschmidt et al., 1968).

The inactivation of pig heart TPN-dependent isocitrate dehydrogenase by DTNB occurs after reaction with five sulfhydryl groups (Colman, 1969). Inactivation can be blocked by isocitrate in the presence of manganous ion or by TPNH. The DTNB-modified enzyme absorbs between 310 and 325 mμ, where the native enzyme is transparent. The enzyme–thionitrobenzoate mixed disulfide groups have an average extinction coefficient of $2.6 \times 10^3\ M^{-1}\ \text{cm}^{-1}$ at 310 mμ, similar to that of the mixed β-mercaptoethanol–nitrobenzoate disulfide ($2.45 \times 10^3\ M^{-1}\ cm^{-1}$ at 323 mμ) (Colman, 1969). DTNB itself has an absorption peak at 323 mμ with an extinction coefficient of $1.66 \times 10^4\ M^{-1}\ cm^{-1}$.

DTNB-modified isocitrate dehydrogenase can be partially reactivated by treatment with β-mercaptoethanol. Upon storage, however, a secondary change takes place, releasing thionitrobenzoate ion and eliminating the ability to be reactivated. This secondary inactivation appears to result from disulfide interchange with unreacted sulfhydryl groups in the partially modified enzyme. The possibility of disulfide interchange in the presence of DTNB was first postulated by Fernandez-Diez et al. (1964), who pointed out that the formation of thionitrobenzoate ion, upon which quantitation of reaction is normally based, would be unaffected by such reactions (Equation 3).

Four sulfhydryl groups in lobster muscle glyceraldehyde-3-phosphate dehydrogenase (i.e., one in each peptide chain) react very rapidly with DTNB, bringing about complete loss of catalytic activity. The subsequent reaction of additional sulfhydryl groups (i.e., a total of five in each peptide chain) is postulated to follow the formation of an intramolecular disulfide linkage between this active-site cysteine residue 148 and the nearby cysteine residue 152 (Wassarman and Major, 1969). Formation of a similar intramolecular disulfide bond is thought to account for the inactivation of guinea pig liver transglutaminase by DTNB (Connellan and Folk,

1969). Catalytic activity of this DTNB-inactivated enzyme could be readily restored by treatment with dithiothreitol.

$$\text{P}-\text{SH} + \underset{\substack{\text{4,4'-dithiodipyridine} \\ \lambda_{max} = 247 \text{ m}\mu \\ \varepsilon_{247} = 1.63 \times 10^4 \, M^{-1} \text{cm}^{-1}}}{\text{Py-S-S-Py}} \longrightarrow \text{P}-\text{S-S-Py} + \underset{\substack{\text{4-thiopyridone} \\ \lambda_{max} = 324 \text{ m}\mu \\ \varepsilon_{324} = 1.98 \times 10^4 \, M^{-1} \text{cm}^{-1}}}{\text{thiopyridone}} \quad (4)$$

Dithiopyridine reacts with —SH groups in a manner similar to DTNB, liberating a molecule of thiopyridone that can be assayed spectrophotometrically. This reagent offers some advantage over DTNB, in the greater pH range over which it gives useful results (Grassetti and Murray, 1967).

Dihydroxydinaphthyl disulfide has been used with fixed proteins to form insoluble disulfide complexes which, after reaction with a diazonium salt, can be used for quantitation of –SH groups in tissue sections (Barka and Anderson, 1963). Its reaction with rabbit muscle heavy meromyosin was accompanied by formation of an absorption peak at 375 mμ (Onodera and Yagi, 1969). Binding of 2 moles greatly enhanced the protein's ATPase activity.

8-4 o-IODOSOBENZOATE

$$\text{P}\underset{\text{SH}}{\overset{\text{SH}}{<}} + \text{(o-COO}^-\text{)C}_6\text{H}_4\text{-IO} \longrightarrow \text{P}\underset{\text{S}}{\overset{\text{S}}{<}} + \text{(o-COO}^-\text{)C}_6\text{H}_4\text{-I} + \text{H}_2\text{O} \quad (1)$$

Since Hellerman et al. (1941) employed o-iodosobenzoate to estimate protein —SH groups, it has become a widely used enzyme inhibitor. Its principal known reaction with proteins is the oxidation of —SH groups to disulfide bonds (Equation 1). Sulfenic acids may be important intermediates in the reaction. Reactions with o-iodosobenzoate are normally carried out near neutral pH, but little is known about the reaction's pH dependence. At low pH, the reagent is quite insoluble in aqueous solutions.

Oxidation of methionine residues has not been detected, but this may deserve reinvestigation. Such oxidations, in the past, have been difficult to detect and frequently were overlooked (see Sections 8-6 and 8-7). For example, the inactivation of pancreatic ribonuclease by o-iodosobenzoate (Ledoux, 1954), since no sulfhydryl groups are present, must result from the oxidation of some other group(s), most likely methionine.

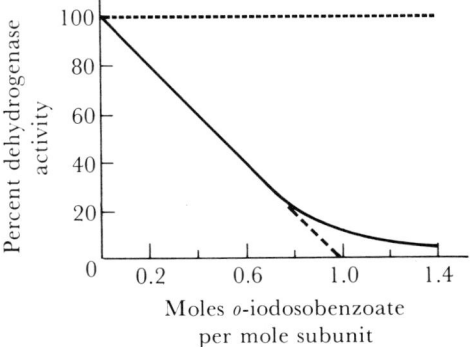

FIGURE 8-1   o-Iodosobenzoate inactivation of pig muscle glyceraldehyde-3-phosphate dehydrogenase incubated at pH 7.5 with increasing concentrations of o-iodosobenzoate at 25° for 10 minutes, at which time samples were diluted in 0.05 M sodium pyrophosphate, pH 8.5, with (dotted line) and without (solid line) 0.01 M dithiothreitol. (Adapted from Parker and Allison, 1969.)

Glyceraldehyde-3-phosphate dehydrogenase from yeast is inactivated by o-iodosobenzoate and, under certain conditions, can be reactivated with thiols (Rafter, 1957). Inactivation of the corresponding enzyme from pig muscle occurs by a stoichiometric reaction with one —SH group per enzyme subunit and, like the yeast enzyme, it can be reactivated by thiols (Parker and Allison, 1969) (Figure 8-1). The catalytically important —SH group in this inactive form of the enzyme is thought to be in the oxidation state of a sulfenic acid (Parker and Allison, 1969; Ehring and Colowick, 1969). Such preparations, although inactive as dehydrogenases, acquire a marked ability to catalyze the hydrolysis of acetyl phosphate (Ehring and Colowick, 1969). This presumed sulfenic derivative is converted by incubation at 37°C into a derivative lacking both dehydrogenase and acetyl phosphatase activity. The transformation presumably involves the formation of a single disulfide bond, producing a conformationally altered form of the protein from which catalytic activity cannot be regained (Equation 2).

$$P(SH)(SH) \underset{RSH}{\rightleftharpoons} P(S-OH)(SH) \xrightarrow{37°C} P\begin{pmatrix}S\\S\end{pmatrix} + H_2O \qquad (2)$$

8-5 TETRATHIONATE

$$S_4O_6^= + P(SH)(SH) \rightleftharpoons P(S-S_2O_3^-)(SH) + S_2O_3^= + H^+ \qquad (1)$$

$$P(S-S_2O_3^-)(SH) \longrightarrow P\begin{pmatrix}S\\S\end{pmatrix} + S_2O_3^= + H^+ \qquad (2)$$

Sodium tetrathionate ($Na_2S_4O_6$) can be either an oxidizing or a reducing agent, depending upon the conditions. It rapidly oxidizes simple thiols to corresponding disulfides. With sulfhydryl proteins, rather stable sulfenylthiosulfate intermediates have been observed (Equation 1) (Pihl and Lange, 1962). These appear to react further with other —SH groups to form disulfides (Equation 2). The dihydrate of sodium tetrathionate may be prepared by iodine oxidation of sodium thiosulfate in 95% ethanol. It is unstable and cannot be stored for more than a few weeks. It can be crystallized from ethanol-water to remove thiosulfate and other impurities.

The reversible inactivation of pig muscle glyceraldehyde-3-phosphate dehydrogenase by ^{35}S-tetrathionate is accompanied by the binding of one equivalent of thiosulfate and the disappearance of one —SH group per enzyme subunit (Parker and Allison, 1969). This sulfenylthiosulfate group is relatively stable, apparently due to structural characteristics of the enzyme which limit its contact with other —SH groups. By treatment with thiols, the thiosulfate group can be displaced and the enzyme reactivated. If, however, the sulfenylthiosulfate derivative of the enzyme is warmed for a short period or is subjected to mildly denaturing conditions, thiosulfate ion is expelled and the enzyme becomes irreversibly inactivated. This irreversible inactivation is thought to be caused by a conformational change accompanying the formation of an intramolecular disulfide bond with a second cysteine residue four residues removed in linear sequence (see Figure 8-2).

Streptococcal proteinase has a single —SH group and reacts with sodium tetrathionate to give an inactive sulfenylthiosulfate derivative (Liu, 1967). Blocking of the reactive —SH group in this way introduces an additional negative charge at the

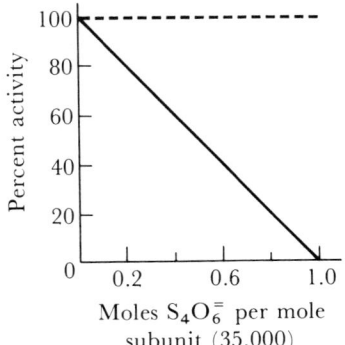

FIGURE 8-2 Tetrathionate inactivation of pig muscle glyceraldehyde-3-phosphate dehydrogenase. The enzyme, 0.01 μmole, was incubated at 0° in 1.0 ml of 0.05 M sodium pyrophosphate, pH 8.5, with increasing concentrations of $Na_2S_4O_6$ at 0°. After 2 minutes, samples were diluted in 0.05 M sodium pyrophosphate, pH 8.5, at 0°, and 0.05 M sodium pyrophosphate, pH 8.5, at 25°, which contained 0.01 M dithiothreitol. Samples which did not contain dithiothreitol were assayed immediately (solid line); samples in 0.01 M dithiothreitol were assayed after 20 minutes at 25° (dashed line). (Adapted from Parker and Allison, 1969.)

active center and increases the susceptibility of an active-center histidine residue to alkylation by the cationic substratelike reagent α-N-bromoacetylarginine methyl ester. If it is not alkylated, the inactive sulfenylthiosulfate enzyme can be reactivated by thiol treatment. Alkylation without prior treatment with tetrathionate gives principally S-alkyl products.

8-6 PERFORMIC ACID

$$\text{(P)}-SH + 3\,H-C\underset{O-OH}{\overset{O}{\diagup}} \longrightarrow \text{(P)}-SO_3H + 3\,H-C\underset{OH}{\overset{O}{\diagup}} \quad (1)$$

$$\text{(P)}-S-S-\text{(P)} + 5\,H-C\underset{O-OH}{\overset{O}{\diagup}} + H_2O \longrightarrow$$

$$2\,\text{(P)}-SO_3H + 5\,H-C\underset{OH}{\overset{O}{\diagup}} \quad (2)$$

$$\text{(P)}-SCH_3 + 2\,H-C\overset{O}{\underset{O-OH}{\diagdown}} \longrightarrow \text{(P)}-\overset{\overset{O}{\uparrow}}{\underset{\underset{O}{\downarrow}}{S}}-CH_3 + 2\,H-C\overset{O}{\underset{OH}{\diagdown}} \qquad (3)$$

Performic acid is an extremely powerful oxidant. It oxidizes both cysteine and cystine to cysteic acid (Equations 1 and 2) and methionine to methionine sulfone (Equation 3). Tryptophan is degraded to formylkynurenine and several other products. Destruction of tryptophan is a major drawback to the use of performic acid for cleaving disulfide bonds and separating peptide chains. Tyrosine, serine, and threonine are also oxidized, but very much more slowly and usually only under relatively drastic conditions. Performic acid oxidizes halide ions to the corresponding halogens, which may then attack tyrosine and histidine residues to form both mono- and dihalogen derivatives.

$$H-C\overset{O}{\underset{OH}{\diagdown}} + H_2O_2 \rightleftharpoons H-C\overset{O}{\underset{O-OH}{\diagdown}} + H_2O \qquad (4)$$

(excess)

Performic acid can be prepared from formic acid and hydrogen peroxide (Equation 4). Sanger (1949) prepared performic acid *in situ* to cleave the A and B chains of insulin in his determination of insulin's primary structure. Other workers add preformed reagent directly to the cooled reaction solution. The oxidation of cysteine and methionine residues converts them into more stable products which are easily quantitated by conventional methods. The rapid elution of the strongly acidic cysteic acid from cation-exchange resins causes some difficulty during amino acid analysis, particularly in preparations containing other ninhydrin-positive materials which elute early in the chromatography. Under controlled conditions, the conversion of methionine to its sulfone is nearly quantitative and the conversion of cysteine into cysteic acid is usually $95 \pm \sim 3\%$.

The conditions required to assure complete oxidation vary somewhat with different proteins but usually require at least a tenfold excess of oxidant. Side reactions are minimized at low temperature ($\sim -10°C$), using a time which just allows complete reaction of cysteine and methionine (Hirs, 1956, 1967). Addition of a small amount of methanol will prevent the solution from freezing. Excessive treatment results in losses of both cysteic acid and methionine sulfone. Reagent remaining after the reaction is completed can be removed by the addition of HBr (HBr is oxidized to Br_2 and can be removed by aspiration), but results in the destruction of tyrosine and histidine (Moore, 1963). Sanger (1949) used a large excess of acetone to precipitate the oxidized chains of insulin, and thereby prevented further reaction. Dilution with a large volume of cold water and immediate freeze-drying is another commonly used procedure (Hirs, 1967).

8-7 HYDROGEN PEROXIDE

$$\boxed{P}-SH + 3\,H_2O_2 \longrightarrow \boxed{P}-SO_3H + 3\,H_2O \tag{1}$$

$$\boxed{P}-SCH_3 + H_2O_2 \longrightarrow \boxed{P}-\underset{O}{\overset{\|}{S}}-CH_3 + H_2O \tag{2}$$

Hydrogen peroxide is a potent oxidant of many organic compounds. In proteins, it attacks principally the sulfur-containing amino acids cysteine and methionine. In the presence of certain metal ions, organic acids, or ethers, it may also attack cystine, tryptophan, and tyrosine.

By judicious use, hydrogen peroxide can be a specific oxidant of methionine in non-SH proteins. Active monosulfoxide derivatives of pancreatic ribonuclease (Neumann et al., 1962), α-chymotrypsin (Koshland et al., 1962; Schachter et al., 1963), and Kunitz trypsin inhibitor (Kassell, 1964) have been prepared using mild hydrogen peroxide treatment at low pH (pH 1 to 3). Monosulfoxide derivatives of adrenocorticotropic and parathyroid hormones were inactive (Dedman et al., 1961; Tashjian et al., 1964). Oxidation of a second methionine in either pancreatic ribonuclease or Kunitz trypsin inhibitor caused a loss of biological activity. Oxidation by hydrogen peroxide of a single methionine residue in subtilisin at pH 8.8 occurs concurrent with changes in kinetic parameters of the enzyme. Although immediately adjacent to the active-site serine, oxidation of this residue does not abolish enzymatic activity (Stauffer and Etson, 1969). Four other methionine residues in the enzyme were not oxidized by this treatment. The single oxidized methionine of α-chymotrypsin monosulfoxide is only three residues from the "active" serine, yet only the enzyme's affinity for substrates appears to be affected (Koshland et al., 1962; Ray and Koshland, 1962; Schachter and Dixon, 1964). Oxidation of a second methionine by hydrogen peroxide occurred after unfolding the enzyme in urea, and further reduced its substrate affinity without decreasing its inherent catalytic ability. Similar results were obtained with a disulfoxide derivative of trypsin (Holeysovsky and Lazdunski, 1968).

The rates of oxidation of methionine and cysteine are affected differently by pH. Oxidation of cysteine decreases in velocity at low pH, whereas the oxidation of methionine increases slightly in going from pH 5 to 1 (Toennies and Callan, 1939).

Cystine is relatively less susceptible to oxidation than either cysteine or methionine. While mild oxidation of simple sulfhydryl compounds readily produces the corresponding disulfides, similar treatment of large polypeptides and proteins usually results in the formation of other oxidation products (Equation 3).

$$\text{(P)}-\text{SH} \xrightarrow{[O]} \begin{bmatrix} \text{(P)}-\text{S}-\text{S}-\text{(P)} \xrightarrow{[O]} \text{(P)}-\text{S}-\overset{\overset{\displaystyle O}{\|}}{\underset{\underset{\displaystyle O}{\|}}{\text{S}}}-\text{(P)} \\ \text{(P)}-\text{SOH} \xrightarrow{[O]} \text{(P)}-\text{SO}_2\text{H} \end{bmatrix} \xrightarrow{[O]} \text{(P)}-\text{SO}_3\text{H} \quad (3)$$

Inactivation of rabbit muscle glyceraldehyde-3-phosphate dehydrogenase by hydrogen peroxide has been shown to result from sulfhydryl-group modification (Little and O'Brien, 1969). Only 3.5 sulfhydryl groups per tetrameric enzyme were readily oxidizable and responsible for the observed inactivation. They appeared to proceed to sulfenic acids rather than to disulfides. The effects of such oxidation could be reversed by treatment with excess of low-molecular-weight thiol or sodium arsenite. Full recovery could be obtained by immediate reduction, but, if delayed, recovery became progressively less successful. Inability to reactivate did not appear solely attributable to disulfide formation.

Hydrogen peroxide reacts with certain organic acids to form corresponding acylperoxides or peracids (i.e., performic, peracetic, etc.), which are much more potent and, therefore, less selective oxidants than hydrogen peroxide (see Section 8-6). The oxidation of both cystine (—S—S—) and tryptophan in α-chymotrypsin by hydrogen peroxide in the presence of its substrate has been attributed to the formation from an acyl enzyme intermediate of an acylperoxide (Schachter et al., 1963). Cytochrome c has been shown to enhance the rate of oxidation of glyceraldehyde-3-phosphate dehydrogenase (Little and O'Brien, 1969).

A dioxane peroxide has been postulated to account for the rapid oxidation of tryptophan by hydrogen peroxide in sodium bicarbonate solutions containing 10% dioxane (Hachimori et al., 1964). Such conditions have been used to determine the accessibility of tryptophan residues in proteins. Several heavy-metal ions promote the breakdown of hydrogen peroxide to free radicals. Ferrous ions and hydrogen peroxide, for example, have been shown to bring about the formation of HO· radicals and have been used to mimic the effects of gamma irradiation on proteins (Slobodian et al., 1968). Halide ions (except F^-) are oxidized by hydrogen peroxide to the corresponding halogens. Both mono- and dichlorotyrosine thus often result from peroxide treatment and can be conveniently avoided only by taking special precautions to exclude chloride ions.

Methionine sulfoxide can be reduced by thiols to methionine. Sulfoxide derivatives of parathyroid hormone (Tashjian et al., 1964), ribonuclease-S-peptide

$$\text{(P)}-\underset{\underset{O}{\|}}{S}CH_3 + 2\,RSH \longrightarrow \text{(P)}-SCH_3 + R-S-S-R + H_2O \qquad (4)$$

(Kenkare and Richards, 1966), ribonuclease (Jori et al., 1968a), and lysozyme (Jori et al., 1968b) have been restored to nearly full biological activity by mild thiol treatment.

In hot hydrochloric acid under conditions commonly used to hydrolyze proteins (6 N HCl, 110°C, 18 hours), methionine sulfoxide decomposes to methionine in about 85% yield:

$$-CH_2-\underset{\underset{}{\overset{\overset{O}{\|}}{S}}}{}-CH_3 \xrightarrow[110°C,\,18\,hr]{6\,N\,HCl} -CH_2-S-CH_3 \qquad (5)$$
$$(85\%)$$

Homocystine, homoserine, and homolanthionine are formed in small amounts (Morihara, 1964). This reformation of methionine in comparatively high yield accounts for the relatively late discovery of methionine sulfoxide as an important oxidation product of proteins (Ray and Koshland, 1960, 1962). Methionine sulfoxide is not destroyed by alkaline hydrolysis and can usually be recovered in about 90% yield. Alternately, it can be determined indirectly after successive treatment with excess iodoacetic and performic acids, followed by normal acid hydrolysis (Neumann et al., 1962; Vithayathil and Richards, 1960). In this treatment, iodoacetic acid does not affect sulfoxide residues but converts methionine to its carboxymethyl-sulfonium derivative. This is resistant to performic acid (Equation 6),

$$-CH_2-S-CH_3 \xrightarrow{ICH_2COO^-} -CH_2\overset{+}{S}\!\!\begin{array}{c}CH_3\\ CH_2COO^-\end{array} \xrightarrow{HC\overset{\overset{O}{\diagup}}{\diagdown}O-OH} \text{no reaction}$$
$$(6)$$

$$-CH_2-\underset{\underset{}{\overset{\overset{O}{\|}}{S}}}{}-CH_3 \xrightarrow{ICH_2COO^-} \text{no reaction} \xrightarrow{HC\overset{\overset{O}{\diagup}}{\diagdown}O-OH} -CH_2-\underset{\underset{O}{\|}}{\overset{\overset{O}{\|}}{S}}-CH_3$$
$$(7)$$

which is then used to oxidize the methionine sulfoxide to methionine sulfone (Equation 7). The sulfone is stable to acid hydrolysis and can be quantitated by amino acid analysis and used as a measure of the sulfoxide (plus sulfone) present at the time of carboxymethylation (Equation 8). The carboxymethylsulfonium derivative is unstable to the hydrolytic conditions, giving rise to several products (Section 6-1) and is not determined (Equation 9).

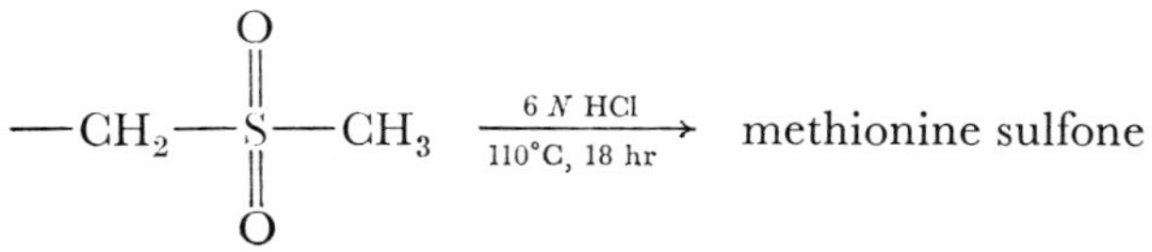

$$-CH_2-\overset{\overset{O}{\|}}{\underset{\underset{O}{\|}}{S}}-CH_3 \xrightarrow[110°C,\ 18\ hr]{6\ N\ HCl} \text{methionine sulfone} \tag{8}$$

$$-CH_2-\overset{+}{S}\overset{CH_3}{\underset{CH_2-COO^-}{\diagdown}} \xrightarrow[110°C,\ 18\ hr]{6\ N\ HCl} \begin{array}{l} \text{methionine} \\ \text{S-carboxymethylhomocystine} \\ \text{homoserine lactone} \end{array} \tag{9}$$

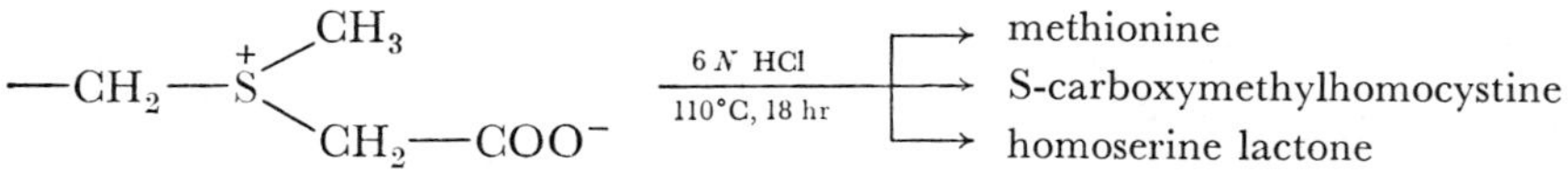

8-8 PHOTOOXIDATION

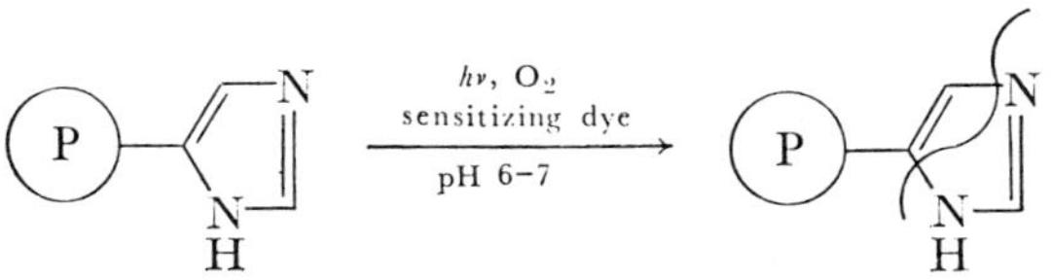

(1)

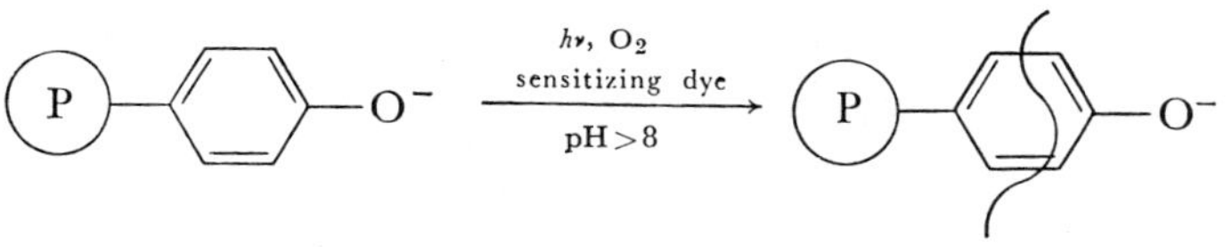

(2)

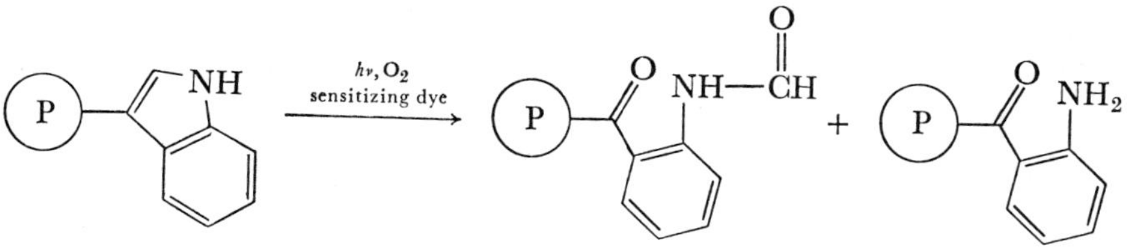

(3)

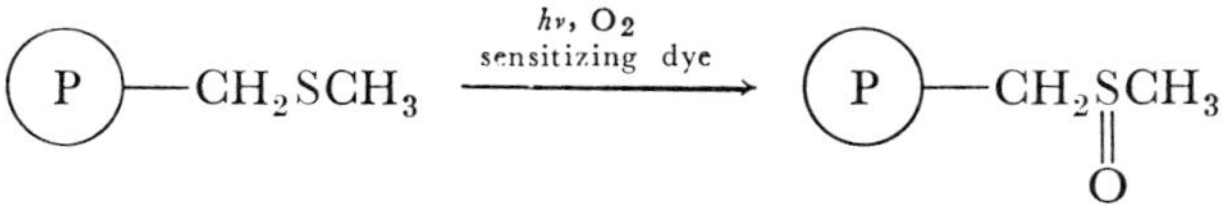

(4)

Raab (1900) first reported the lethal effects of visible light on microorganisms treated with certain dyes. This photodynamic effect was investigated by Weil and co-workers (Weil et al., 1951; Weil and Buchert, 1951; Weil and Seibles, 1955; Weil,

1965) as a means to selectively oxidize proteins and amino acids. Histidine, tyrosine, tryptophan, methionine, and cystine were shown to react, whereas other common amino acids did not. Under the same conditions, cysteine is oxidized to cystine without a sensitizing dye.

Dyes vary greatly in their efficiency as photosensitizers. All effective photosensitizing dyes appear to act via the formation of long-lived excited states, probably triplets, and are capable of being photoreduced (Oster et al., 1959). Sensitized photooxidations have been shown to proceed via several different mechanisms depending upon the dye and its concentration, the concentration of oxygen ($\sim 10^{-3} M$ for aqueous solution exposed to air), the nature of the oxidizable substrate, and a number of additional factors. The reaction may proceed either by transferring the excited-state energy of the sensitizer to oxygen or by a direct interaction of the excited dye with the substrate (Equation 5). Methylene-blue-sensitized oxidations of histidine and methionine, for example, appear to be of the former type, whereas reaction with tyrosine appears to involve a direct electron transfer from tyrosine to the excited sensitizer. The photosensitized oxidation of tryptophan appears to follow a complex or mixed type mechanism (Weil, 1965).

$$\text{dye} \xrightarrow{h\nu} \text{dye}^* \begin{cases} \xrightarrow{O_2} \text{dye} + O_2^* \xrightarrow{\text{substr}} [\text{substr } O_2] \longrightarrow \text{substr}_{ox} \\ \xrightarrow{\text{substr}} \text{dye}_{red} + \text{substr}_{ox} \end{cases}$$

(5)

The photooxidation products of the aromatic amino acids have not been well characterized, but the aromatic rings are known to be the main points of attack. Peptide bonds are not broken. Kynurenine and N-formylkynurenine are major degradation products of tryptophan (Equation 3). Histidine is photooxidized to aspartic acid and urea via several intermediate compounds (Tomita et al., 1969). Whether, in proteins, the reaction proceeds to completion or only to one of these intermediates is not known. Methionine is usually oxidized to methionine sulfoxide plus, in some cases, a small amount of methionine sulfone.

Two dyes, methylene blue and rose bengal, have been most widely used to photosensitize the oxidation of proteins. Riboflavin, proflavin, fluorescein, and eosin are among those which have been used effectively with amino acids (Bellin and Yankus, 1968; Grossweiner and Zwicker, 1961). Methylene blue has been and remains the most widely used photosensitizing dye for proteins. However, recent evidence has shown rose bengal to be somewhat more selective for histidine (Westhead, 1965; Bellin and Yankus, 1968). Being anionic is thought to facilitate the latter's specificity by favoring the formation of short-lived dye-imidazole complexes.

Photooxidation proceeds rapidly even at low temperature. Its specificity for the various amino acid side chains is partially determined by pH. Oxidation of histidine, for example, is usually the fastest reaction at neutral pH but is quite slow at low pH. Tyrosine is most reactive at higher pH's, where it is anionic (Sluyterman, 1962; Weil, 1965; DasGupta and Boroff, 1965). Tryptophan and methionine are the only amino acids readily oxidized below pH 4. Inactivation of an enzyme resulting from the destruction of histidine is often indicated by a characteristic pH dependence, similar to the photooxidative action spectrum of histidine (Figure 8-3) (Freude, 1968; Westhead, 1965; Martinez-Carrion et al., 1967).

The effects of oxygen concentration on the photodynamic inactivation of several enzymes and on the destruction of several model compounds have been investigated and found to differ with different dyes (Weil and Buchert, 1951; Hodgson et al., 1969). In general, quantum yields were found to increase as oxygen concentrations

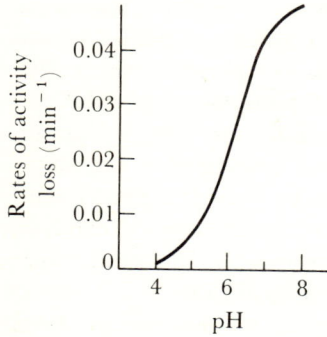

FIGURE 8-3   Rate of activity loss of aspartate aminotransferase as a function of pH at 9° in the presence of methylene blue exposed to incandescent light. (From Martinez-Carrion et al., 1967.)

of the saturating atmosphere increased to 20% and, thereafter, at higher concentrations to remain relatively constant. Bellin and Yankus (1968) have described a decrease in quantum yield for the rose-bengal-sensitized oxidation of histidine at oxygen concentrations greater than 40%.

In acidic solution, the methylene-blue-photosensitized oxidation of bovine pancreatic ribonuclease decreases that enzyme's catalytic activity and converts both of its methionine residues into methionine sulfoxide (Jori et al., 1968a). At higher pH, extensive inactivation is accompanied by the destruction of histidine residues (Weil and Seibles, 1955). Ribonuclease contains no tryptophan. Reduction with thioglycolic acid has been shown to completely restore methionine and, in the former case, catalytic activity.

Using proflavin to sensitize the photooxidation of lysozyme in aqueous formic acid resulted in the oxidation of 1.9 moles of methionine and a quantitative conversion of all six tryptophan residues into kynurenine (Benassi et al., 1967; Galiazzo et al., 1968). Reduction of this modified enzyme with β-mercaptoethanol restored the methionine but not the catalytic activity. It is not clear, however, whether the failure to regain activity was due only to the absence of tryptophan in the partially restored enzyme or whether its absence also interfered with correct refolding of the peptide chain. Similar photooxidation, sensitized by methylene blue in 84% acetic acid, resulted in the oxidation of both methionines but not tryptophan. This latter product retained low catalytic activity and could be restored to near full activity by reduction with thiols (Jori et al., 1968b).

Photooxidation of sheep heart phosphofructokinase in the presence of methylene blue and either ATP or fructose-6-phosphate has been shown to render it insensitive to ATP inhibition (Ahlfors and Mansour, 1969). Kinetic studies at pH 8.2 of this photooxidized enzyme revealed no significant change as compared to the native enzyme in the K_m values for the two substrates, fructose-6-phosphate and ATP. At pH 6.9, lower K_m's than the native enzyme were obtained for both substrates. Homotropic interaction between the fructose-6-phosphate binding sites was eliminated, as were the inhibition by citrate and the activation by 5'-AMP. Photooxidation in the absence of both substrates led to complete destruction of catalytic activity. The groups oxidized in each case were not determined. The results were taken as evidence for separate catalytic and regulatory sites for ATP.

In most cases, loss of enzymatic activity following photooxidation has been attributed to the destruction of critical histidine residues. Thus, for example, the inactivations of rabbit muscle aldolase (Hoffee et al., 1967), pig heart aspartate amino transferase (Martinez-Carrion et al., 1967; Martinez-Carrion et al., 1970), cytochrome c (Nakatani, 1960), yeast enolase (Westhead, 1965), bovine carboxypeptidase A (Freude, 1968), and rennin (Hill and Laing, 1965) have all been attributed to destruction of histidine residues. The inactivation of ribonuclease A (Weil and Seibles, 1955), mentioned above, presumably resulted from the destruction of both histidine and methionine, as did the inactivation of phosphoglucomutase (Ray and Koshland, 1962).

Cresol red and crystal violet have been shown to selectively photosensitize the oxidation of cysteine, both free and in proteins (Jori et al., 1969). In acidic solutions, cysteine can be quantitatively converted to cysteic acid, while at higher pH reaction in the dark brings about its oxidation to cystine.

Pyridoxal-5-phosphate binds specifically to 6-phosphogluconate dehydrogenase and in that state has been used as a specific photosensitizer for the oxidation of adjacent histidine residues (Rippa and Pontremoli, 1969) (see Section 6-9). Irradiation of proteins containing a covalently linked sensitizer in a known position has been shown in the case of 41-dinitrophenyllysylribonuclease to give information on the three-dimensional structure (Scoffone et al., 1970). Irradiation of such products should result in the selective modification of photooxidizable residues adjacent to the sensitizer, and in this way can be used to identify such groups.

Because methionine sulfoxide reverts, in part, to methionine upon acid hydrolysis (see Section 8-7) and tryptophan is destroyed, special procedures are required to ascertain their presence. Alkaline hydrolysis largely preserves both. The alkaline hydrolysis and analysis procedure of Ray and Koshland (1962) typically accounts for 85% of the methionine sulfoxide. Methods for the quantitation of tryptophan have been described in Chapter 4. The indirect procedure of Neumann et al. (1962) for the determination of methionine and methionine sulfoxide by sequential treatment with iodoacetate and performic acid has been described in the preceding section. Problems associated with the detection of methionine sulfoxide have been discussed by Neumann (1967). Photooxidation is at this time probably the best general procedure for selectively modifying the histidine residues of proteins. Under somewhat different conditions, the same basic procedure can be used to selectively oxidize methionine, tryptophan, and cysteine residues. More information on photooxidation is available in two recent reviews by Spikes and Straight (1967) and Foote (1968). Photooxidation using methylene blue has been reviewed by Ray (1967).

8-9 N-BROMOSUCCINIMIDE

Witkop and coworkers (Patchornik et al., 1958; Ramachandran and Witkop, 1959; Viswanatha et al., 1960) have shown N-bromosuccinimide (NBS) to be a useful reagent for selectively cleaving tryptophanyl peptide bonds in peptides and proteins. It also cleaves tyrosyl peptide bonds, but more slowly than those of tryptophan (Ramachandran and Witkop, 1967). Histidine residues undergo a similar reaction, but very much more slowly (Shaltiel and Patchornik, 1963). NBS is a potent oxidizing agent. It oxidizes —SH groups more rapidly than its reaction with tryptophan. Methionine and cystine can also be oxidized. The reactions of NBS with proteins have been thoroughly reviewed (Witkop, 1961; Ramachandran and Witkop, 1967; Spande and Witkop, 1967; Spande et al., 1970).

NBS and N-bromoacetamide, which is chemically similar, are stable, readily obtainable crystalline compounds. N-Bromoacetamide has been found superior to NBS for cleaving certain small peptides (Patchornik et al., 1958). It reacts with proteins somewhat more slowly than does NBS. Both compounds are highly reactive sources of very electrophilic bromonium ions (Br^+). With each, the initial reaction involves addition of bromine to the γ-δ carbon-carbon double bonds of tryptophan, tyrosine, or histidine residues. Scission of the respective peptide bonds occurs subsequently, as shown in Equations 1 and 2.

$$\text{[peptide]}-NH-\underset{\underset{O}{\|}}{C}-\underset{\underset{NH}{|}}{CH}-CH_2-\text{[indole]} + 1\tfrac{1}{2}\,\text{[NBS]} \xrightarrow{pH\sim 4}$$

$$\text{[support]}-\overset{+}{NH_3} \quad NH-\underset{CH_2}{CH}\underset{\underset{O}{\|}}{\overset{C-O}{\diagup}}\text{[oxindole]}-NH + 1\tfrac{1}{2}\,\text{[succinimide]}-NH + 1\tfrac{1}{2}\,Br^- \qquad (1)$$

$$\text{[peptide]}-NH-\underset{\underset{O}{\|}}{C}-\underset{\underset{NH}{|}}{CH}-CH_2-\text{[phenyl]}-OH + 3\,\text{[NBS]} \xrightarrow{pH\sim 4}$$

$$\text{[support]}-\overset{+}{NH_3} \quad NH-\underset{CH_2}{CH}\underset{}{\overset{C-O}{\diagup}}\text{[dibromocyclohexadienone]}=O + 3\,\text{[succinimide]}-NH + Br^- \qquad (2)$$

Of the aromatic amino acids, tryptophan is most susceptible to NBS oxidation. The reaction is best performed in acidic media, usually in acetate or formate buffers at pH 4 or below. In small peptides or model compounds, only 1.5 equivalents of NBS are required to effect the cleavage of a tryptophan residue, whereas in proteins, due to competing side reactions, usually 3 to 6 equivalents are necessary for maximal cleavage. Yields from 50–90% of the theoretical number of cleavages can be obtained with peptides and other model compounds, but with proteins yields normally are between 10% and 50%. The reaction is accompanied by a decrease in 280-mμ absorbance and the appearance of a much stronger absorbance, centered at 261 mμ

($\varepsilon = 10,300\ M^{-1}\ cm^{-1}$). These changes in absorption have been used to quantitatively determine tryptophan contents of proteins (Spande and Witkop, 1967).

Reaction with NBS can be used to probe the accessibility and reactivity of tryptophan residues in proteins (Section A-8) and to assess the involvement of such residues in various properties. Careful treatment of chicken egg white lysozyme with NBS, for example, gives a progressive destruction of tryptophan, accompanied by loss of catalytic activity (Rao and Ramachandran, 1962; Bernier and Jolles, 1961). All six of its tryptophans are accessible to, and can be oxidized by, NBS at pH 4. The catalytic inactivation correlates, however, with the destruction of just one of these so-called exposed tryptophan residues (Hayashi et al., 1963) and constituted some of the early evidence for tryptophan in the active center of lysozyme. Under slightly different conditions, considerable oxidation of tyrosine and histidine was found to precede complete inactivation (Kronman et al., 1967). Treatment with NBS in 8 M urea gives rise to N-bromourea. This less reactive intermediate gives relatively greater specificity for tryptophan (i.e., to the exclusion of tyrosine) (Funatsu et al., 1964).

Tyrosyl peptides react with 3 equivalents of NBS to form a dibromodienone spirolactone carboxy-terminal fragment ($\varepsilon_{260} = 10,000$ to $11,000\ M^{-1}\ cm^{-1}$) and a new amino-terminal fragment (Equation 2) (Schmir et al., 1959; Corey and Haefele, 1959; Witkop, 1961; Ramachandran and Witkop, 1967). Reaction is slower than with tryptophan and has not been widely used with proteins. The reaction of NBS with histidine is very similar to its reaction with tryptophan and tyrosine but takes place very slowly even at elevated temperatures. It has been used to cleave peptides but appears to be of little use for intact proteins.

REFERENCES

Ahlfors, C. E., and T. E. Mansour (1969): *J. Biol. Chem.*, **244**, 1247.
Bailey, J. L., and R. D. Cole (1959): *J. Biol. Chem.*, **234**, 1733.
Barka, T., and P. J. Anderson (1963): *Histochemistry*, Harper & Row, New York.
Bellin, J. S., and C. A. Yankus (1968): *Arch. Biochem. Biophys.*, **123**, 18.
Benassi, C. A., E. Scoffone, G. Galiazzo, and G. Jori (1967): *Photochem. Photobiol.*, **6**, 857.
Bernier, I., and P. Jolles (1961): *Compt. Rend. Acad. Sci. (Paris)*, **253**, 745.
Bewley, T. A., J. S. Dixon, and C. H. Li (1968): *Biochim. Biophys. Acta*, **154**, 420.
Bewley, T. A., and C. H. Li (1969): *Int. J. Protein Res.*, **1**, 117.
Butterworth, P. H. W., H. Baum, and J. W. Porter (1967): *Arch. Biochem. Biophys.*, **118**, 716.
Cecil, R. (1963): in *The Proteins*, edited by H. Neurath, vol. 1, p. 379.
Cecil, R., and U. E. Loening (1960): *Biochem. J.*, **76**, 146.
Cecil, R., and J. R. McPhee (1955): *Biochem. J.*, **60**, 496.
Cecil, R., and R. G. Wake (1962): *Biochem. J.*, **82**, 401.
Cecil, R., and R. D. J. Weitzman (1964): *Biochem. J.*, **93**, 1.
Chan, W. W. C. (1968): *Biochemistry*, **7**, 4247.
Chibnall, A. C., and M. W. Rees (1958): *Biochem. J.*, **68**, 105.
Chignell, D. A., W. B. Gratzer, and R. C. Valentine (1968): *Biochemistry*, **7**, 1082.

Cleland, W. W. (1964): *Biochemistry*, **3**, 480.
Cole, R. D. (1967): *Methods Enzymol.*, **11**, 206.
Colman, R. F. (1969): *Biochemistry*, **8**, 888.
Connellan, J. M., and J. E. Folk (1969): *J. Biol. Chem.*, **244**, 3173.
Corey, E. J., and L. F. Haefele (1959): *J. Am. Chem. Soc.*, **81**, 2225.
Crestfield, A. M., J. Skupin, S. Moore, and W. H. Stein (1960): *Fed. Proc.*, **19**, 341.
Damjanovich, S., and K. Kleppe (1966): *Biochim. Biophys. Acta*, **122**, 145.
DasGupta, B. R., and D. A. Boroff (1965): *Biochim. Biophys. Acta*, **97**, 157.
Dedman, M. L., T. H. Farmer, and C. J. O. R. Morris (1961): *Biochem. J.*, **78**, 348.
Dixon, G. H., and A. C. Wardlaw (1960): *Nature*, **188**, 721.
Ehring, R., and S. P. Colowick (1969): *J. Biol. Chem.*, **244**, 4589.
Eldjarn, L., and A. Pihl (1957): *J. Biol. Chem.*, **225**, 499.
Ellman, G. L. (1959): *Arch. Biochem. Biophys.*, **82**, 70.
Fava, A., A. Iliceto, and E. Camera (1957): *J. Am. Chem. Soc.*, **79**, 833.
Fernandez-Diez, M. J., D. T. Osuga, and R. E. Feeney (1964): *Arch. Biochem. Biophys*, **107**, 449.
Foote, C. S. (1968): *Science*, **162**, 963.
Fraenkel-Conrat, H., A. Mohammad, E. D. Ducay, and D. K. Mecham (1951): *J. Am. Chem. Soc.*, **73**, 625.
Freude, K. A. (1968): *Biochim. Biophys. Acta*, **167**, 485.
Funatsu, M., N. M. Green, and B. Witkop (1964): *J. Am. Chem. Soc.*, **86**, 1846.
Galiazzo, G., G. Jori, and E. Scoffone (1968): *Biochem. Biophys. Res. Commun.*, **31**, 158.
Grassetti, D. R., and J. F. Murray (1967): *Arch. Biochem. Biophys.*, **119**, 41.
Grossweiner, L. I., and E. F. Zwicker (1961): *J. Chem. Phys.*, **34**, 1411.
Gutte, B., and R. B. Merrifield (1969): *J. Am. Chem. Soc.*, **91**, 501.
Hachimori, Y., H. Horinishi, K. Kurihara, and K. Shibata (1964): *Biochim. Biophys. Acta*, **93**, 346.
Hayashi, K., T. Imoto, and M. Funatsu (1963): *J. Biochem. (Tokyo)*, **54**, 381.
Hellerman, L., F. P. Chinard, and P. A. Ramsdell (1941): *J. Am. Chem. Soc.*, **63**, 2551.
Hill, R. D., and R. R. Laing (1965): *Biochim. Biophys. Acta*, **99**, 352.
Hirs, C. H. W. (1956): *J. Biol. Chem.*, **219**, 611.
Hirs, C. H. W. (1967): *Methods Enzymol.*, **11**, 59.
Hodgson, C. F., E. B. McVey, and J. D. Spikes (1969): *Experientia*, **25**, 1021.
Hoffee, P., C. Y. Lai, E. L. Pugh, and B. L. Horecker (1967): *Proc. Nat. Acad. Sci.*, **57**, 107.
Holeysovsky, V., and M. Lazdunski (1968): *Biochim. Biophys. Acta*, **154**, 457.
Jori, G., G. Galiazzo, A. Marzotto, and E. Scoffone (1968a): *Biochim. Biophys. Acta*, **154**, 1.
Jori, G., G. Galiazzo, A. Marzotto, and E. Scoffone (1968b): *J. Biol. Chem.*, **243**, 4272.
Jori, G., G. Galiazzo, and E. Scoffone (1969): *Int. J. Protein Res.*, **1**, 289.
Kassell, B. (1964): *Biochemistry*, **3**, 152.
Kastenschmidt, L. L., J. Kastenschmidt, and E. Helmreich (1968): *Biochemistry*, **7**, 3590.
Kenkare, U. W., and F. M. Richards (1966): *J. Biol. Chem.*, **241**, 3197.
Kolthoff, I. M., A. Anastasi, and B. H. Tan (1958): *J. Am. Chem. Soc.*, **80**, 3235.
Kolthoff, I. M., and W. Stricks (1951): *Anal. Chem.*, **23**, 763.
Koshland, D. E., D. N. Strumeyer, and W. J. Ray (1962): *Brookhaven Symp. Biol.*, **15**, 101.
Kress, L. F., and M. Laskowski (1967): *J. Biol. Chem.*, **242**, 4925.
Kronman, M. J., F. M. Robbins, and R. E. Andreotti (1967): *Biochim. Biophys. Acta*, **147**, 462.
Kung, Y. T., et al., (1965): *Scientia Sinica*, **14**, 1710.
Ledoux, L. (1954): *Biochim. Biophys. Acta*, **13**, 121.
Light, A., and N. K. Sinha (1967): *J. Biol. Chem.*, **242**, 1358.
Little, C., and P. J. O'Brien (1969): *Eur. J. Biochem.*, **10**, 533.

Liu, T. Y. (1967): *J. Biol. Chem.*, **242,** 4029.
Liu, W. K., and J. Meienhofer (1968): *Biochem. Biophys. Res. Commun.*, **31,** 467.
McPhee, J. R. (1956): *Biochem. J.*, **64,** 22.
Madsen, N. B., and C. F. Cori (1956): *J. Biol. Chem.*, **223,** 1055.
Markus, G. (1964): *J. Biol. Chem.*, **239,** 4163.
Martinez-Carrion, M., R. Kuczenski, D. C. Tiemeier, and D. L. Peterson (1970): *J. Biol. Chem.*, **245,** 799.
Martinez-Carrion, M., C. Turano, F. Riva, and P. Fasella (1967): *J. Biol. Chem.*, **242,** 1426.
Moore, S. (1963): *J. Biol. Chem.*, **238,** 235.
Morihara, K. (1964): *Bull. Chem. Soc. (Japan)*, **37,** 1781.
Nakatani, M. (1960): *J. Biochem. (Tokyo)*, **48,** 633.
Neumann, H., R. F. Goldberger, and M. Sela (1964): *J. Biol. Chem.*, **239,** 1536.
Neumann, H., M. Shinitzky, and R. A. Smith (1967b): *Biochemistry*, **6,** 1421.
Neumann, H., and R. A. Smith (1967): *Arch. Biochem. Biophys.*, **122,** 354.
Neumann, H., I. Z. Steinberg, J. R. Brown, R. F. Goldberger, and M. Sela (1967a): *Eur. J. Biochem.*, **3,** 171.
Neumann, H., I. Z. Steinberg, and E. Katchalski (1965): *J. Am. Chem. Soc.*, **87,** 3841.
Neumann, N. P. (1967): *Methods Enzymol.*, **11,** 485.
Neumann, N. P., S. Moore, and W. H. Stein (1962): *Biochemistry*, **1,** 68.
Onodera, M., and K. Yagi (1969): *J. Biochem. (Tokyo)*, **66,** 751.
Oster, G., J. S. Bellin, R. W. Kimball, and M. E. Schrader (1959): *J. Am. Chem. Soc.*, **81,** 5095.
Parker, D. J., and W. S. Allison (1969): *J. Biol. Chem.*, **244,** 180.
Patchornik, A., W. B. Lawson, and B. Witkop (1958): *J. Am. Chem. Soc.*, **80,** 4748.
Pihl, A., and R. Lange (1962): *J. Biol. Chem.*, **237,** 1356.
Raab, O. (1900): *Z. Biol.*, **39,** 524.
Rafter, G. W. (1957): *Arch. Biochem. Biophys.*, **67,** 267.
Ramachandran, L. K., and B. Witkop (1959): *J. Am. Chem. Soc.*, **81,** 4028.
Ramachandran, L. K., and B. Witkop (1967): *Methods Enzymol.*, **11,** 283.
Rao, G. J. S., and L. K. Ramachandran (1962): *Biochim. Biophys. Acta*, **59,** 507.
Ray, W. J. (1967): *Methods Enzymol.*, **11,** 490.
Ray, W. J., and D. E. Koshland (1960): *Brookhaven Symp. Biol.*, **13,** 135.
Ray, W. J., and D. E. Koshland (1962): *J. Biol. Chem.*, **237,** 2493.
Richmond, V. (1966): *Biochim. Biophys. Acta*, **127,** 499.
Rippa, M., and S. Pontremoli (1969): *Arch. Biochem. Biophys.*, **133,** 112.
Rosnati, L. (1945): *Gass. Chim. Ital.*, **75,** 225.
Sanger, F. (1949): *Biochem. J.*, **44,** 126.
Schachter, H., and G. H. Dixon (1964): *J. Biol. Chem.*, **239,** 813.
Schachter, H., K. A. Halliday, and G. H. Dixon (1963): *J. Biol. Chem.*, **238,** PC3134.
Schmir, G., L. A. Cohen, and B. Witkop (1959): *J. Am. Chem. Soc.*, **81,** 2228.
Scoffone, E., G. Galiazzo, and G. Jori (1970): *Biochem. Biophys. Res. Commun.*, **38,** 16.
Seon, B. K. (1967): *J. Biochem. (Tokyo)*, **61,** 606.
Shaltiel, S., and A. Patchornik (1963): *J. Am. Chem. Soc.*, **85,** 2799.
Shapira, E., and R. Arnon (1969): *J. Biol. Chem.*, **244,** 1026.
Slobodian, E., C. J. Delgado, O. Slywka, and S. Rubenfeld (1968): *156th Meeting of the American Chemical Society, Biological Chemistry Division*, Abstract #166.
Sluyterman, L. A. E. (1962): *Biochim. Biophys. Acta*, **60,** 557.
Spande, T. F., and B. Witkop (1967): *Methods Enzymol.*, **11,** 498.
Spande, T. F., B. Witkop, Y. Degani, and A. Patchornik (1970): *Adv. Protein Chem.*, **24,** 97.
Spikes, J. D., and R. Straight (1967): *Ann. Rev. Phys. Chem.*, **18,** 409.

Stauffer, C. E., and D. Etson (1969): *J. Biol. Chem.*, **244,** 5333.
Stockmayer, W. H., D. W. Rice, and C. C. Stephenson (1955): *J. Am. Chem. Soc.*, **77,** 1980.
Swan, J. M. (1957): *Nature*, **180,** 643.
Tashjian, A. H., D. A. Ontjes, and P. L. Munson (1964): *Biochemistry*, **3,** 1175.
Toennies, G., and T. P. Callan (1939): *J. Biol. Chem.*, **129,** 481.
Tomita, M., M. Irie, and T. Ukita (1969): *Biochemistry*, **8,** 5149.
Van Rensburg, N. J. J., and O. A. Swanepoel (1967): *Arch. Biochem. Biophys.*, **121,** 729.
Viswanatha, T., W. B. Lawson, and B. Witkop (1960): *Biochim. Biophys. Acta*, **40,** 216.
Vithayathil, P. J., and F. M. Richards (1960): *J. Biol. Chem.*, **235,** 2343.
Wassarman, P. M., and J. P. Major (1969): *Biochemistry*, **8,** 1076.
Weil, L. (1965): *Arch. Biochem. Biophys.*, **110,** 57.
Weil, L., and A. R. Buchert (1951): *Arch. Biochem. Biophys.*, **34,** 1.
Weil, L., W. G. Gordon, and A. R. Buchert (1951): *Arch. Biochem. Biophys.*, **33,** 90.
Weil, L., and T. S. Seibles (1955): *Arch. Biochem. Biophys.*, **54,** 368.
Weil, L., and T. S. Seibles (1961): *Arch. Biochem. Biophys.*, **95,** 470.
Westhead, E. W. (1965): *Biochemistry*, **4,** 2139.
Witkop, B. (1961): *Adv. Protein Chem.*, **16,** 221.

9
electrophilic reagents

9-1 IODINE

$$P\text{-}C_6H_4\text{-}O^- + I_3^- \longrightarrow P\text{-}C_6H_3(I)\text{-}O^- + H^+ + 2I^- \xrightarrow{I_3^-}$$

$$\longrightarrow P\text{-}C_6H_2(I)_2\text{-}O^- \quad (1)$$

$$P\text{-(imidazole)} + I_3^- \longrightarrow P\text{-(4-I-imidazole)} + H^+ + 2I^- \xrightarrow{I_3^-}$$

$$\longrightarrow P\text{-(2,4-diiodoimidazole)} \quad (2)$$

$$P\text{-SH} + I_3^- \longrightarrow P\text{-SI} + H^+ + 2I^- \xrightarrow{H_2O}$$

$$P\text{-SOH} \longrightarrow \begin{cases} \xrightarrow{P\text{-SH}} P\text{-S-S-P} \\ \xrightarrow{[O]} P\text{-SO}_3^- \end{cases} \quad (3)$$

Under mild conditions, the reaction of iodine with proteins affects only tyrosyl, histidyl, and cysteinyl residues. Under harsh conditions, tryptophanyl and methionyl residues may also be affected (Koshland et al., 1963). Mono- and diiodo derivatives of both tyrosine and histidine are formed. Cysteinyl residues can be oxidized to several different products, depending upon the conditions.

$$\text{I} + I_3^- \rightleftharpoons \text{II} + 2I^- \longrightarrow \text{III} + H^+ \qquad (4)$$

The rate-limiting step in the iodination of phenols appears to be loss of a hydrogen ion from the quinoidlike intermediate II. The reaction is subject to a large kinetic isotope effect (Grovenstein and Kilby, 1957) and general base catalysis (Mayberry and Bertoli, 1965). Its rate is inversely proportional to the square of the iodide-ion concentration under conditions where titratable iodine is in the form of triiodide (Berliner, 1951). Kinetic data indicate the reactive species to be molecular iodine and phenolate anion (Berliner, 1951; Mayberry and Bertoli, 1965). Iodination of tyrosines at neutral or slightly acidic pH results predominantly in formation of diiodotyrosine. Monoiodination is rate-limiting under these conditions (Li, 1942; Roche et al., 1951). Above pH 8, substitution of the second iodine is slower than the first, and appreciable amounts of monoiodotyrosine may accumulate (Figure 9-1). The energy of activation for substitution of the second iodine is greater by approximately 4 kcal than that for the first (Table 9-1). A nonpolar environment appears to favor diiodination of the "buried" tyrosyl residues in human serum albumin (Perlman and Edelhoch, 1967). Decreased solvent polarity promotes the preferential conversion of tyrosine to diiodotyrosine (Mayberry and Hockert, 1970). The inactivation of pancreatic ribonuclease upon iodination correlates with the incorporation of the second iodine into tyrosine residue 115 (Friedman et al., 1966).

Reaction of iodine with peptides or proteins usually differs from that with the free amino acid, due, presumably, to the different electronic and/or steric environments of the reactive centers. These factors produce considerable variation in the susceptibility of individual residues to iodination. Rate constants for iodination of several low-molecular-weight tyrosine derivatives are given in Table 9-1. A method for determining the relative rates of iodination of various tyrosyl residues in proteins has been described based on a double-labeling procedure (Roholt and Pressman, 1967). Two batches of protein are separately labeled to different extents with the two radioactive isotopes ^{125}I and ^{131}I, respectively. After mixing, the proteins are enzymatically digested and the resulting peptides separated. The relative reactivity

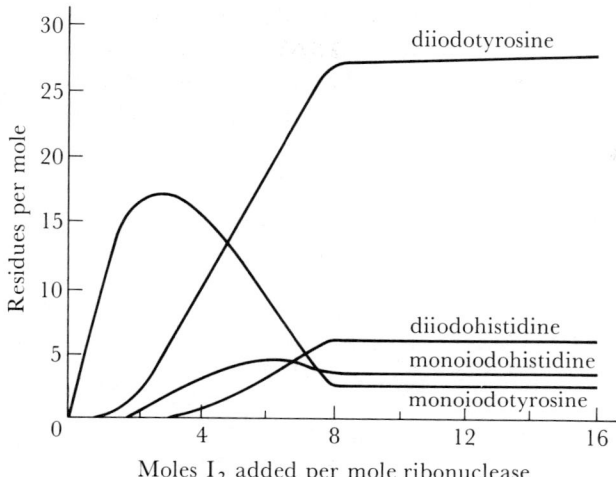

FIGURE 9-1 Incorporation of iodine into ribonuclease at 0°C, pH 8.54, in 0.2 M Tris. (From Covelli and Wolff, 1966a.)

of each tyrosine is then determined from the ratio of the two isotopes in each tyrosyl-containing peptide. The procedure precludes the need for quantitative recovery of each peptide. Relative reactivities of the four tyrosyl residues in α-chymotrypsin by this method are in the order 146 > 94 > 171 > 229. The least reactive tyrosine, namely 229, is less than 10% as reactive as the next least reactive tyrosine and about

TABLE 9-1 Second-order rate constants (25°C) and energies of activation for the iodination of tyrosine and derivatives in pH 9.6 carbonate buffer (from Mayberry et al., 1965)

	K_{obs} (liters/mole sec)	E_a (kcal/mole)
L-Tyrosine	47.7	17.4
Glycyl-L-tyrosine	71.8	16.6
N-Acetyl-L-tyrosine	38.9	17.7
L-Tyrosine methyl ester	45.2	16.3
N-Acetyltyrosine methyl ester	36.6	16.4
L-Tyrosyl-L-tyrosine	90.7	17.4
N-Acetyl-3-iodo-L-tyrosine	5.2	21.5
O-Methyltyrosine	0	—

2% as reactive as the most reactive (Dube et al., 1966). To avoid diiodination and the difficulty thereby introduced, very low levels of iodine should be used.

Iodination of tyrosyl residues does not always halt with the formation of diiodinated derivative (Reineke and Turner, 1945). Iodination of casein, for example, has been shown to result under some circumstances in the formation of small amounts

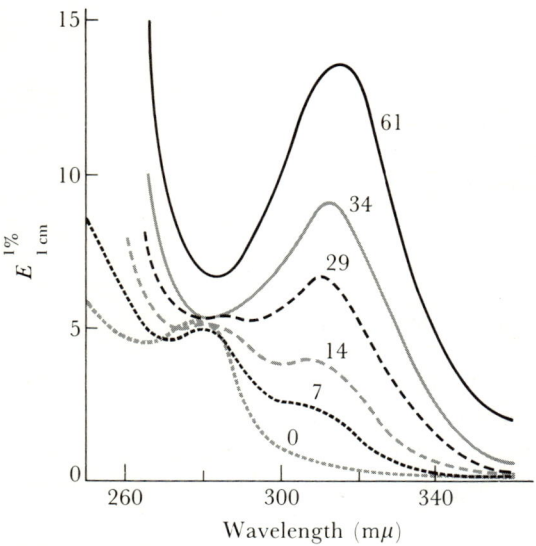

FIGURE 9-2 Ultraviolet absorption spectra of human serum albumin at pH 10.75 as a function of iodine content. Numbers indicate the number of atoms of iodine per mole of protein. (From Hughes and Straessle, 1950.)

of thyroxin (Ludwig and von Mutzenbecker, 1939). This product is apparently formed via the intermediate formation of 3,5-diiodo-4-hydroxybenzaldehyde (Pitt-Rivers, 1948).

Tyrosine, mono- and diiodotyrosines, and thyroxin are spectrally distinguishable (Edelhoch, 1962). A marked bathochromic shift accompanies the addition of each iodine. This is illustrated in Figure 9-2 by the spectral changes which parallel the introduction of iodine into human serum albumin. The wavelength of maximum absorption in this case, determined at pH 10.75, changes from 275 to 312 mμ, as shown. A considerable increase in extinction coefficient also accompanies the reaction. Absorption maxima and extinction coefficients for several tyrosine derivatives are given in Table 9-2. The spectral properties of protein-bound iodotyrosines

TABLE 9-2 Spectral properties of tyrosine and several iodinated derivatives (from Edelhoch, 1962)

	In 0.1 M HCl		In 0.1 M NaOH		
	$\varepsilon \times 10^{-3}$	λ_{max}	$\varepsilon \times 10^{-3}$	λ_{max}	pK
Tyrosine	1.40	274.5	2.40	293	10.1
Monoiodotyrosine	2.75	283	4.10	305	8.2
Diiodotyrosine	2.75	287	6.25	311	6.4
Thyroxine	4.20	298	6.18	325	6.7

are similar to those of the free iodoamino acids. An anomalous low value by approximately 10% in absorptivity has been observed for the diiodotyrosines in iodinated ribonuclease (Gruen et al., 1959).

The pK of the phenolic hydroxyl of tyrosine is decreased by the ortho substitution of iodine for hydrogen. pK values of several iodinated tyrosine derivatives are given in Table 9-2. The estimated pK's of tyrosine and its mono- and diiodo derivatives in thyroglobulin, insulin, and lysozyme are given in Table 9-3. An intrinsic pK of 7.7 has been demonstrated for polydiiodotyrosine (Katchalski and Sela, 1953). A similar value for diiodotyrosine was observed in iodinated thyroglobulin (Edelhoch, 1962). Perturbation of the monomer's pK to this higher value is indicative of either the hydrophobic environment, the electronegative field, or a combination of these effects. An increase of 1.29 pH units in the pK of monoiodotyrosine is obtained in 50% methanol as compared to water (Mayberry and Hockert, 1970).

In a study of the effect of tyrosine iodination on tobacco mosaic virus coat protein, methyl mercuric nitrate was shown to retard reaction with the protein's sulfhydryl group. Only partial protection was obtained, however, even with a large excess of mercurial (Fraenkel-Conrat and Sherwood, 1967). Using low levels of

TABLE 9-3 Apparent pK's of tyrosine, monoiodotyrosine, and diiodotyrosine in proteins (Wolff and Covelli, 1966)

	Isoelectric point	Apparent pK		
		Tyrosine	Monoiodotyrosine	Diiodotyrosine
Thyroglobulin	4.7	11.35	~9.3	~7.5
Insulin	5.3	—	—	7.9
Lysozyme	11.1	10.5–12.8	7.9	6.0

iodine, selective modification of a small number of tyrosyl residues in rabbit muscle aldolase is possible without affecting sulfhydryl groups (Wassarman and Kaplan, 1968). Inactivation is apparently brought about mainly by substitution at the carboxy-terminal tyrosyl groups, and not through oxidation of sulfhydryl groups.

Of the common amino acids, cysteine, not tyrosine, is the most susceptible to attack by iodine. Although the reaction conditions are mild, the method is seldom used to selectively modify sulfhydryl groups of proteins due, perhaps, to the abundance of other reagents for the same purpose. The possible pathways for the modification of sulfhydryl groups by iodine are indicated in Equation 3. Iodination of cysteine (—SH) under most conditions gives cystine (—S—S—), while the same treatment of proteins very often gives other derivatives. The initial reaction of KI_3 with tobacco mosaic virus transforms the single sulfhydryl group of each coat protein subunit into a sulfenyl iodide (—SI) (Fraenkel-Conrat, 1955). In contrast to the lability of low-molecular-weight sulfenyl iodides, this group is completely stable in the intact virus. Upon denaturation of the virus, the sulfenyl iodide group decomposes, releasing iodide ion. Denaturation in the presence of cysteine gives rise to the formation of a disulfide linkage, as shown in Equation 6.

$$\boxed{P}\!-\!SI + H_2O \rightleftharpoons \boxed{P}\!-\!SOH + H^+ + I^- \tag{5}$$

$$\boxed{P}\!-\!SI + RSH \rightleftharpoons \boxed{P}\!-\!S\!-\!SR + H^+ + I^- \tag{6}$$
$$(OH^-) \qquad\qquad\qquad\qquad\qquad (OH^-)$$

Under similar conditions, rabbit muscle glyceraldehyde-3-phosphate dehydrogenase is inactivated by iodine monochloride (ICl). Inactivation correlates with reaction of the initial 2 equivalents of iodine monochloride. The 2 equivalents of iodine bound under these conditions are eliminated upon denaturation of the protein or upon treatment with dithiothreitol. Inactivation has been attributed to the formation of a sulfenyl iodide derivative with the enzyme's active-site sulfhydryl groups, and not to the modification of tyrosine or histidine residues (Parker and Allison, 1969). Reactivation to 60% or 70% of the initial catalytic activity was obtained upon treatment with dithiothreitol.

The reaction of iodine (i.e., KI_3) with the sulfhydryl groups of rabbit muscle creatine kinase effects a loss of catalytic activity similar to that with rabbit muscle glyceraldehyde-3-phosphate dehydrogenase upon treatment with iodine monochloride (ICl), but without a significant amount of iodine being bound (Trundel and Cunningham, 1969). Restoration of catalytic activity upon treatment with thiols was also possible. The stoichiometry of the reaction, its reversal by thiols, and the absence of iodine in the thiol-treated protein indicated the probable product of the reaction to be a sulfenic acid. The conversion of sulfhydryl groups to sulfenic acids in the presence of iodine presumably comes about as indicated in Equation 5 via the formation of a transient sulfenyl iodide. Reactivation of the enzymatic activity upon treatment with thiols presumably involves the formation of a mixed-disulfide intermediate (Equation 6), followed by its scission by a second equivalent of thiol.

Both sulfenyl iodides and sulfenic acids are labile groups except when, in some cases, they are protected by a protein's structure. When these protective structures are not present, a variety of higher oxidation products can result (Equation 3).

The reaction of histidine with iodine is only slightly slower than that of tyrosine. As with tyrosine, both a mono- and diiodinated product may result (Covelli and Wolff, 1966). Initial substitution takes place principally at carbon 5 in the imidazole ring, and subsequent addition of the second takes place at carbon 2 (Brunings, 1947; Holloway et al., 1967; Bensusan and Naidu, 1967). Both iodinated derivatives break down during normal acid hydrolysis (Cha and Scheraga, 1963; Fraenkel-Conrat, 1950; Koshland et al., 1963), making their detection difficult by usual procedures.

5-iodohistidine   2,5-diiodohistidine

It is well established but not generally appreciated that both mono- and di-iodohistidine are formed under the usual conditions for iodination of proteins. The presence of iodohistidines has been demonstrated after iodine treatment of many proteins (Wolff and Covelli, 1969). The ratio of tyrosine to histidine iodination, however, is subject to large variation, depending upon the protein and the conditions employed. An inverse relationship has been observed between the tryptophan contents of several proteins and the proportion of iodine reacting with histidine (Wolff and Covelli, 1969). This was most extreme in the cases of trypsin and α-chymotrypsin, where no iodohistidine was formed even after treatment with a considerable excess of iodine.

Relatively less reaction with histidine as compared to that with tyrosine is obtained at pH 8.5 than at higher pH's. The iodination of histidine appears to require dissociation of the imidazole ring proton, suggesting that the reaction occurs by an attack of iodine on the imidazole anion (Wolff and Covelli, 1969). An alternate proposal suggests initial formation of an N–I intermediate to be mandatory in the reaction (Brunings, 1947).

Oxidation of tryptophanyl residues usually does not occur under conditions of mild iodination (Hughes and Straessle, 1950) but during extensive treatment (Koshland et al., 1963). The oxidation by iodine of a single tryptophan in lysozyme, however, has been accomplished without apparent modification of any other residues (Hartdegen and Rupley, 1964). Methionyl residues may be oxidized to both the sulfoxide and sulfone by iodine, but only under relatively vigorous conditions (Koshland et al., 1963). Histidinyl, methionyl, and tryptophanyl residues are, in general, more easily oxidized as free amino acids than as components of a protein or polypeptide chain (Li, 1945). Both mono- and dibromotyrosyl residues are attacked by chymotrypsin (Vaslow and Doherty, 1953). Iodotyrosines are presumably also susceptible.

It is often desirable, particularly when radioactive iodine is used, to generate the iodinating agent in the reaction mixture. Iodine can thus be generated in potassium iodide solution (hence KI_3) by addition of a stronger oxidant such as potassium iodate (KIO_3) (McFarlane, 1956; Wassarman and Kaplan, 1968), hydrogen peroxide (Gruen et al., 1959; Marchalonis, 1969), or chloramine T (Hunter and Greenwood, 1962). Iodine monochloride is a commercially available reddish-brown liquid that has been used as an iodinating agent (Parker and Allison, 1969). The relative advantages of the various iodination agents for the production of radioactive, ^{131}I- and ^{125}I-labeled antibodies and peptide hormones have been discussed (McFarlane and Greenwood, 1966).

Triiodide ion is the most frequently employed source of iodine for the iodination of proteins. It is formed upon addition of iodine to solutions of sodium iodide. In this form, iodine is made water soluble. Reaction of the reddish-brown solution with proteins appears to involve prior formation of iodine

$$I_3^- \underset{k_2}{\overset{k_1}{\rightleftarrows}} I_2 + I^-$$

a reaction which in rare cases can be rate-limiting. The equilibrium between iodine and triiodide has been investigated by Davies and Gwynne (1952). Rates for the iodination of tyrosine as previously mentioned decrease with the square of the iodide concentration.

Reaction can be followed from the loss of triiodide color at 355 mμ (Cunningham and Nuenke, 1961), iodometrically with thiosulfate (Herriott, 1937), or with other agents (Hughes and Straessle, 1950). Covalently bound iodine can be determined by the use of radioactively labeled iodine or via fusion with an alkali metal and subsequent analysis (Stimmel and McCullagh, 1936; Herriott, 1937). Oxidative side reactions can be ascertained indirectly by the method of Herriott (1937). Iodotyrosines are converted back to tyrosine during acid hydrolysis (Herriott, 1937). Contents of both mono- and diiodotyrosine can be determined from their characteristic absorptions by the method of Edelhoch (1962).

9-2 TETRANITROMETHANE

$$\text{P}-\text{C}_6\text{H}_4-\text{OH} + (\text{NO}_2)_4\text{C} \xrightarrow{\text{pH}>8}$$

$$\text{P}-\text{C}_6\text{H}_3(\text{NO}_2)-\text{O}^- + (\text{NO}_2)_3\text{C}^- + 2\text{H}^+ \quad (1)$$

The usefulness of tetranitromethane (TNM) as a nitrating agent for proteins was first demonstrated by Wormall (1930). More recently, the major product of its reaction with nonsulfhydryl proteins has been identified as 3-nitrotyrosine (Sokolovsky et al., 1966; Riordan et al., 1966). Nitration of tyrosine residues occurs readily under mild conditions, with only a few important side reactions.

Nitration of phenols with TNM was shown to be first order in both TNM and phenoxide ion (rate = k[TNM][PhO$^-$]) (Bruice et al., 1968). The reaction rate is markedly enhanced at high pH and exceedingly slow below pH 7. A charge transfer complex has been proposed as an intermediate in the reaction (Bruice et al., 1968).

The mildly oxidative conditions lead to a rapid oxidation of sulfhydryl groups which may (Sokolovsky et al., 1966) or may not (Riordan and Christen, 1968) later be reversed by treatment with an excess of low-molecular-weight thiol. Oxidation of sulfhydryl groups in aldolase appeared to give other than disulfides (Riordan and Christen, 1968). Potassium tetrathionate has been used to reversibly block sulfhydryl groups of lobster muscle arginine kinase to prevent their oxidation by TNM (Kassab et al., 1970) (see Section 8-5). Oxidation of methionine to methionine sulfoxide was observed when carbonic anhydrase was treated with TNM (Nilsson and Lindskog, 1967), although no such oxidation was observed with methionine itself (Riordan et al., 1967). Tetranitromethane has been observed to react with tryptophan in several proteins (Mühlrad et al., 1968; Cuatrecasas et al., 1968). The product of its reaction with α,N-acetyltryptophan has been identified as α,N-acetyl-7-nitrotryptophan (Morihara and Nagami, 1969). No other amino acids have been shown to be affected by TNM under mild conditions. An intermediate enzyme-substrate complex of aldolase and dihydroxyacetone phosphate, however, reacts with TNM (Christen and Riordan, 1968). TNM has been used to nitrate the side chain vinyl groups of ferriheme (Atassi, 1969).

Treatment with TNM promotes aggregation of both rabbit γ-globulin and collagen, presumably through the formation of intermolecular cross-linkages (Doyle et al., 1968). The apparent formation of cross-linkages has been observed in insulin (Boesel and Carpenter, 1970), trypsinogen, trypsin, ribonuclease, lysozyme, and Kunitz trypsin inhibitor (Vincent et al., 1970). Decreased amounts of tyrosine in excess of that converted to 3-nitrotyrosine suggest the cross-linkages to be between tyrosine residues.

In alkaline solution 3-nitrotyrosine has a strong absorption band centered at 428 mμ ($\varepsilon = 4100$ M^{-1} cm^{-1}) which upon acidification is shifted to 360 mμ ($\varepsilon = 2790$ M^{-1} cm^{-1}) (Figure 9-3). An isosbestic point at 381 mμ ($\varepsilon = 2200$ M^{-1} cm^{-1}) may be used for quantitative purposes at intermediate pH values. The strongly colored ($\varepsilon_{350} = 14{,}400$ M^{-1} cm^{-1}) nitroformate ion formed during the nitration of tyrosine must be removed prior to such quantitation. The formation of this colored

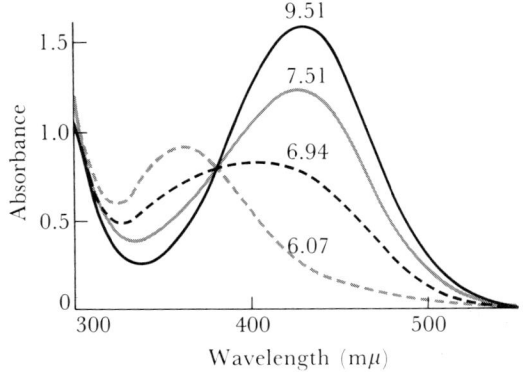

FIGURE 9-3 Absorption spectra of α-N-acetyl-3-nitrotyrosine (2.5×10^{-4} M) in a solution containing 0.2 M Tris, 0.2 M acetate, and 0.5 M NaCl at the pH values indicated. (From Riordan et al., 1967.)

anion by side reactions precludes its use as a measure of the rate of tyrosyl nitration. Dialysis, gel filtration, or extraction with tributylphosphate (Beaven and Gratzer, 1968) will remove the nitroform and permit spectrophotometric quantitation. Nitrotyrosine is stable in hot acid and may be quantitated by amino acid analysis after acid hydrolysis (Sokolovsky et al., 1967).

Nitrotyrosine has a pK between 6.8 and 7.0 and is therefore relatively more ionized than tyrosine at neutral pH. In proteins, somewhat higher values have usually been found. In carbonic anhydrase and glutamic dehydrogenase, for example, the pK of nitrotyrosyl residues was near 8.0 (Nilsson and Lindskog, 1967; Price and Radda, 1969). Values of 6.9 and 7.0 for the pK of nitrotyrosine were found in trypsinogen (Vincent et al., 1970) and arginine kinase (Kassab et al., 1970), respectively.

Nitration of proteins may be carried out at pH 8 and room temperature by adding aliquots of TNM in 95% ethanol to a buffered solution containing approximately 10 mg/ml protein. Nitration of carboxypeptidase under these conditions, using a fourfold excess of TNM per mole of enzyme and a 45-minute incubation, resulted in the nitration of 1.2 tyrosyl residues out of a possible 19 (Riordan et al., 1967). Using a 64-fold excess and a 4-hour incubation, 6.7 tyrosines were nitrated (Sokolovsky et al., 1966). The enzyme was found quite resistant to further nitration. Accessibility of tyrosine residues appears to be an important factor limiting the extent of reaction with most proteins. Nitration with TNM thus appears to be one of the more useful procedures for determining numbers of exposed tyrosyl groups (see Section 1-6). Upon nitration of trypsin or trypsinogen, for example, only five or six of the ten residues were found to react (Kenner et al., 1968), while in sperm whale myoglobin all three tyrosyls were accessible (Atassi, 1968). With most proteins which have been studied, only a relatively small fraction of the tyrosine residues react (Beaven and Gratzer, 1968; Sokolovsky et al., 1966; Verpoorte and Lindblow, 1968; Cuatrecasas et al., 1968; Nilsson and Lindskog, 1967).

Nitration of the bacterial protease subtilisin (Carlsberg strain) effects as much as a six- to sevenfold stimulation of its activity toward a single substrate, clupein (Johansen et al., 1967). Its activity toward other substrates was changed less. Nitration of a different strain of the enzyme (i.e., subtilisin novo) had little effect on enzymatic activity. Nitration of a single tyrosine residue in glutamic dehydrogenase reduced its inhibition by GTP. The presence of GTP during modification protected this residue (Price and Radda, 1969).

$$\text{P}-\text{C}_6\text{H}_3(\text{NO}_2)-\text{O}^- \xrightarrow{[\text{H}]} \text{P}-\text{C}_6\text{H}_3(\text{NH}_2)-\text{O}^- \quad (2)$$

Treatment of 3-nitrotyrosine with sodium hydrosulfite ($Na_2S_2O_4$) results in its rapid conversion into 3-aminotyrosine (Sokolovsky et al., 1967). Under the mild conditions used, no side reactions were observed. Aminotyrosine has absorption maxima at 288 mμ ($\varepsilon = 2800\ M^{-1}\ cm^{-1}$) near neutrality, at 275 mμ ($\varepsilon = 1600\ M^{-1}\ cm^{-1}$) in acid, and at 302 m$\mu$ ($\varepsilon = 4200\ M^{-1}\ cm^{-1}$) in alkali. These absorption maxima are characteristic of aminotyrosine (I) and its two ionized forms (II and III), respectively.

Aminotyrosine has two nucleophilic centers both reversibly ionizable and susceptible to chemical attack. Acylation of either or both of these centers may be used to modify the ionic state of such residues (Sokolovsky et al., 1967). With pK's near 4.8, these amino groups are susceptible to a number of reactions at pH's lower than those where other protein groups are reactive. The chelation of heavy-metal ions between these two ligands has been proposed to be experimentally feasible (Sokolovsky et al., 1967). Aminotyrosine is stable under the conditions of protein hydrolysis and may be quantitated by amino acid analysis.

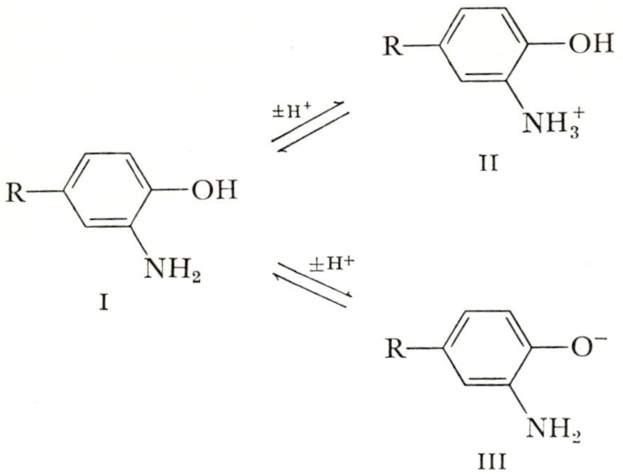

[Structures I, II, III showing equilibria of aminophenol with $\pm H^+$]

(3)

Reduction of nitrotyrosine to aminotyrosine in lysozyme was accomplished in 0.05 M Tris buffer, pH 8.0, by adding a 32-fold molar excess of sodium hydrosulfite ($Na_2S_2O_4$) to the enzyme. Reaction was completed within 10 minutes and resulted in a near-quantitative conversion of nitrotyrosine to aminotyrosine (Sokolovsky et al., 1967).

Similar reduction of two tetranitromethane-treated staphylococcal nuclease derivatives gave aminotyrosyl derivatives of that enzyme (Cuatrecasas et al., 1968). The low pK of the aminotyrosyl groups was employed to prepare additional selectively modified derivatives. Treatment of either with the fluorescent dye dimethylamino-naphthalenesulfonyl chloride (Section 5-4) at pH 4.5 to 5 gave derivatives specifically labeled at only the aminotyrosyl group (Cuatrecasas et al., 1968). Treatment of the same nuclease derivatives with either of two bifunctional reagents, p,p'-difluoro-m,m'-dinitrodiphenyl sulfone or 1,5-difluoro-2,4-dinitrobenzene, gave specific intramolecularly cross-linked derivatives, where one end of the cross-linking agent was bound to an aminotryosine group (Cuatrecasas et al., 1969).

9-3 DIAZONIUM IONS

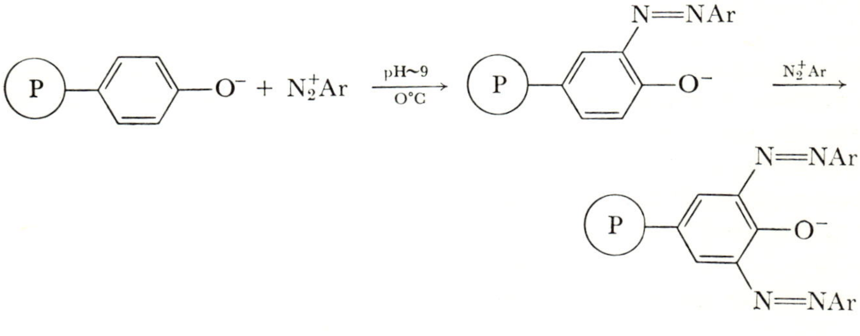

[Reaction scheme: P–C6H4–O⁻ + N2⁺Ar →(pH~9, 0°C) mono-azo product →(N2⁺Ar) bis-azo product]

(1)

Electrophilic Reagents

$$P\text{-imidazole-H} + N_2^+Ar \xrightarrow[0^\circ C]{pH \sim 9} P\text{-imidazole(H)-N=N-Ar} \xrightarrow{N_2^+Ar}$$

$$P\text{-imidazole(H)(N=N-Ar)(N=N-Ar)} \quad (2)$$

$$P\text{-}NH_2 + N_2^+Ar \longrightarrow P\text{-}NH\text{-}N{=}NAr \xrightarrow{N_2^+Ar}$$

$$P\text{-}N(N{=}NAr)(N{=}NAr) \quad (3)$$

In 1904, Pauly reported the formation of a brightly colored material upon treating protein solutions with diazoatized arsanilic acid, and showed the color to be a direct result of the presence of histidine and tyrosine in the protein. The observed stoichiometry indicated bis coupling to each residue (Pauly, 1915). Later workers (Higgins and Harrington, 1959; Howard and Wild, 1957) demonstrated similar reactions with the amino, guanidino, and indole moieties. Reaction with amino groups proceeds rapidly under mild conditions, forming colorless products which are often undetected. In mildly alkaline solutions, the last reaction may be more rapid than with tyrosine or histidine residues.

In spite of the heterogenous nature of the reaction products, diazoatization of proteins has seen widespread use. Immunologists have taken particular advantage of these reagents to augment the antigenic properties of proteins and for the production of particular antigenic determinants. Diazoatization is exceedingly useful in this respect. The availability of many different arylamines has allowed considerable versatility in the choice of such determinants. Diazonium salts have been extensively used for the affinity labeling of antibodies (Metzger et al., 1964) and for the selective modification of several enzymes (Kagan and Vallee, 1969).

The colored product obtained when histidine or related compounds are treated with a diazonium reagent is the basis of several colorimetric procedures for determining these compounds. Such colorimetric procedures have not received widespread acceptance for quantitating histidine or tyrosine in proteins because a large excess of reagent is required to insure complete reaction, resulting in high blank values. Decomposition to phenols, which then react with remaining diazonium ion to produce brightly colored azophenols (III), accounts for the high blank values (Equation 4).

$$X-\text{C}_6\text{H}_4-N_2^+ \xrightarrow{OH^-} X-\text{C}_6\text{H}_4-OH + N_2\uparrow \xrightarrow{X-\text{C}_6\text{H}_4-N_2^+}$$

$$X-\text{C}_6\text{H}_4-N=N-\text{C}_6\text{H}_3(OH)(X) \quad (4)$$

I, II, III

This problem is not encountered using diazonium-1-H-tetrazole (DHT) (Horinishi et al., 1964), because the corresponding hydrolysis product 5-hydroxy-1-H-tetrazole does not react with the remaining DHT. This has made possible the development of more accurate colorimetric procedures for the quantitation of both histidine and tyrosine in proteins (Horinishi et al., 1964; Sokolovsky and Vallee, 1966; Takenaka et al., 1969).

amino-1-H-tetrazole $\xrightarrow{HNO_2, HCl, 0°C}$ diazonium-1-H-tetrazole (5)

Diazonium reagents are usually prepared at low temperature in the presence of several equivalents of strong acid. The excess acid is necessary to prevent undesirable side reactions but should be neutralized before the reagent is used. Excess nitrous acid can be detected with starch-iodide paper and eliminated by the addition of urea. Diazonium ions are rather unstable and, in most cases, should be prepared just before their use. Fluoroborate salts of many diazonium ions are stable enough to be sold commercially. The preparation of such compounds has been described by Roe (1957) and by Traylor and Singer (1967). Diazonium ions may be bound to cation-exchange resins and, in this form, are relatively stable and convenient to use (Phillips et al., 1965). Due to its instability and high nitrogen content, diazonium-1-H-tetrazole is potentially very explosive and should be handled with care. Concentrations exceeding 0.2 M are reportedly dangerous (Sokolovsky and Vallee, 1966).

The rates of reaction of the different amino acid residues histidine, tyrosine, and lysine all increase with increasing pH. The optimum reaction rate for each is near

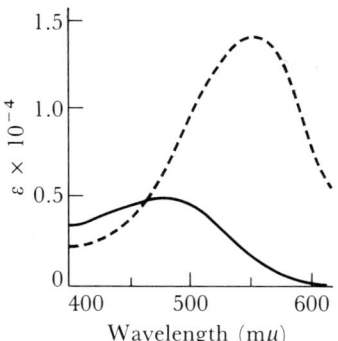

FIGURE 9-4   Absorption spectra of monoazotetrazole-N-acetyltyrosine (solid curve) and bisazotetrazole-N-acetyltyrosine (dashed curve), both approximately $1 \times 10^{-4}\ M$ at pH 8.8. (From Sokolovsky and Vallee, 1966.)

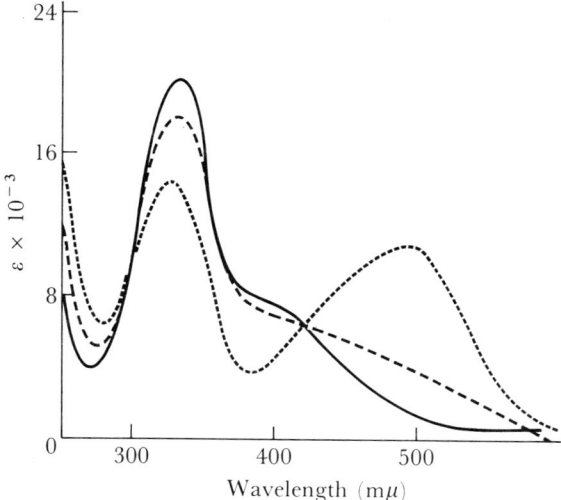

FIGURE 9-5   Effect of pH on the absorption spectrum of mono(p-azobenzenearsonic acid)-chloroacetyltyrosine; $0.1\ M$ HCl (solid curve), pH 8.7 (dashed curve), $0.1\ M$ NaOH (dotted curve). (Adapted from Tabachnick and Sobotka, 1959.)

pH 9. At higher pH's, the reaction is slowed by the competitive formation of azohydroxides and their corresponding salts (Equation 6).

$$Ar-N\equiv N^+ + OH^- \rightleftharpoons Ar-N=N-OH \rightleftharpoons Ar-N=N-O^- + H^+ \quad (6)$$

Tyrosine residues react with diazonium ions, forming monoazo derivatives or, with excess, bisazo derivatives. The spectra of these compounds are pH-dependent (see Figures 9-4 and 9-5). The pK of the phenolic hydroxyl of monoazotyrosine is approximately 8.8; that of bisazotyrosine is near 8.2. Both mono- and bisazotyrosine decompose when heated strongly in acid. They may be quantitated by amino acid analysis from the decrease in tyrosine as compared to the unmodified protein. Horinishi et al. (1964), Sokolovsky and Vallee (1966), and Takenaka et al. (1969) have described colorimetric procedures for the determination of azotetrazole derivatives of both tyrosine and histidine. Absorption maxima and extinction coefficients of several such compounds are given in Table 9-4.

Histidine and tyrosine residues in proteins react with diazonium reagents at similar rates. Bisazohistidyl residues are usually formed at lower molar ratios of diazonium to protein than are bisazotyrosyls. The pK of monoazohistidine is approximately 7.6. Azohistidines, like azotyrosines, are unstable in hot concentrated acids and can be determined after acid hydrolysis only by comparison with the unmodified protein. Extinction coefficients and absorption maxima of several such compounds are given in Table 9-4, and the spectra of mono- and bisazotetrazole-N-acetylhistidine are shown in Figure 9-6.

TABLE 9-4 Spectral properties of azo derivatives of amino acids

Compound	λ_{max} (mμ)	$\varepsilon \times 10^{-3}$
Azotetrazole-N-acetylhistidine[a]	360	11.5
Azobenzenearsonic acid-N-acetylhistidine[b]	420	22.3
Bisazotetrazole-N-acetylhistidine[a]	480	21.0
Azotetrazole-N-acetyltyrosine[a]	478	5.2
Azobenzenearsonic acid-N-chloroacetyltyrosine[b]	490	11.0
Bisazotetrazole-N-acetyltyrosine	548	14.1
Bisazobenzenearsonic acid-N-chloroacetyltyrosine[b]	545	17.5
Bisazobenzenearsonic acid-ε-aminocaproic acid[b]	378	30.8

[a] Measured at pH 8.8.   [b] In alkaline solution.

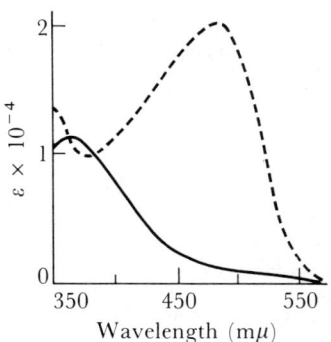

FIGURE 9-6 Absorption spectra of monoazotetrazole-N-acetylhistidine (solid curve) and bisazotetrazole-N-acetylhistidine (dashed curve). Both approximately $1 \times 10^{-4}\,M$ in pH 8.8 bicarbonate. (From Sokolovsky and Vallee, 1966.)

Amino groups react with 2 moles of diazonium to give disubstituted compounds known as triazens (Equation 3). These compounds absorb maximally near 365 mμ but have no appreciable visible absorption. Decomposition of triazens is hastened at low pH and such decomposition is occasioned by an evolution of nitrogen.

Other protein groups also react with diazonium ions but quite slowly. The products of such reactions with tryptophanyl, cysteinyl, and arginyl residues have not been well characterized and will not be discussed here. Further information can be obtained from the original literature (Howard and Wild, 1957; Higgins and Harrington, 1959).

REFERENCES

Atassi, M. Z. (1968): *Biochemistry*, **7**, 3078.
Atassi, M. Z. (1969): *Biochim. Biophys. Acta*, **177**, 663.
Beaven, G. H., and W. B. Gratzer (1968): *Biochim. Biophys. Acta*, **168**, 456.
Bensusan, H. B., and M. S. R. Naidu (1967): *Biochemistry*, **6**, 12.
Berliner, E. (1951): *J. Am. Chem. Soc.*, **73**, 4307.
Boesel, R. W., and F. H. Carpenter (1970): *Biochem. Biophys. Res. Commun.*, **38**, 678.
Bruice, T. C., M. J. Gregory, and S. L. Walters (1968): *J. Am. Chem. Soc.*, **90**, 1612.
Brunings, K. J. (1947): *J. Am. Chem. Soc.*, **69**, 205.
Cha, C. Y., and H. A. Scheraga (1963): *J. Biol. Chem.*, **238**, 2958.
Christen, P., and J. F. Riordan (1968): *Biochemistry*, **7**, 1531.
Covelli, I., and J. Wolff (1966): *J. Biol. Chem.*, **241**, 4444.
Cuatrecasas, P., S. Fuchs, and C. B. Anfinsen (1968): *J. Biol. Chem.*, **243**, 4787.
Cuatrecasas, P., S. Fuchs, and C. B. Anfinsen (1969): *J. Biol. Chem.*, **244**, 406.
Cunningham, L. W., and B. J. Nuenke (1961) *J. Biol. Chem.*, **236**, 1716.

Davies, M., and E. Gwynne (1952): *J. Am. Chem. Soc.*, **74,** 2748.
Doyle, R. J., J. Bello, and O. A. Roholt (1968): *Biochim. Biophys. Acta*, **160,** 274.
Dube, S. K., O. A. Roholt, and D. Pressman (1966): *J. Biol. Chem.*, **241,** 4665.
Edelhoch, H. (1962): *J. Biol. Chem.*, **237,** 2778.
Fraenkel-Conrat, H. (1950): *Arch. Biochem. Biophys.*, **27,** 109.
Fraenkel-Conrat, H. (1955): *J. Biol. Chem.*, **217,** 373.
Fraenkel-Conrat, H., and M. Sherwood (1967): *Arch. Biochem. Biophys.*, **120,** 571.
Friedman, M. E., H. A. Scheraga, and R. F. Goldberger (1966): *Biochemistry*, **5,** 3770.
Grovenstein, E., and D. C. Kilby (1957): *J. Am. Chem. Soc.*, **79,** 2972.
Gruen, L., M. Laskowski, and H. A. Scheraga (1959): *J. Biol. Chem.*, **234,** 2050.
Hartdegen, F. J., and J. A. Rupley (1964): *Biochim. Biophys. Acta*, **92,** 625.
Herriott, R. (1937): *J. Gen. Physiol.*, **20,** 335.
Higgins, H. G., and K. J. Harrington (1959): *Arch. Biochem. Biophys.*, **85,** 409.
Holloway, C. T., R. P. M. Bond, I. G. Knight, and R. B. Beechey (1967): *Biochemistry*, **6,** 19.
Horinishi, H., Y. Hachimori, K. Kurihara, and K. Shibata (1964): *Biochim. Biophys. Acta*, **86,** 477.
Howard, A. N., and F. Wild (1957): *Biochem. J.*, **65,** 651.
Hughes, W. L., and R. Straessle (1950): *J. Am. Chem. Soc.*, **72,** 452.
Hunter, W. M., and F. C. Greenwood (1962): *Nature*, **194,** 495.
Johansen, J. T., M. Ottesen, and I. Svendsen (1967): *Biochim. Biophys. Acta*, **139,** 211.
Kagan, H. M., and B. L. Vallee (1969): *Biochemistry*, **8,** 4223.
Kassab, R., A. Fattoum, and L. A. Pradel (1970): *Eur. J. Biochem.*, **12,** 264.
Katchalski, E., and M. Sela (1953): *J. Am. Chem. Soc.*, **75,** 5284.
Kenner, R. A., K. A. Walsh, and H. Neurath (1968): *Biochem. Biophys. Res. Commun.*, **33,** 353.
Koshland, M. E., F. M. Englberger, M. J. Erwin, and S. M. Gaddone (1963); *J. Biol. Chem.*, **238,** 1343.
Li, C. H. (1942): *J. Am. Chem. Soc.*, **64,** 1147.
Li, C. H. (1945): *J. Am. Chem. Soc.*, **67,** 1065.
Ludwig, W., and P. von Mutzenbecker (1939): *Hoppe-Seyl. Z. Physiol. Chem.*, **258,** 195.
Marchalonis, J. J. (1969): *Biochem. J.*, **113,** 299.
Mayberry, W. E., and D. Bertoli (1965): *J. Org. Chem.*, **30,** 2029.
Mayberry, W. E., and T. J. Hockert (1970): *J. Biol. Chem.*, **245,** 697.
Mayberry, W. E., J. E. Rall, and D. Bertoli (1965): *Biochemistry*, **4,** 2606.
McFarlane, A. S. (1956): *Biochem J.*, **62,** 135.
McFarlane, A. S., and F. C. Greenwood (1966): *Nature*, **210,** 18.
Metzger, H., L. Wofsy, and S. J. Singer (1964): *Proc. Nat. Acad. Sci.*, **51,** 612.
Morihara, K., and K. Nagami (1969): *J. Biochem. (Tokyo)*, **65,** 321.
Mühlrad, A., A. Corsi, and A. L. Granata (1968): *Biochim. Biophys. Acta*, **162,** 435.
Nilsson, A., and S. Lindskog (1967): *Eur. J. Biochem.*, **2,** 309.
Parker, D. J., and W. S. Allison (1969): *J. Biol. Chem.*, **244,** 180.
Pauly, H. (1904): *Hoppe-Seyl. Z. Physiol. Chem.*, **42,** 508.
Pauly, H. (1915): *Hoppe-Seyl. Z. Physiol. Chem.*, **94,** 284.
Perlman, R. L., and H. Edelhoch (1967): *J. Biol. Chem.*, **242,** 2416.
Phillips, J. H., S. A. Robrish, and C. Bates (1965): *J. Biol. Chem.*, **240,** 699.
Pitt-Rivers, R. (1948): *Biochem. J.*, **43,** 223.
Price, N. C., and G. K. Radda (1969): *Biochem. J.*, **114,** 419.
Reineke, E. P., and C. W. Turner (1945): *J. Biol. Chem.*, **161,** 613.
Riordan, J. F., and P. Christen (1968): *Biochemistry*, **7,** 1525.
Riordan, J. F., M. Sokolovsky, and B. L. Vallee (1966): *J. Am. Chem. Soc.*, **88,** 4104.

Riordan, J. F., M. Sokolovsky, and B. L. Vallee (1967): *Biochemistry*, **6,** 358.
Roche, J., S. Lissitzky, O. Michel, and R. Michel (1951): *Biochim. Biophys. Acta*, **7,** 439.
Roe, A. (1957): in *Organic Reactions*, vol. 5, edited by R. Adams, Wiley, New York, p. 204.
Roholt, O. A., and D. Pressman (1967): *Biochim. Biophys. Acta*, **147,** 1.
Sokolovsky, M., J. F. Riordan, and B. L. Vallee (1966): *Biochemistry*, **5,** 3582.
Sokolovsky, M., J. F. Riordan, and B. L. Vallee (1967): *Biochem. Biophys. Res. Commun.*, **27,** 20.
Sokolovsky, M., and B. L. Vallee (1966): *Biochemistry*, **5,** 3574.
Stimmel, B. F., and D. R. McCullagh (1936): *J. Biol. Chem.*, **116,** 21.
Tabachnick, M., and H. Sobotka (1959): *J. Biol. Chem.*, **234,** 1726.
Takenaka, A., T. Suzuki, O. Takenaka, H. Horinishi, and K. Shibata (1969): *Biochim. Biophys. Acta*, **194,** 293.
Traylor, P. S., and S. J. Singer (1967): *Biochemistry*, **6,** 881.
Trundle, D., and L. W. Cunningham (1969): *Biochemistry*, **8,** 1919.
Vaslow, F., and D. G. Doherty (1953): *J. Am. Chem. Soc.*, **75,** 928.
Verpoorte, J. A., and C. Lindblow (1968): *J. Biol. Chem.*, **243,** 5993.
Vincent, J. P., M. Lazdunski, and H. Delaage (1970): *Eur. J. Biochem.*, **12,** 250.
Wassarman, P. M., and N. O. Kaplan (1968): *J. Biol. Chem.*, **243,** 720.
Wolff, J., and I. Covelli (1966): *Biochemistry*, **5,** 867.
Wolff, J., and I. Covelli (1969): *Eur. J. Biochem.*, **9,** 371.
Wormall, A. (1930): *J. Exp. Med.*, **51,** 295.

10
miscellaneous reagents

10-1 DICARBONYLS

10-1.1 1,2-Cyclohexanedione

Cyclohexanedione (Toi et al., 1967), benzil (Itano and Gottlieb, 1963), and certain other 1,2-diketones (Yamada and Itano, 1966) react in alkaline solution with the guanidino moieties of proteins. Cyclohexanedione appears to be one of the most useful of these reagents. In strongly alkaline solution (i.e., 0.2 M NaOH), its reaction with arginine results in the formation of a single product, N^5-(4-oxo-1,3-diazospiro[4.4]non-2-ylidene)-L-ornithine (Equation 1). In 0.1 to 0.05 M sodium hydroxide two additional products are formed (Toi et al., 1967). Treatment of proteins with cyclohexanedione in even less alkaline solutions (i.e., pH 11 to 12) results in the modification of both guanidino and amino groups (Liu et al., 1968). The unknown reaction with amino groups is indicated by formation of a yellow color with a broad absorption maximum near 440 mμ. A similar product can be formed from cyclohexanedione and lysine (Liu et al., 1968).

Modification of several protein inhibitors of trypsin with cyclohexanedione has been employed to show the essentiality of arginine residues for their trypsin-inhibitory activities. Thus, chicken ovomucoid and soybean trypsin inhibitor lost almost all antitrypsin activity upon such treatment whereas several other ovomucoids, which were not arginine-type trypsin inhibitors, were relatively unaffected (Liu et al., 1968).

Modification of arginine and lysine residues of proteins with cyclohexanedione can be quantitated by determination of the residues remaining either in the intact proteins or by amino acid analysis after hydrolysis (see Table 4-1).

A quantitative procedure for the determination of arginine, arginine residues in proteins, and other guanidinium compounds has been developed which employs the related diketone phenanthroquinone. Reaction with this compound converts the respective guanidinium groups into intensely fluorescent substances which can be quantitatively determined even at very low levels (Yamada and Itano, 1966).

10-1.2 Glyoxal

Glyoxal has been shown to react with arginine residues in proteins in aqueous solutions between pH 8 and 9, but the nature of the derivative formed is unknown (Nakaya et al., 1967; Kotaki et al., 1966; Nakaya et al., 1969). Reaction can be detected and quantitated by the decrease in arginine found by amino acid analysis after acid hydrolysis. Lysine residues react under the same conditions, also giving an unknown product. Treatment with glyoxal inactivates D-amino acid oxidase concomitant with the modification of 18 of 28 arginine residues. Benzoate ion which binds to the enzyme protects it from inactivation.

10-1.3 Phenylglyoxal

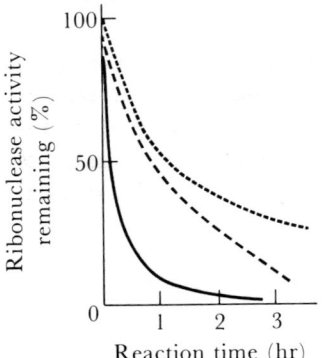

FIGURE 10-1   Rates of inactivation of ribonuclease A by reaction with phenylglyoxal (solid curve), methylglyoxal (dashed curve), and glyoxal (dotted curve) at pH 8.0 and 25°. Activity was measured against RNA as substrate. Protein, 0.5%; phenylglyoxal hydrate, 1.5%; methylglyoxal, 1.5%; glyoxal hydrate, 1.5%. (Adapted from Takahashi, 1968.)

At neutral pH, phenylglyoxal reacts with guanidino groups of proteins, giving derivatives containing two phenylglyoxal moieties per guanidino group (Equation 2). Under mildly acidic conditions (i.e., below pH 4), these derivatives are sufficiently stable to permit the isolation of peptides so labeled after enzymatic or chemical cleavage of the peptide chain. The derivative slowly decomposes at neutral or alkaline pH's, and incubation under such conditions in the absence of excess phenylglyoxal can regenerate most of the initial guanidino groups (Takahashi, 1968).

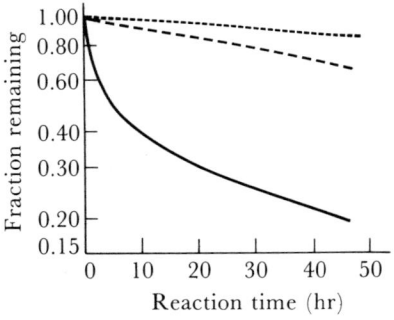

FIGURE 10-2   The loss of arginine (solid curve), lysine (dashed curve), and histidine (dotted curve) from BSA exposed to diacetyl trimer (0.4 M) at 25°C in pH 7 phosphate. (From Yankeelov et al., 1968.)

Treatment at pH 8.0 and 25°C with 1.5% phenylglyoxal destroys 80–90% of the catalytic activity of pancreatic ribonuclease within 30 minutes. Analysis of proteolytic digests of the modified protein showed the reaction to occur primarily with only two of four possible arginine residues (i.e., residues 39 and 85). Deamination of the NH_2-terminal amino group occurred during the treatment, giving a corresponding α-keto group (Takashashi, 1968) (see Figure 10-1).

Phenylglyoxal appears to be useful for the modification of accessible arginine residues of proteins. Its blockage of these groups is potentially reversible. Deamination of NH_2-terminal residues has been found to occur during such treatment and has been proposed as a method for subtractively identifying such groups. Upon extensive treatment with phenylglyoxal, ε-amino groups may also react.

10-1.4 DIACETYL TRIMER

$$\text{(3)}$$

Diacetyl (2,3-butanedione) condenses in mildly alkaline aqueous solution (pH 8.2) (Yankeelov et al., 1966) or in the presence of powdered glass (Yankeelov et al., 1968) to a bicyclic trimer, which may be isolated and purified as a stable crystalline product. Treatment of proteins with this reagent at neutral pH results in the modification primarily of their guanidino groups (Figure 10-2). A much slower reaction with amino groups is accompanied by the development of a strong reddish color ($\lambda_{max} \sim 530$ mμ) (Yankeelov et al., 1966; Grossberg and Pressman, 1968). Neither reaction product is stable in strong acid. Consequently, both are destroyed during acid hydrolysis. The extent of modification may be determined from the loss of arginine and lysine.

10-1.5 MALONALDEHYDE

$$\text{(4)}$$

A 1,3-dialdehyde, malonaldehyde, reacts with arginine in strong acid (10 N HCl) to give a crystalline product absorbing maximally at 315 mμ (King, 1966). These reaction conditions are probably too severe for even the most stable protein, but such modification may be useful in sequencing work by limiting tryptic digestion to lysyl bonds. The product can be quantitated on the basis of its absorbance at 315 mμ ($\varepsilon = 3490\ M^{-1}\ cm^{-1}$) at pH 2.0. It is unstable to the usual conditions for acid hydrolysis and reverts upon such treatment to ornithine ($\sim$66% yield) and arginine ($\sim$22% yield).

Malonaldehyde is formed during the oxidation of unsaturated lipids, and under such relatively mild conditions has been shown to react with amino groups of proteins (Kwon and Olcott, 1966; Crawford et al., 1967; Buttkus, 1967; Chio and Tappel, 1969). The reaction of glycine with malonaldehyde involves a 1,4-addition of the nucleophilic nitrogen atom of glycine to the α,β-unsaturated carbonyl system of the malonaldehyde enol (Crawford et al., 1967). More recently the stoichiometry has been found to involve 2 moles of amino acid to give a 1-amino-3-iminopropene derivative, R—N=CH—CH=CHNHR (Chio and Tappel, 1969). Inactivation of pancreatic ribonuclease by malonaldehyde under the same conditions is thought to involve a similar reaction to form intramolecular cross-links of the type

$$\text{P}\begin{array}{c}\text{NH—CH}\\ \\ \text{N=CH}\end{array}\text{CH}$$

These adducts have a characteristic intense fluorescence around 470 mμ when excited at 395 mμ (Chio and Tappel, 1969). Malonaldehyde, a product of lipid peroxidation, has been linked to aging processes in a variety of biological systems by virtue of such reactions (Chio and Tappel, 1969).

The modification of arginine residues of proteins with cyclohexanedione, diacetyl trimer, glyoxal, and phenylglyoxal can be accomplished under relatively mild, slightly alkaline conditions, where many proteins are able to maintain a relatively native conformation. A principle side reaction involving the modification of amino groups has been described for each reagent. In contrast, modification of arginine residues with malonaldehyde requires strongly acidic conditions. Its use as a selective reagent for the modification of arginine residues in proteins appears limited by these required harsh conditions. The relative virtues of each of the above arginine reagents are not sufficiently distinct at this time to make one obviously preferable to the others.

10-2 MERCURIALS

$$\text{P}\!-\!S^- + {}^+\!HgR \rightleftharpoons \text{P}\!-\!SHgR \qquad (1)$$

Mercurials react rapidly and specifically with the sulfhydryl groups of proteins. No other groups are usually affected under normal conditions. Mercurials are widely used to quantitate and to determine the effect of substitution on sulfhydryl groups.

The most commonly used inorganic mercurial is mercuric chloride ($HgCl_2$). It is moderately soluble in both water and organic solvents (6.8 g/100 ml water, 26 g/100 ml ethanol at 25°C). Organic mercurials are less soluble in water and more soluble in organic solvents. Low aqueous solubility can be a problem if high concentrations are required. The solubility of mercurials in lipid solvents may be important in considering their relative reactivity with sulfhydryl groups in different environments, and in directing their specificity to different sites in multisulfhydryl proteins (Robinson et al., 1967). Reactions of mercurials with proteins have been thoroughly reviewed by Webb (1966).

Mercuric salts in aqueous solutions tend to exist as a series of tetrahedral coordination complexes, the predominant form depending upon the conditions, particularly upon the pH and the availability of various anionic ligands. Mercuric chloride in aqueous solution, depending upon the pH and chloride concentrations, exists in a series of complexes like the series indicated in Equation 2. Similar complexes are formed with other anions. As these ligands affect the availability of the mercurial, they can have a profound effect on its reactivity. Organic mercurials form similar complexes with the organic group occupying one corner of the tetrahedron. The various mercurials are normally obtained as chloride, hydroxide, nitrate, or acetate salts, but once dissolved they assume a condition determined by that medium. Since the particular salt employed is of secondary concern, it is most reasonable to refer to these reagents by the name of the cation. Thus, both p-hydroxymercuribenzoate and p-chloromercuribenzoate will, when no distinction is necessary, be referred to as p-mercuribenzoate, and a similar convention will be followed with other mercurials.

$$HgCl_2 \underset{2\,Cl^-}{\overset{2\,H_2O}{\rightleftarrows}} \begin{array}{c} HgCl_2(H_2O)_2 \\ \\ HgCl_4^= \end{array} \underset{Cl^-}{\overset{OH^-}{\rightleftarrows}} \begin{array}{c} \overset{H^+}{\rightleftarrows} HgCl_2(H_2O)(OH)^- \\ Cl^- \big\Vert H_2O \\ HgCl_3(OH)^= \end{array} \underset{Cl^-}{\overset{OH^-}{\rightleftarrows}} HgCl_2(OH)_2^= \quad (2)$$

Being divalent, mercuric ion can react with 1 or 2 sulfhydryl equivalents. The uncertainty resulting from this possibility constitutes a serious drawback to its quantitative use. This divalency, however, has been used in a particularly interesting way by Hughes (1947, 1949) to form linkages between molecules of mercaptalbumin. Mercaptalbumin, having a single sulfhydryl group, as indicated below, is converted in the presence of mercuric ion initially into a simple mercaptide (Equation 3). When the ratio of mercury to protein is less than 1, this is followed by a second reaction wherein dimers are formed (Equation 4). If additional mercuric ion is then added, the mercury-linked mercaptalbumin dimer can react further to form a mercury monomer (Equation 5). At ratios of mercury to protein near 0.5, a crystalline fraction of the mercaptalbumin mercuric dimer can be isolated. The second step (Equation 4) is slow and first order in each of the two different macromolecules and is the rate-determining step in dimer formation.

$$\text{(P)}-SH + Hg^{++} \rightleftharpoons \text{(P)}-SHg^+ + H^+ \quad (3)$$

$$\text{(P)}-SHg^+ + \text{(P)}-SH \rightleftharpoons \text{(P)}-S-Hg-S-\text{(P)} + H^+ \quad (4)$$

$$\text{(P)}-S-Hg-S-\text{(P)} + Hg^{++} \rightleftharpoons 2\,\text{(P)}-SHg^+ \quad (5)$$

More rapid dimer formation was obtained using the divalent mercurial compound 3,6-bis(mercurimethyl)dioxane,

$$^+Hg-CH_2-CH\begin{array}{c}O-CH_2\\ \\CH_2-O\end{array}CH-CH_2-Hg^+$$

which presumably does not require so close an approach of the two protein molecules (Edsall et al., 1954).

E. coli DNA polymerase has a single exposed sulfhydryl group which has been shown to react with mercuric ion to give both a simple mercaptide and a mercury-linked dimer. Both forms have full polymerase and exonuclease activity (Jovin et al., 1969). The radioactive isotope ^{203}Hg was used to provide a sensitive label for physical studies of the two forms.

Intramolecular mercury bridges have been formed in papain (Arnon and Shapira, 1969) and bovine pancreatic ribonuclease (Sperling et al., 1969), following specific reduction of single disulfide bridges in each. The mercuric ion in these cases binds together the two newly formed sulfhydryl groups and the attached polypeptide chains in positions similar to those before reduction (Equation 6).

$$\{-S-S-\} \xrightarrow{[H]} \{-SH \quad HS-\} \xrightarrow{Hg^{2+}} \{-SHgS-\} \quad (6)$$

The optimum rate of reaction of mercurials with proteins is usually near pH 5. While higher pH's increase the concentration of the more reactive anionic form ⓟ–S⁻, greater reactivity may not result, one reason being the increased competition of OH⁻ ligands for the mercurial. Reaction rates of sulfhydryl groups in different proteins and of different sulfhydryl groups within any one protein can vary greatly but approach those of low-molecular-weight thiols in urea or other denaturants. Some sulfhydryl groups in proteins react with mercurials only after treatment with

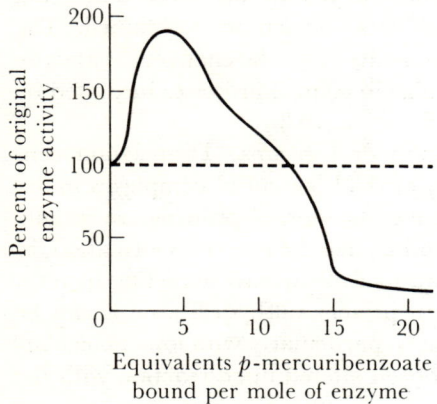

FIGURE 10-3 Effect of p-mercuribenzoate on the phosphatase activity of rabbit liver fructose-1,6-diphosphatase. (From Pontremoli et al., 1965.)

such agents. Because of such differences in reactivity, it is often possible to selectively modify a fraction of the total sulfhydryl groups in a protein. In some cases, it is possible to attribute changes in catalytic properties to reaction with a fraction of the total thiol groups. The reaction of rabbit liver fructose diphosphatase with p-mercuribenzoate, for example, effects a marked stimulation of catalytic activity with the first 2 equivalents bound, and thereafter decreases activity as the remaining 18 groups react (Figure 10-3) (Pontremoli et al., 1965).

$^+$Hg—⟨ ⟩—COO⁻ $^+$HgCH$_3$

p-mercuribenzoate   methylmercuric ion

Mercaptide formation often initiates changes in the noncovalent structure of proteins—for example, in an increased aggregation of 3-phosphoglyceraldehyde dehydrogenase (Elödi, 1960) and in the dissociation of muscle enolase (Malmström, 1962), muscle phosphorylase (Madsen and Cori, 1955, 1956), and myosin ATPase (Kominz, 1961). A number of sulfhydryl-containing enzymes fail to react with mercurials or react without noticeable changes in catalytic activity. Increased catalytic activity upon treatment with mercurials has been observed with many enzymes (see Figure 10-3). Incubation of *E. coli* aspartate transcarbamylase with mercurials, for example, brings about its dissociation into separate catalytic and regulatory subunits, loss of its sensitivity to CTP inhibition, and an increase in its maximal catalytic velocity (Gerhart and Pardee, 1962). When three to four rapidly reacting sulfhydryl groups in pig heart mitochondrial malate dehydrogenase react with p-mercuribenzoate, significant stimulation of catalytic activity is obtained. The sulfhydryl groups involved appear to be at allosteric sites, and the changes in catalytic activity upon mercaptide formation may be similar to control processes mediated by naturally occurring activators (Silverstein and Sulebele, 1970).

Mercurials combine very strongly with sulfhydryl groups. Dissociation constants (i.e., $K_{eq} = [RS^-][R'-Hg^+]/[R'HgSR]$) of thiol-mercurial complexes under conditions such as those normally used for the modification of proteins are usually 10^{-20} or less. The equilibrium amount of mercaptide, like the rate of its formation, is, in part, determined by the pH and the concentrations of various competing ligands. Removal of mercurials from their complex with protein sulfhydryl groups can be accomplished by treatment with competing ligands, particularly with low-molecular-weight thiols. Where physiological activity has been altered upon reaction with the mercurial, very often it can be recovered at least in part by such a treatment.

Different mercurials typically have different characteristic rates and extents of reaction with particular proteins. Differences in reactivity can be attributed to differences in size, ionic charge, and hydrophilicity of the reagent, in combination with effects of the protein environment near sulfhydryl groups. Organic mercurials, being larger than mercuric ion, are often less reactive, apparently due to steric effects limiting their approach to, and hence impeding their reaction with, many sulfhydryl groups. The presence or absence of an additional ionic charge can increase or decrease the reactivity with sulfhydryl groups, in particular ionic or nonionic environments. It is often found, for example, that p-mercuribenzoate gives more complete reaction than phenylmercuric ion with many proteins. It is difficult to know why this should be, but it is apparent that in the former the two opposite charges cancel, making it a neutral reagent, whereas the latter reagent is cationic. Sulfhydryl groups buried in hydrophobic centers of proteins react best with more hydrophobic mercurials. It is sometimes possible to show with a series of related mercurials a regular increase in reactivity with increasing hydrophobicity. Thus, the inhibition of fumarase by a series of alkylmercury nitrates has been found to increase in proportion to the number of methylene groups in the alkyl side chains (Robinson et al., 1967).

Organic mercurials absorb strongly in the ultraviolet (Figure 10-4). The extinction coefficient of the p-mercuribenzoate ion, at its 233-mμ absorption

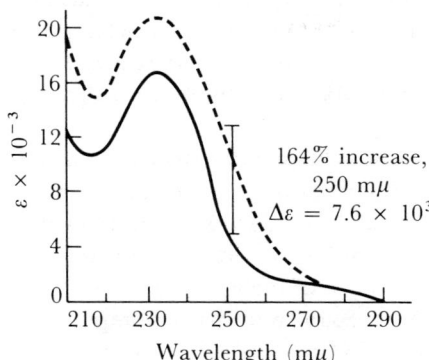

FIGURE 10-4 Absorbancy of p-mercuribenzoate (solid curve) and its mercaptide with cysteine (dashed curve) in 0.05 M phosphate, pH 7.0. (From Boyer, 1954.)

maximum, is $1.7 \times 10^4 \, M^{-1} \, cm^{-1}$. This increases to $2.2 \times 10^4 \, M^{-1} \, cm^{-1}$ upon mercaptide formation (Riordan and Vallee, 1967). The difference at this wavelength, however, is less than that at 250 to 255 mμ, where mercaptide formation is usually followed (Boyer, 1954). Both mercurial and protein absorb strongly at these wavelengths, and appropriate blanks must be used. The increments of increased absorption differ with different proteins and with different conditions. Determinations are therefore done by a spectrophotometric titration as illustrated in Figure 10-5. The endpoint, at the equivalence of p-mercuribenzoate and sulfhydryl groups is indicated by a change in slope.

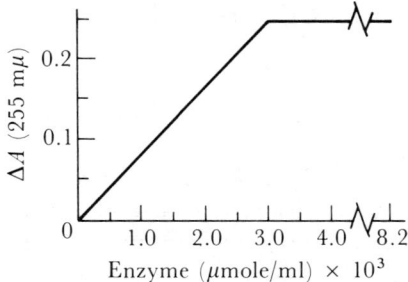

FIGURE 10-5 Spectrophotometric titration of 3-phosphoglyceraldehyde dehydrogenase (0.03 μmole/ml) with p-mercuribenzoate at pH 4.6. (Adapted from Boyer and Segal, 1954.)

The azomercurial dye, 4-(*p*-dimethylaminobenzenazo)phenylmercuric acetate, has been used to measure sulfhydryl group contents of proteins using changes in absorption accompanying its combination with them (Horowitz and Klotz, 1956). *p*-Mercuriphenylazo-β-naphthol has been similarly used to determine sulfhydryl contents of tissues and tissue homogenates (Bennett, 1951).

The affinity of mercurials for sulfhydryl groups is far greater than it is for any other common protein group. Under normal conditions, this should allow specific modification of these groups in proteins. However, when used in excess, as they often are, secondary binding to other groups should be considered. It is interesting how little evidence pertains to the occurrence or the absence of such reactions.

When hemerythrin is treated with *p*-mercuribenzoate, a transient complex with other than sulfhydryl groups has been observed (Klapper, 1970). Mercurial is slowly displaced from this less stable complex in favor of that with sulfhydryl groups.

Carboxypeptidase A contains 1 zinc atom per mole, which can be removed by dialysis at low pH or against phenanthroline. Catalytic activity can be restored to the apoenzyme by addition of zinc or certain other divalent metal ions (i.e., Mn^{++}, Fe^{++}, Ni^{++}, and Co^{++}). The apoenzyme also combines strongly with *p*-mercuribenzoate. The latter complex, however, has no catalytic activity (Vallee et al., 1960). The stability constant of this complex is similar to that of *p*-mercuribenzoate with sulfhydryl groups. Recent studies have shown, however, that carboxypeptidase has no sulfhydryl groups (Lipscomb, 1967). The strong binding of *p*-mercuribenzoate by this protein without benefit of sulfhydryl groups illustrates the possibility of similar complexes in other proteins.

Silver salts are stable, readily available, and combine with sulfhydryl groups of proteins even more firmly than do mercurials. Silver ion (Ag^+) is univalent, similar to organic mercurials. Complexes of silver in ammonia, $Ag(NH_3)_2^+$, and tris-(hydroxymethyl)aminomethane, $Ag(Tris)_2^+$, have been used for amperometric titration of protein sulfhydryl groups (Benesch et al., 1955; Benesch and Benesch, 1948). A tendency of sulfhydryl groups to combine with more than one silver ion detracts from the usefulness of these methods for quantitation of sulfhydryl groups.

10-3 CYANOGEN BROMIDE

Cyanogen bromide is used to selectively cleave peptide bonds of methionine residues in proteins and polypeptides (Gross and Witkop, 1962). Cleavage involves the initial formation of a methionine cyanosulfonium derivative (II), which, under acidic conditions, is immediately converted into the transient homoserine iminolactone (III). The reaction goes spontaneously to completion with the formation of a new NH_2-terminal peptide and COOH-terminal homoserine lactone peptide fragment (IV). Under normal conditions, the reaction goes rapidly to completion with neither intermediate being isolated.

Miscellaneous Reagents

[Structures I, II, III, IV depicting the mechanism of cyanogen bromide cleavage at methionine residues]

Cyanogen bromide is presently the most specific and most useful reagent for the chemical cleavage of peptide bonds. Other than methionine, only cysteine is known to react slowly under the usual acidic reaction conditions (70% formic acid). These rather harsh conditions, however, preclude its use with many biologically active proteins. By a somewhat milder treatment, cyanogen bromide has been used to cleave some more stable biologically active proteins. At pH 6, for example, cleavage of about one-half the methionyl residues in rabbit immunoglobulin, type G (IgG), molecular weight 150,000 grams, gives rise to a fragment of 94,800 grams molecular weight lacking some biological properties of the unmodified protein but having retained the capacity to precipitate antigen (Cahnmann et al., 1966). The role of methionine residues in several protein inhibitors of trypsin has been investigated using cyanogen bromide. For example, cleavage of both methionyl groups failed to significantly decrease the ability of turkey ovomucoid to inhibit trypsin and chymotrypsin (Feinstein and Feeney, 1967). The cyanogen bromide–methionine reaction is extremely useful as an adjunct to enzymatic methods for cleaving peptide bonds and determining sequences of proteins and peptides. Its uses as a protein reagent have been well covered by several recent reviews (Gross, 1967; Witkop, 1968; Spande et al., 1970).

10-4 SULFENYL HALIDES

$$\text{P}-\text{indole-NH} + \text{ClS}-\text{C}_6\text{H}_4-\text{NO}_2 \xrightarrow{\text{20-60\% acetic acid}}$$

$$\text{P}-\text{(2-(4-nitrophenylthio)indole)-NH} + \text{H}^+ + \text{Cl}^- \quad (1)$$

$$\text{P}-\text{SH} + \text{ClS}-\text{C}_6\text{H}_3(\text{NO}_2)_2 \xrightarrow{\text{20-60\% acetic acid}}$$

$$\text{P}-\text{S}-\text{S}-\text{C}_6\text{H}_3(\text{NO}_2)_2 + \text{H}^+ + \text{Cl}^- \quad (2)$$

Sulfenyl halides have been shown to react selectively with tryptophan and cysteine residues in proteins, giving a 2-thioether derivative and a mixed disulfide, respectively (Equations 1 and 2) (Scoffone et al., 1968; Fontana et al., 1968). Both reactions take place in aqueous solutions in the presence of high concentrations of an organic acid. For proteins able to withstand these conditions without loss of biological activity, these reagents can be used to examine the effect of substitution on tryptophan and cysteine residues. An important use for these reagents appears to be for quantitative determination of these residues.

The reaction of sulfenyl halides with tryptophan produces a chromophore that is readily quantitated spectrophotometrically. The mixed disulfide formed from cysteinyl residues can be cleaved to give the corresponding nitrothiophenolate ion by 0.1 M NaOH and removed to prevent its interference with the quantitation of tryptophan. The nitrothiophenolate ion can also be determined and used to quantitate amounts of cysteine originally present (Boccú et al., 1970). The 4-nitrophenylthioether derivative of tryptophan has an absorption maximum at 328 mμ and the 4-nitrothiophenolate ion at 450 mμ.

Due to the harsh conditions necessary for reaction, sulfenyl halides are most useful as modifying agents for small, stable, or not-too-complex proteins and polypeptides. Modification of the single tryptophan residue in the peptide hormones

ACTH and human growth hormone, for example, has been used to investigate the importance of these residues for biological activity. Treatment of ACTH with 2-nitrophenylsulfenyl chloride results in the loss of its lipolytic-stimulatory activity, and the modified hormone blocks the lipolytic activity of the unmodified ACTH (Ramachandran and Lee, 1970). In contrast, when the single tryptophan residue of growth hormone is substituted with 2-nitrophenylsulfenyl chloride, its physical-chemical properties remain similar to those of the unmodified hormone, and full growth-promoting potency is retained (Brovetto-Cruz and Li, 1969).

Modification of all six tryptophan residues in lysozyme with 2-nitrophenylsulfenyl chloride drastically alters that enzyme's physical-chemical properties and completely eliminates catalytic activity (Habeeb and Atassi, 1969).

10-5 NITROUS ACID

Nitrous acid was used by Van Slyke (1929) to deaminate and, from the nitrogen liberated, to quantitatively determine α-amino acids. It is a weak acid ($pK \sim 3.3$) and in aqueous solution gives rise to several species (Equation 1).

$$NO_2^- \underset{}{\overset{H^+}{\rightleftharpoons}} HONO \underset{}{\overset{H^+}{\rightleftharpoons}} H_2^+ONO \overset{NO_2^-}{\rightleftharpoons} N_2O_3 + H_2O \tag{1}$$

$$H_2^+ONO \rightleftharpoons NO^+ + H_2O$$

$$N_2O_3 \rightarrow NO\cdot + NO_2\cdot$$

Deamination rates of amino acids vary with the square of the concentration of nitrous acid, suggesting dinitrogen trioxide (N_2O_3) to be an important reactive species. The initial step in the reaction appears to be formation of a nitrosoamine by the nucleophilic displacement of nitrite ion from dinitrogen trioxide (Equation 2). Tautomerization leads to diazohydroxide which, in turn, ionizes to a diazonium salt (Equation 2).

$$R-\underset{H}{\overset{H}{N}}:\overset{O}{\overset{\|}{N}}-ONO \xrightarrow{-NO_2^-} R-\underset{H}{\overset{H}{\overset{+}{N}}}-N=O \longrightarrow$$

$$R-\underset{H}{\overset{+}{N}}-N=\overset{+}{O}H \xrightarrow{-H^+} R-N=N-OH \xrightarrow{-OH^-} R\overset{+}{N}\equiv N \tag{2}$$

Aliphatic diazonium ions are quite unstable. The adjacent carbonyl group present upon reaction with α-amino acids presumably confers a stabilizing effect on the intermediate diazonium ion (see Section 7-1), but it does not appear to alter the ultimate course of reaction. Diazonium ions can react with a variety of nucleophiles or can be converted by β-elimination of nitrogen into corresponding unsaturated compounds (Equations 3 and 4). These two alternate reactions do not account completely for the products formed, however, since none of those expected is obtained in large amount. One product, ε-hydroxy-α-aminocaproic acid, can be found in hydrolysates at a position near glycine upon amino acid analysis (Shields et al., 1959). In deaminated papain, however, only about 20% of the deaminated lysine appeared as this product (Shields et al., 1959). A second component observed by amino acid analysis eluting just prior to tyrosine was also in low amount.

$$\begin{array}{c} CH_2 \\ | \\ H-C-N_2^+ \\ | \\ C=O \\ | \end{array} \quad \begin{array}{c} \xrightarrow{H_2O,\ -N_2} \\ \\ \xrightarrow{X^-,\ -N_2} \\ \\ \xrightarrow{-N_2} \end{array} \quad \begin{array}{c} CH_2 \\ | \\ H-C-OH\ +\ H^+ \\ | \\ C=O \\ | \end{array} \quad (3a)$$

$$\begin{array}{c} CH_2 \\ | \\ H-C-X \\ | \\ C=O \\ | \end{array} \quad (3b)$$

$$\begin{array}{c} CH \\ \| \\ H-C \\ | \\ C=O \\ | \end{array}\ +\ H^+ \quad (3c)$$

α-Amino groups, due to their lower pK values, react with nitrous acid very much more rapidly than ε-amino groups (Van Slyke, 1929; Belenkii and Orestova, 1965; Maeda and Ishida, 1967). Nitrous acid can thus be used in some cases to specifically modify terminal α-amino groups without affecting ε-amino groups. Modification of tryptophan and tyrosine residues is an important side reaction. Reaction with tryptophan is indicated by the formation of a yellow color, but the compound(s) responsible have not been identified. Tyrosine can apparently be converted into 3-nitrosotyrosine. In *E. coli* L-asparaginase, an absorption maximum

$$\begin{array}{c}\text{N}_2^+ \\ | \\ \text{CH}_2 \\ | \\ \text{CH}_2 \\ | \\ \text{CH}_2 \\ | \\ \text{CH}_2 \\ | \\ \text{CH}-\text{NH}_2 \\ | \\ \text{C}=\text{O} \\ | \end{array}$$

$$\xrightarrow[-N_2]{H_2O} \begin{array}{c} \text{OH} \\ | \\ \text{CH}_2 \\ | \\ \text{CH}_2 + \text{H}^+ \\ | \\ \text{CH}_2 \\ | \end{array} \quad (4a)$$

$$\xrightarrow[-N_2]{H_2O} \begin{array}{c} \text{CH}_3 \\ | \\ \text{CH}-\text{OH} + \text{H}^+ \\ | \\ \text{CH}_2 \\ | \end{array} \xrightarrow[110°C,\,18\,hr]{6\,N\,HCl} \text{CH}_3\text{-} \underset{\text{(lactone ring)}}{\bigcirc} \text{-NH}_3^+ + \text{H}_2\text{O} \quad (4b)$$

$$\xrightarrow[-N_2]{} \begin{array}{c} \text{CH}_2 \\ \| \\ \text{CH} + \text{H}^+ \\ | \\ \text{CH}_2 \\ | \end{array} \quad (4c)$$

at 710 mμ appeared after deamination with nitrous acid but could be avoided by the inclusion of a small amount of phenol or salicylic acid in the reaction solution (Wagner et al., 1969).

Deamination of the NH$_2$-terminal amino group of pancreatic ribonuclease greatly decreased its ability to precipitate with rabbit antiribonuclease antibody without greatly altering its enzymatic activity (Van Vunakis et al., 1960). At pH 5 and low ionic strength, the deaminated enzyme had an increased ability to cleave yeast RNA, but at higher pH or ionic strength, its ability was similar to that of the unmodified enzyme.

The inactivation of porcine elastase by nitrous acid is dependent upon the enzyme's conformational state. Above pH 4.2 the enzymatically important α-amino group is inaccessible to the reagent, but at lower pH (pH 3.0) a reversible change in the enzyme's conformation exposes this group and increases its reactivity (Gertler and Hofmann, 1967). The inactivation of trypsin by nitrous acid is similar in most respects to that of elastase but is independent of pH between pH 4.2 and 3.0 (Scrimger and Hofmann, 1967). The NH$_2$-terminus of trypsin is as reactive at pH 4.2 as is that of elastase at pH 3.0. Substrates of trypsin protect it from inactivation by nitrous acid.

E. coli L-asparaginase, upon treatment with nitrous acid, experiences a loss of its single NH_2-terminal amino group and one of its 13 ε-amino groups (Wagner et al., 1969). This doubly deaminated derivative has catalytic activity similar to the untreated enzyme. The NH_2-terminus of the antitumor and antimicrobial peptide Neocarzinostatin can be selectively removed by nitrous acid without affecting its ε-amino group. The desamino peptide appears to have increased activity against certain tumors (Maeda and Ishida, 1967).

REFERENCES

Arnon, R., and E. Shapira (1969): *J. Biol. Chem.*, **244**, 1033.
Belenkii, B. G., and V. A. Orestova (1965): *Biokhimiya*, **30**, 878.
Benesch, R., and R. E. Benesch (1948): *Arch. Biochem. Biophys.*, **19**, 35.
Benesch, R. E., H. A. Lardy, and R. Benesch (1955): *J. Biol. Chem.*, **216**, 663.
Bennett, H. S. (1951): *Anat. Rec.*, **110**, 231.
Boccú, E., F. M. Veronese, A. Fontana, and C. A. Benassi (1970): *Eur. J. Biochem.*, **13**, 188.
Boyer, P. D. (1954): *J. Am. Chem. Soc.*, **76**, 4331.
Boyer, P. D., and H. L. Segal (1954): in *A Symposium on the Mechanism of Enzyme Action*, edited by W. D. McElroy and H. B. Glass, Johns Hopkins Press, Baltimore, p. 520.
Brovetto-Cruz, J., and C. H. Li (1969): *Biochemistry*, **8**, 4695.
Buttkus, H. (1967): *J. Food Sci.*, **32**, 432.
Cahnmann, H. J., R. Arnon, and M. Sela (1966): *J. Biol. Chem.*, **241**, 3247.
Chio, K. S., and A. L. Tappel (1969): *Biochemistry*, **8**, 2821.
Crawford, D. L., T. C. Yu, and R. O. Sinnhuber (1967): *J. Food Sci.*, **32**, 332.
Edsall, J. T., R. H. Maybury, R. B. Simpson, and R. Straessle (1954): *J. Am. Chem. Soc.*, **76**, 3131.
Elödi, P. (1960): *Biochim. Biophys. Acta*, **40**, 272.
Feinstein, G., and R. E. Feeney (1967): *Biochim. Biophys. Acta*, **140**, 55.
Fontana, A., E. Scoffone, and C. A. Benassi (1968); *Biochemistry*, **7**, 980.
Gerhart, J. C., and A. B. Pardee (1962): *J. Biol. Chem.*, **237**, 891.
Gertler, A., and T. Hofmann (1967): *J. Biol. Chem.*, **242**, 2522.
Gross, E. (1967): *Methods Enzymol.*, **11**, 238.
Gross, E., and B. Witkop (1962): *J. Biol. Chem.*, **237**, 1856.
Grossberg, A. L., and D. Pressman (1968): *Biochemistry*, **7**, 272.
Habeeb, A. F. S. A., and M. Z. Atassi (1969): *Immunochemistry*, **6**, 555.
Horowitz, M. G., and I. M. Klotz (1956): *Arch. Biochem. Biophys.*, **63**, 77.
Hughes, W. L. (1947): *J. Am. Chem. Soc.*, **69**, 1836.
Hughes, W. L. (1949): *Cold Spring Harbor Symp. Quant. Biol.*, **14**, 79.
Itano, H. A., and A. J. Gottlieb (1963): *Biochem. Biophys. Res. Commun.*, **12**, 405.
Jovin, T. M., P. T. Eglund, and A. Kornberg (1969): *J. Biol. Chem.*, **244**, 3009.
King, T. P. (1966): *Biochemistry*, **5**, 3454.
Klapper, M. H. (1970): *Biochem. Biophys. Res. Commun.*, **38**, 172.
Kominz, D. R. (1961): *Biochim. Biophys. Acta*, **51**, 456.
Kotaki, A., M. Harada, and K. Yagi (1966): *J. Biochem. (Tokyo)*, **60**, 592.
Kwon, T. W., and H. S. Olcott (1966): *Biochim. Biophys. Acta*, **130**, 528.
Lipscomb, W. N. (1967): *Am. Chem. Soc. Abstr.*, **155**, 2-M.
Liu, W. H., G. Feinstein, D. T. Osuga, R. Haynes, and R. E. Feeney (1968): *Biochemistry*, **7**, 2886.

Madsen, N. B., and C. F. Cori (1955): *Biochim. Biophys. Acta*, **18,** 156.
Madsen, N. B., and C. F. Cori (1956): *J. Biol. Chem.*, **223,** 1055.
Maeda, H., and N. Ishida (1967): *Biochim. Biophys. Acta*, **147,** 597.
Malmström, B. G. (1962): *Arch. Biochem. Biophys. Suppl.*, 247.
Nakaya, K., H. Horinishi, and K. Shibata (1967): *J. Biochem. (Tokyo)*, **61,** 345.
Nakaya, K., T. Suzuki, O. Takenaka, and K. Shibata (1969): *Biochim. Biophys. Acta*, **194,** 301.
Pontremoli, S., B. Luppis, S. Traniello, M. Rippa, and B. L. Horecker (1965): *Arch. Biochem. Biophys.*, **112,** 7.
Ramachandran, J., and V. Lee (1970): *Biochem. Biophys. Res. Commun.*, **38,** 507.
Riordan, J. F., and B. L. Vallee (1967): *Methods Enzymol.*, **11,** 541.
Robinson, G. W., R. A. Bradshaw, L. Kanarek, and R. L. Hill (1967): *J. Biol. Chem.*, **242,** 2709.
Scoffone, E., A. Fontana, and R. Rocchi (1968): *Biochemistry*, **7,** 971.
Scrimger, S. T., and T. Hofmann (1967): *J. Biol. Chem.*, **242,** 2528.
Shields, G. S., R. L. Hill, and E. L. Smith (1959): *J. Biol. Chem.*, **234,** 1747.
Silverstein, E., and G. Sulebele (1970): *Biochemistry*, **9,** 274.
Spande, T. F., B. Witkop, Y. Degani, and A. Patchornik (1970): *Adv. Protein Chem.*, **24,** 97.
Sperling, R., Y. Burstein, and I. Z. Steinberg (1969): *Biochemistry*, **8,** 3810.
Takahashi, K. (1968): *J. Biol. Chem.*, **243,** 6171.
Toi, K., E. Bynum, E. Norris, and H. A. Itano (1967): *J. Biol. Chem.*, **242,** 1036.
Vallee, B. L., T. L. Coombs, and F. L. Hoch (1960): *J. Biol. Chem.*, **235,** PC45.
Van Slyke, D. D. (1929): *J. Biol. Chem.*, **83,** 425.
Van Vunakis, H., E. Leikhim, R. Delaney, L. Levine, and R. K. Brown (1960): *J. Biol. Chem.*, **235,** 3430.
Wagner, O., E. Irion, A. Arens, and K. Bauer (1969): *Biochem. Biophys. Res. Commun.*, **37,** 383.
Webb, J. L. (1966): *Enzyme and Metabolic Inhibitors*, vol. 2, Academic Press, New York, p. 729.
Witkop, B. (1968): *Science*, **162,** 318.
Yamada, S., and H. A. Itano (1966): *Biochim. Biophys. Acta*, **130,** 538.
Yankeelov, J., M. Kochert, J. Page, and A. Westphal (1966): *Fed. Proc.*, **25,** 590.
Yankeelov, J. A., C. D. Mitchell, and T. H. Crawford (1968): *J. Am. Chem. Soc.*, **90,** 1664.

III

III
appendix

selected techniques for the modification of protein side chains

A-1 AMINO GROUPS

Amino groups of proteins are basic groups (pK values near 8 for α-NH$_2$ and 9.5 for ε-NH$_2$) and are positively charged except at high pH. Only the uncharged form, that which predominates at pH values higher than their pK, is reactive as a nucleophile. Higher pH thus usually enhances their reactivity with most reagents.

Acetic Anhydride. Acetylation with acetic anhydride is probably the most frequently used procedure for the modification of protein amino groups. The preferred method for most work is that first described by Fraenkel-Conrat (1957) as follows:

> To the protein solution, preferably but not necessarily of high concentration (2–10%), is added an equal volume of saturated solution of sodium acetate; the solution (or suspension) is cooled in an ice bath and then treated with a total amount of acetic anhydride approximately equal to the weight of protein used, distributed over three to six additions in the course of 1 hour at 0° (e.g., 10 mg of protein in 0.1 ml of H$_2$O, 0.1 ml of sodium acetate, five times 2 μl of acetic anhydride). Dialysis or gel filtration can be used to separate the modified protein from unwanted components of the reaction solution.

The high concentration of sodium acetate serves as a buffer and also helps to direct the selectivity of modification to amino groups (see Section 5-1). Acetylation in the absence of high sodium acetate concentrations is sometimes desirable, in which case the same general procedure can be employed substituting a more desirable buffer or maintaining the pH by periodic addition of alkali. More extensive acetylation of tyrosine residues should be expected under such conditions.

Succinic Anhydride. Succinylation of protein amino groups using succinic anhydride proceeds under mildly alkaline conditions similar to those suitable for acetylation with acetic anhydride (see Section 5-1). Variations of the procedure described by Habeeb et al. (1958) are commonly used, such as the following for the succinylation of staphylococcal enterotoxin B (Chu et al., 1969).

In a typical experiment, 180 mg of enterotoxin B was dissolved in 15 ml of 1.0 M carbonate buffer at pH 8.0; 20 mg of solid succinic anhydride was added to the stirred solution every 10 minutes for 70 minutes. The reaction solution was maintained at pH 8.0 with 0.1 N NaOH by pH-stat titration. The reaction was stopped at either 1 or 2 hours by dilution to 50 ml with distilled water and then dialyzed. Such treatment resulted in the succinylation of approximately half the total amino groups. The extent of succinylation can be expected to vary with different proteins, and the molar excess of reagent employed (i.e., frequently 10- to 20-fold) should be adjusted to achieve the desired extent of modification. Particularly with small amounts of protein, where a correspondingly small amount of succinic anhydride is used, it is frequently convenient to employ the reagent dissolved in a solution of dry dioxane.

Maleic Anhydride. Modification of proteins with maleic anhydride proceeds similarly as in the procedure of Butler et al. (1969) for the preparation of maleyl-chymotrypsinogen:

> Bovine chymotrypsinogen A (20 mg) is dissolved in 2.0 ml of 0.1 M sodium pyrophosphate buffer, pH 9.0, and treated at 2°C with 300 μl of 1.0 M maleic anhydride in redistilled dioxane. The maleic anhydride solution is added in six additions and the pH of the mixture is maintained at 9.0 by the addition of 0.1 M NaOH. When the reaction is complete, the maleyl-chymotrypsinogen is desalted by passing it through a column (40 × 3 cm) of Sephadex G-25 in 0.01 M NH$_3$, and the fractions that contain the protein are pooled to give a solution with a protein concentration of about 0.6 mg/ml. More than 90% of the amino groups are blocked.

> For unblocking of maleyl-chymotrypsinogen, the solution of maleyl-chymotrypsinogen is adjusted to pH 3.5 with formic acid and aqueous NH$_3$. The solution is then incubated at 37°C for 30 hours after which the reaction is stopped by addition of alkali to raise the pH above neutrality. More than 90% of the maleyl groups are removed by this procedure.

Reaction with 2-methylmaleic anhydride (citraconic anhydride) proceeds under the same conditions. The deacylation procedure, however, is considerably more rapid (Dixon and Perham, 1968; Gibbons and Perham, 1970). Where mild conditions or more rapid deacylation is desirable the latter reagent appears preferable. These reagents are discussed in Section 5-1.

Cyanate (Carbamylation). Cyanate reacts with amines at pH 5 and above to give substituted ureas (Section 5-2). For the modification of protein amino groups, variations of the procedure of Stark et al. (1960) are employed. Reaction rates vary relatively little from pH 5 to 1 unit below the pK of the group being modified. A buffer or some other means for controlling the pH of the reaction solution such as a pH-stat should, however, be used to prevent its rise due to hydrolysis of the reagent. Modification typically involves a concentration of sodium or potassium cyanate not greater than 1 M, at approximately neutral pH for a time, and a temperature sufficient to give the desired extent of reaction. The preparation of carbamylated rabbit γ-globulin (Chen et al., 1962) is given as a typical example.

Rabbit γ-globulin (20 to 30 mg/ml) in 1 M KCNO at 38° in pH 8 borate buffer gave after 4.3 and 9 hours, respectively, 76% and 84% carbamylation of the amino groups. The reactions were stopped and excess reagent removed by dialysis.

O-Methylisourea (Guanidination). O-Methylisourea reacts with ε-amino groups of proteins converting lysine residues into homoarginine residues (see Section 5-3). The reaction is strongly pH-dependent, giving greater modification at higher pH. The procedure of Klee and Richards (1957) for the guanidination of pancreatic ribonuclease follows:

> The pH of a solution of O-methylisourea hydrochloride is adjusted with 2.5 M NaOH to the desired value between 8.5 and 11 as measured at room temperature. This solution is cooled to 2° and added to the sample of ribonuclease to give a final concentration of 0.5 M O-methylisourea and 0.5% protein. The reagent is a good buffer in the pH region 9 to 11, and only very small changes in pH are expected from solution of the protein in the adjusted reagent or from the subsequent reaction.
>
> The reaction is allowed to proceed at 2° for periods of time from a few hours to 3 or 4 days. Samples are taken at appropriate intervals and placed in sufficient acetate buffer, pH 5.0, to bring the resulting pH below 7. The extent of reaction obtained at the various pH values and times is shown in Figure 5-8. A pH of 10 or greater is required for extensive reaction.

Ethyl Acetimidate (Amidination). Amidination of proteins with ethyl acetimidate occurs in aqueous solutions between pH 7 and 10. The reaction is strongly pH-dependent, being more rapid at higher pH. A procedure similar to that of Wofsy and Singer (1963) for bovine serum albumin is applicable to many proteins.

> To achieve 62-65% amidination, 0.14 g ethyl acetimidate hydrochloride and 0.1 ml of 5 M NaOH are mixed and added to 5 ml of an approximately 1% protein solution in borate buffer, pH 8.5, $\mu = 0.1$. (The reagent is about 0.2 M in a total volume of 6 ml. The amount of 5 M NaOH mixed with the reagent prior to addition of the protein solution is somewhat less than that required for neutralization; the preliminary mixing avoids subjecting the protein to extremes of pH.) The pH is adjusted and maintained at 8.3 to 8.6 for 2 hours, the reaction solution being stirred and kept at 0°. After 2 hours, the solution is dialyzed to stop the reaction.
>
> To attain 85% amidination, the same procedure can be used but with an increase in reagent concentration to 1 M. For exhaustive amidination, the combined effect of more reagent, higher pH and temperature, and longer reaction time can be used (Wofsy and Singer, 1963). Amidination is discussed in Section 5-3.

Reductive Alkylation. Reduction of the Schiff bases resulting from the reaction of various aldehydes and ketones with protein amino groups can be used to produce a large variety of substituted proteins (see Section 6-8). The procedure, known as *reductive alkylation*, can be done between pH 8 and 10 at low temperature (Means and Feeney, 1968).

Reductive methylation of most proteins is best done at 0° in 0.20 M borate buffer (pH 9.0). For each milliliter of solution containing 2.5 to 10.0 mg of protein, approximately 0.5 mg of sodium borohydride is added, followed by five increments (0.5 μl/ml of reaction solution) of 37% aqueous formaldehyde solution over a period of 30 minutes. Such treatment generally results in the methylation of more than 80% of the amino groups. The relative amounts of formaldehyde and sodium borohydride indicated are such that a very slight excess of borohydride should remain at the end of the reaction. This is important in that it prevents other possible reactions between the protein and formaldehyde. More extensive modification is best obtained by a repetition of the same procedure. Reductive ethylation using acetaldehyde in place of the formaldehyde solution can be done similarly, but for equivalent treatment gives less extensively modified proteins.

Reductive isopropylations can be done at 0° in 0.18 M borate buffer (pH 9.0) containing 10% v/v acetone. Several small portions of sodium borohydride are added (0.5 to 1 mg/ml of reaction solution) to each milliliter of solution containing 2.5 to 10.0 mg of protein, until the desired extent of modification is achieved. The samples can be dialyzed or subjected to gel permeation chromatography to separate the protein from unreacted reagents and reaction products.

Trinitrobenzenesulfonic Acid. The reaction of trinitrobenzenesulfonic acid (TNBS) with protein amino groups takes place at pH values near 7 or above, and can be used to study the effect of amino-group substitution and to quantitatively determine amino groups (Section 6-5). The recommended procedure is essentially that first described by Habeeb (1966):

To 1 ml of protein solution (0.6 to 1 mg/ml) is added 1 ml of 4% NaHCO$_3$ (pH 8.5) and 1 ml of 0.1% TNBS in water. The solution is then incubated at 37°–40°C in the dark for a time sufficient to give the desired extent of reaction. Two hours at 40° gives near-quantitative reaction. Lower temperatures give slower and more selective reaction. To quantitate the extent of reaction, 1 ml of 10% sodium dodecyl sulfate should be added to solubilize the protein upon addition of 0.5 ml of 1 M HCl. The absorbance of the solution at 335 to 345 mμ is read against a blank containing 1 ml of water instead of protein solution. An extinction coefficient of 1.4×10^4 M^{-1} cm^{-1} can be used to calculate the number of amino groups present.

A-2 SULFHYDRYL GROUPS

Sulfhydryl groups are very reactive with many reagents. They react readily with most alkylating and arylating agents, combine with many heavy-metal ions, and are easily oxidized by even very mild oxidizing agents. They react readily with most acylating agents, but the resulting thiol esters are quite unstable in an aqueous environment and usually hydrolyze. Four of the more useful reagents for the modification of protein sulfhydryl groups and procedures for their use are given below.

p-Mercuribenzoate. Solutions of *p*-mercuribenzoate can be prepared by adjusting an aqueous suspension of the reagent (*p*-chloro- or *p*-hydroxymercuribenzoate) to pH 10 or 11 to achieve solubilization and then readjusting to a desired lower pH. Insoluble material, if present, can be removed by filtration or centrifugation. The reagent concentration can be determined by measuring the absorption at 232 mμ employing an extinction coefficient of 1.69×10^4 M^{-1} cm^{-1} (Boyer, 1954).

For qualitatively determining the effect of *p*-mercuribenzoate substitution, the desired amount of reagent can be added to the protein in water or a buffer compatible with the protein. The amount used is usually based on moles of sulfhydryl groups present, varying from a stoichiometric equivalence to a several-hundredfold excess. Sufficient time should be allowed for completion of the reaction because of large variation in reaction rates. For most proteins, maximum reaction rates are obtained at pH 4.5 to 5. To determine the kinetics of the reaction of a reagent with sulfhydryl groups, the desired amount of reagent is added and the rate of mercaptide formation determined from the rate of increase in absorbance at 250 to 255 mμ (see Figure 10-4). The increment of absorbance per mercaptide group can be estimated, if one knows the number of sulfhydryl groups, by determining the total increase in the presence of excess reagent. This value varies for different proteins and for the same protein under different conditions (Swenson and Boyer, 1957). Typically, values from 5 to 7.5×10^3 M^{-1} cm^{-1} are found. At pH values near neutrality, the change in absorbance is greatest at 250 mμ, while at pH 4.5 it is greatest near 255 mμ.

For quantitative determination of sulfhydryl groups with *p*-mercuribenzoate, a procedure such as that described by Boyer (1954) is recommended. This involves adding the protein in question to a solution of *p*-mercuribenzoate as in a titration and noting the endpoint as that amount of protein no longer giving a linear increment of increased absorbance (see Figure 10-5). Because both the reagent and the protein absorb strongly at the wavelengths employed, care must be exercised to employ proper blanks. Under some circumstances, for example, when the protein is available only in small amounts or has a very low content of sulfhydryl groups, it is more desirable to reverse the sequence of additions (i.e., adding the reagent to the protein, and observing the change in absorbance and the endpoint in an analogous way).

Iodoacetate. The reactivity of sulfhydryl groups with haloacetates far exceeds that of other protein groups under almost all conditions (Section 6-1). It is therefore possible to employ an extremely wide range of conditions, usually designed primarily to satisfy requirements of the protein being modified, and still obtain selective reaction with sulfhydryl groups. While a pH between approximately 6 and 8.5 is optimal, reaction specificity continues high for a considerable distance on either side of this range. Specific modification can usually be enhanced by using just the theoretical amount of reagent required for reaction with sulfhydryl groups. For quantitative reaction, urea, SDS, or another denaturant is frequently needed. Solutions of haloacetates should be made up just prior to their use, since they are subject to slow hydrolysis ($—CH_2—I \xrightarrow{OH^-} —CH_2—OH + I^-$) under alkaline conditions. The procedure of Crestfield et al. (1963) for the reduction and S-carboxymethylation of proteins is:

> To 5 to 100 mg of protein in a 12-ml screw-cap vial maintained under a nitrogen barrier, add 3.61 g of deionized, crystalline urea, 0.30 ml of EDTA solution (50 mg of disodium EDTA per ml), 3.0 ml of Tris buffer, pH 8.6 (5.23 g of Tris and 9 ml of 1.0 N HCl diluted to 30 ml with water), and finally 0.10 ml of mercaptoethanol. The solution is made up to a 7.5-ml mark with water, and a solution 8 M in urea and 0.2% in EDTA is used to fill the vial completely. A disk of polyethylene is slid over the top in order to exclude air and is held in place with the screw cap. After 4 hours at room temperature (22°–25°), the contents of the vial are transferred to a 25-ml beaker under the nitrogen barrier. A freshly prepared solution of 0.268 g of iodoacetic acid in 1.0 ml of 1.0 N NaOH is added to the reaction mixture. The iodoacetate added is slightly less on a molar basis than the amount of mercaptoethanol. The —SH groups of cysteine residues react most rapidly, and since the excess iodoacetate reacts faster with mercaptoethanol above pH 8 than with thioether sulfur, alkylation of methionine is kept to a minimum. The three equivalents of Tris per equivalent of iodoacetate keep the solution more alkaline than pH 8.3.

Exclusion of light while working up the solution is important in order to prevent the formation of iodine which may react with tyrosine and histidine residues. Fifteen minutes after addition of the iodoacetate, the reaction is stopped by passing the sample through a column of Sephadex G-75 to separate the protein from other components of the reaction solution.

When haloacetates are used to determine the effects of sulfhydryl group substitution on a protein's properties, conditions should be chosen under which the protein is stable. Temperature, pH, and the concentration of reagent can be adjusted to attain the desired extent or rate of reaction. The procedure of Battell et al. (1968) for the carboxymethylation of muscle phosphorylase b is more or less typical:

> The reaction was carried out in 0.13 M KCl, 0.033 M sodium glycerophosphate, 1.5 mM EDTA at pH 7.5 with a protein concentration of ~10 mg/ml, and 10 mM iodoacetamide. Aliquots were taken and the reaction stopped at various times by addition of a 50-fold excess of mercaptoethanol over iodoacetamide. After 8 hours incubation at 30°C approximately 85% of the enzyme's catalytic activity was lost. At this high excess of iodoacetamide (approximately 30-fold) inactivation was pseudo-first order to beyond 95% completion.

N-Ethylmaleimide. N-Ethylmaleimide shows high selectivity for reaction with more exposed protein sulfhydryl groups. At pH values near neutrality its reaction with sulfhydryl groups is sufficiently specific to allow considerable variation in other reaction parameters such as temperature and reagent concentration. It can be used to quantitatively determine sulfhydryl groups (see Section 6-2). However, its more important use is for substitution of sulfhydryl groups to determine the effects of modification on various properties. The reaction of N-ethylmaleimide with isocitrate dehydrogenase (Colman and Chu, 1970) was done, for example, at pH 7.7 in

0.1 M triethanolamine buffer containing 0.3 M Na$_2$SO$_4$, 1 mM EDTA, and 10% glycerol. Enzyme (0.37 mg/ml) was incubated at 30°C with 0.09 mM reagent and the reaction was followed spectrophotometrically from the decrease in absorbance at 302 mμ using $\Delta\varepsilon = 620$ (see Section 6-2). After 60 minutes more than 80% of the dehydrogenase activity and almost all reductive carboxylation activity were lost. In the presence of the enzyme's substrate, isocitrate, and the activating cation, Mn^{++}, negligible activity was lost during this period. Under these conditions of reagent excess (approximately sevenfold based on two sulfhydryl groups per 58,000 g) the changes in catalytic properties appeared pseudo-first order for 60 minutes. Hydrolysis of the reagent was shown to have negligible influence on the observed rates under these conditions (i.e., half-life = 318 minutes).

5,5'-Dithiobis(2-nitrobenzoic acid) (DTNB). DTNB can be used to effect the substitution of sulfhydryl groups in proteins and to quantitatively determine them spectrophotometrically. Variations of the procedure first described by Ellman (1959) can be used for quantitation of sulfhydryl groups. A stock solution of DTNB (39.6 mg in 10 ml of 0.1 M phosphate buffer, pH 7.0) can be prepared and used for several weeks or longer until blank values become too high. To each 3.0-ml sample of protein in 0.1 M phosphate buffer (pH 8.0), 0.02 ml of stock DTNB solution is added; after 2 to 3 minutes, the optical density at 412 mμ is determined against a blank lacking protein. An extinction coefficient of 13,600 M^{-1} cm^{-1} can be used to calculate the concentration of sulfhydryl groups reacting. Denaturants like sodium dodecyl sulfate, urea, or guanidine hydrochloride can be used to increase the reactivity of slow-reacting groups.

To determine the effect of substitution on the properties of a protein, reaction conditions should be chosen to be compatible with the protein. As an example, the procedure of Kastenschmidt et al. (1968) for reaction of DTNB with muscle phosphorylase b follows:

> To 10 mg of phosphorylase b in 0.9 ml of 50 mM sodium glycerophosphate and 2 mM EDTA (pH 6.8) was added 0.1 ml of 10 mM DTNB in the same buffer. The molar concentrations of enzyme and DTNB in the solution were 5.4×10^{-5} M and 1×10^{-3} M, respectively, giving an 18.5-fold excess of DTNB. The reaction solution was incubated for as long as 7 hours at room temperature or until the desired extent of reaction was indicated by the increased absorbance at 412 mμ. To stop the reaction and separate the modified protein from excess reagent and reaction products, the solution was passed through a column packed with Sephadex G-25 preequilibrated with 50 mM glycerophosphate and 2 mM EDTA at pH 6.8. Under these conditions the reaction was biphasic: two sulfhydryl groups per dimer of phosphorylase b react rapidly and two others more slowly. Reaction of the latter two effects changes in quaternary structure and a decrease in catalytic activity. Following a 6-hour incubation period a mixture of products was obtained consisting of catalytically active dimers with one modified sulfhydryl per monomer, an inactive monomer with two modified sulfhydryl groups, and partially active, polydisperse, high-molecular-weight aggregates with between one and two modified groups per monomer.

A-3 DISULFIDE GROUPS

Disulfide bonds in proteins are subject to cleavage by mild reducing agents (Section 8-1), by a few oxidizing agents (Section 8-6), or by nucleophilic displacement (Section 8-2). Of the procedures for cleavage, those involving reduction are probably the mildest and most specific. Many smaller proteins when reduced retain the capacity to correctly refold and reoxidize under appropriate conditions. Complete reduction usually requires the presence of urea or another denaturant; limited and selective cleavage of a few disulfide bonds can frequently be obtained in the absence of such agents. Procedures have been described for limited reduction of disulfide bonds in proteins using dithiothreitol (Bewley and Li, 1969; Liu and Meienhofer, 1968; Sperling et al., 1969), sodium borohydride (Light and Sinha, 1967; Kress and Laskowski, 1967), sodium phosphorothioate (Neumann et al., 1967), and β-mercaptoethanol (Shapira and Arnon, 1969). Complete reduction of disulfide bonds is usually accomplished in 6 M guanidine/HCl or 8 M urea using a procedure such as that described by Anfinsen and Haber (1961) for the reduction and reoxidation of pancreatic ribonuclease.

Reduction. In a typical experiment, 350 mg of ribonuclease were dissolved in 10 ml of a freshly prepared 8 M solution of recrystallized urea, adjusted to pH 8.6 with 5% methylamine. Mercaptoethanol was added at a level of 1 μl per mg of protein, the container was flushed with nitrogen, and the solution was allowed to stand for $4\frac{1}{2}$ hours at room temperature. Methylamine was employed for neutralization to insure the decomposition of any thioglycollides that might be present in the reducing agent. After this period, the pH was adjusted to 3.5 with glacial acetic acid and the entire solution was applied to a column of Sephadex G-25, equilibrated with 0.1 M acetic acid. The column was developed with the same solvent. Titrations with p-mercuribenzoate (Sections 10-2 and A-2) can be used to quantitate the extent of reaction. The same procedure has also been applied to trypsinogen, chymotrypsinogen, and lysozyme, giving fully reduced products that were soluble in the dilute acetic acid solvent.

Conversion of sulfhydryl groups in reduced ribonuclease to disulfide bonds is extremely slow at the hydrogen-ion concentration of the acetic acid solution employed for the Sephadex columns. At room temperature and pH 3.0, a sample of protein containing 7.8 sulfhydryl groups per mole contained, after 24, 72, and 97 hours of exposure to atmospheric oxygen, respectively, 7.5, 6.9, and 4.5 sulfhydryl groups per mole. At icebox temperatures, the sulfhydryl group content is undiminished for two to three days.

Reoxidation. High yields of enzymatically active material and the absence of precipitation is achieved when oxidation is carried out at low protein concentrations and when conditions of surface denaturation (e.g., bubbling with air) are avoided.

Efficient oxidation occurs when dilute solutions, adjusted to pH 8.0 to 8.5, are allowed simply to stand in open vessels at room temperature for approximately 20 hours. Under these conditions, the yields of regenerated, active ribonuclease have been uniformly between 80% and 100%.

Sulfite. Sulfite reacts with disulfide bonds of proteins cleaving them and giving 1 equivalent each of S-sulfonate and a sulfhydryl group. In a dissociating medium (i.e., in high concentrations of urea or guanidine/HCl), and in the presence of an oxidizing agent, the reaction can proceed until all half-cystine residues are converted into S-sulfonates. Complete sulfitolysis can be obtained by stepwise addition of an oxidant such as iodosobenzoate (Section 8-4) or more often in a continuous fashion using cupric ion as the oxidative agent (Swan, 1957). The procedure taken from Cole (1967) is as follows:

> To 1 g of protein in 50 ml of water, add 0.5 M NH$_4$OH to pH 9.0. Add to this 1.0 mmole of CuSO$_4$ dissolved in 40 ml of water adjusted to pH 9.0, followed by 10 ml of 0.7 M Na$_2$SO$_3$. Allow to stand for 2 hours at room temperature and then dialyze against 0.1 M sodium citrate pH 7.0 at 4°C. When all the cupric ion has been removed, dialyze against water to remove other electrolytes. While such a procedure gives complete cleavage of the disulfide bonds in α-lactalbumin and β-lactoglobulin most proteins will require 8 M urea or another denaturing medium in place of water.

A-4 CARBOXYL GROUPS

The δ- and ε-carboxyl groups of aspartic and glutamic acids, respectively, are the principal anionic groups in proteins. They are acidic groups with pK values usually between pH 4.5 and 5.0. They can be esterified under relatively mild conditions by reaction with one of several diazoacetate derivatives (see Section 7-1). However, only a relatively small number of the most reactive groups become modified. A harsher procedure giving more complete reaction requires placing the protein in anhydrous methanol containing a moderate amount of hydrochloric acid. The following procedure is essentially that described by Fraenkel-Conrat and Olcott (1945).

The dry protein is suspended in cold absolute methanol and concentrated hydrochloric acid or hydrogen chloride gas is added to a final acid concentration of 0.02 to 0.10 N. The reaction mixture is maintained between 0° and room temperature for one to several days. The exact time, temperature, and acid concentration may be varied depending upon the extent of esterification desired and will usually have to be chosen on the basis of several trial preparations. At 2° using 0.07 N HCl, for example, nearly 300 hours is required to completely esterify pancreatic ribonuclease whereas eight of its eleven carboxyl groups are modified in about 120 hours. With most proteins using 0.1 N HCl and a temperature of 25° reaction will be completed within 24 hours. Reaction is stopped finally by dilution with a large volume of ice-cold water, and the excess acid and methanol are removed by dialysis against 0.001 N HCl. Dialysis against neutral or slightly alkaline solutions results in slow saponification of the methoxyl groups. In moderate to strongly alkaline solution, or in the presence of hydroxylamine, saponification is quite rapid and can be used to regenerate the unmodified protein. Saponification of ribonuclease methyl ester at pH 10.4 and 25°, for example, is complete in approximately 26 hours (Broomfield et al., 1965).

Estimation of the extent of the reaction is possible by comparison of the modified and unmodified proteins' titration curves between pH 2.0 and ~7.5. The presence of so-called buried or abnormally titrating carboxyl groups in some proteins can be a source of error. The extent of reaction can be determined based on the amount of methanol released during alkaline hydrolysis using distillation to separate it from the hydrolysate and then effecting its oxidation with excess dichromate and back-titrating with ferrous sulfate (Vithayathil and Richards, 1961). A third method involves reduction of the methoxyl groups with lithium borohydride in tetrahydrofurane and determination of the resulting hydroxyamino acid residues (Chibnall et al., 1958; Vithayathil and Richards, 1961).

Carboxyl groups of proteins can also be converted to amides through reaction with one of several amines promoted by a water-soluble carbodiimide (Section 7-2). For quantitative reaction, urea, guanidine/HCl, or another denaturant, can be used as in the following procedure of Hoare and Koshland (1967) for the quantitative determination of protein carboxyl groups. The same procedure, but in the absence of a denaturing agent, can be used to selectively modify proteins affecting up to about half the carboxyl groups.

In a typical reaction, a solution at pH 4.75 containing the protein (13.3 mg/ml), 1.33 M glycine methyl ester, and 7.5 M urea is kept at 25° in a water-jacketed vessel attached to a pH-stat. A solution of 0.40 M carbodiimide [i.e., 1-benzyl-3-(3-dimethylaminopropyl)carbodiimide or 1-ethyl-3-(3-dimethylaminopropyl)carbodi-imide] in 7.5 M urea is then added to a concentration of reagent of 0.1 M, and the pH is maintained by automatic titration with 0.5 M HCl. After standing at 25°, the solution is dialyzed at 0° against 0.001 M HCl. The number of groups modified can be determined by amino acid analysis after acid hydrolysis to detect the increase in the amount of glycine. For proteins having a high glycine content, other amines or radioactive glycine methyl ester can be used. Different amines can also be used to give the modified groups particularly desired properties. Thus, by using amino-ethanesulfonic acid the negative charge of the original carboxyl group can be maintained or, conversely, by using a diamine such as ethylenediamine, it can be replaced by a positive charge.

A-5 GUANIDINO GROUPS

Guanidino groups in proteins are strongly basic ($pK > 12$) and remain protonated except in strongly alkaline solution. They are resistant to most chemical treatments.

In the last few years, several methods have been developed for their modification. These methods are similar in that each involves reaction with a dicarbonyl compound or a derivative of a dicarbonyl. The modified groups are not ionized at neutral pH values. The procedure of Toi et al. (1967) for reaction with 1,2-cyclohexanedione in 0.2 M NaOH is comparatively specific for arginine but too harsh for many proteins. The following procedure using less alkaline conditions is applicable to a greater number of proteins, although it is less specific in that some modification of amino groups also takes place (Liu et al., 1968).

> The protein (10 to 20 mg) and an equal amount of 1,2-cyclohexanedione are dissolved in 5.0 ml of water and adjusted to pH 11 by addition of triethylamine and then stored in the dark at room temperature for 12 to 24 hours. To determine the effects of the high pH a control must be run without 1,2-cyclohexanedione. Extensive, but not quantitative, modification of guanidino groups occurs under these conditions. The extent of reaction can be calculated from the loss of arginine as measured by amino acid analysis after acid hydrolysis. The yellow color of the protein after modification is due to the formation of an unidentified product from reaction with amino groups. This reaction usually involves some (10–40%) of the amino groups and is decreased by higher pH values. The unidentified product absorbs maximally near 440 mμ.

Modification of guanidino groups can be accomplished under less alkaline conditions by reaction with a condensation trimer of diacetyl. Preparation of the reagent, and its use for the modification of bovine serum albumin, has been described by Yankeelov et al. (1968).

The reagent (trimer) is prepared by mixing 196 g of diacetyl (2,3-butanedione) with 415 g of clean, alkali-washed (0.1 N NaOH, H_2O), dry, powdered glass and by incubating with occasional mixing for five days at 25°C. The product is extracted from the hardened mass with diethyl ether and dried with anhydrous magnesium sulfate. Evaporation of the ether and storage overnight in the cold produces approximately 76 g (39%) of product which can be recrystallized from ether as white needles [m.p. 112.5°–114°C; $\lambda_{max} = 286$ mμ (in 1 mM HCl)]. A crude preparation of the reagent can also be prepared by incubating a 15% solution of diacetyl for 24 to 48 hours in Tris, borate, or phosphate buffer adjusted to pH 8.8. The pH of the solution slowly decreases and must periodically be readjusted. The reagent is usable when upon 15-fold dilution its absorption at 400 mμ reaches 0.45 (Grossberg and Pressman, 1968).

In a typical experiment, 6 mg/ml of bovine plasma albumin was allowed to react with 0.4 M reagent in 0.5 M phosphate buffer (pH 7.0) at 25°. After 40 hours, less than 20% of the initial arginine and about 70% of the initial lysine remained (see Figure 10-2). The decreased arginine and lysine contents can be determined by amino acid analysis after acid hydrolysis. The red color ($\lambda_{max} = 530$ mμ) which forms during the reaction is presumably due to reaction with amino groups.

Phenylglyoxal is another useful reagent which reacts with guanidino groups of proteins near neutral pH (Takahashi, 1968). The reaction is reversible and the guanidino groups can be partially recovered after modification by incubation at neutral or alkaline pH values in the absence of the reagent. The procedure given below is taken from Takahashi (1968) for the modification of bovine pancreatic ribonuclease.

The reaction mixture is prepared by adding a 0.3–3% solution of phenylglyoxal hydrate dissolved in 0.2 M N-ethylmorpholine acetate buffer (pH 8.0) to an equal volume of a 1–6% solution of ribonuclease dissolved in the same buffer. The solution is incubated at 25°. At suitable intervals, samples can be removed for assay of enzymatic activity. With 0.5% ribonuclease and 1.5% phenylglyoxal, the decrease in enzymatic activity is complete in about 2 hours. The modified protein can be freed from the excess reagent by passing it through a column of Sephadex G-25 pre-equilibrated with 0.01 M ammonium acetate at 4°. It is stable below pH 4 but decomposes at neutral or alkaline pH values, regenerating the original guanidino groups ($\sim$80%).

A-6 PHENOLIC GROUPS

Phenolic groups of tyrosine residues are important structural and functional components of most proteins. They are subject to a large number of chemical modification procedures. The phenolic hydroxyl group is susceptible to modification by many acylating and alkylating agents. The aromatic ring is susceptible to attack and substitution by a number of electrophilic reagents. Both types of reactions become more rapid at higher pH's, where the phenolic group is increasingly more ionized. In most cases, modification of phenolic groups is readily quantitated, because most modifications result in a distinctive change in absorbance.

Reaction with N-Acetylimidazole. The most useful procedure for modification of phenolic hydroxyl groups is that first described by Riordan et al. (1965) using N-acetylimidazole to effect their acetylation (see Section 5-1):

The protein in 0.05 M sodium borate, or 0.02 M veronal, buffer at pH 7.5 is treated with approximately a 60-fold molar excess of N-acetylimidazole. The solution after conclusion of the reaction is dialyzed or subjected to gel filtration to separate the protein from low-molecular-weight components of the solution. To avoid deacetylation of the phenolic groups, the pH of the solution should not exceed 8. The extent of reaction can be determined from the change in absorbance using $\Delta\varepsilon_{278} = 1160\ M^{-1}\ cm^{-1}$, or by the hydroxamate procedure of Balls and Wood (1956). Acetylation of amino groups, which also frequently occurs upon such treatment, can be detected by determining the number of such groups remaining as compared to unmodified protein (see Section 4-1).

Iodination of phenolic groups has been one of the most useful protein modification techniques. The reaction conditions as well as the reagent itself are subject to great variation (see Section 9-1). The following procedure of Azari and Feeney (1961) for the iodination of chicken conalbumin with KI_3 is suitable for extensive iodination of most proteins (for less extensive reaction, less reagent should be used):

> The protein is dissolved in 0.1 M borate buffer (pH 9.5) and chilled to 1°–2°C. A chilled 0.05 M solution of iodine in 0.24 M KI is added, and the solution is incubated for about 15 minutes. The amount of iodine added is that required to convert all tyrosine to diiodotyrosine (tyrosine + 2 I_2 → diiodotyrosine + 2 I^-). The progress of the reaction can be followed from the decrease in intensity of the reddish-brown to yellow color of the reagent, and a few drops of 1.0 M $NaHSO_3$ can be added to stop it at any time. The modification of tyrosine can be estimated by the spectrophotometric procedure of Edelhoch (1962). Total iodine incorporated can be determined by standard iodide analysis following alkali-metal fusion.

Nitration with Tetranitromethane. An important recently developed procedure for ring substitution of phenolic groups is that of Sokolovsky et al. (1966) for nitration with tetranitromethane:

> The reaction is normally carried out in 0.05 M Tris buffer (pH 8.0) at room temperature. Aliquots of the reagent, 0.84 M in 95% ethanol, are added and the reaction allowed to proceed for about 30 minutes. The amount of reagent required will vary with different proteins and with extent of modification desired. A 64-fold excess of tetranitromethane over carboxypeptidase A gave modification of 6.7 of the 19 phenolic groups. Gel filtration or dialysis can be used to separate the protein from the reaction solution. The nitrotyrosyl residues are stable and can be readily quantitated from their absorbance (see Section 9-2) or by amino acid analysis after acid hydrolysis. A useful further modification is the reduction of the nitro groups to amino groups (see Section 9-2).

A-7 IMIDAZOLE GROUPS

Imidazole groups of histidine residues are involved in the catalytic function of many proteins. They are weakly basic groups (p$K \sim 6.5$) and subject to a large number of modification reactions. They react readily with most acylating, alkylating, and many electrophilic reagents. Acylation of imidazole groups is seldom observed since most acylimidazoles rapidly hydrolyze in aqueous solutions. Dye-sensitized photochemical oxidation of imidazole groups has been the most generally useful procedure for their modification and for determining their importance to various properties of proteins (see Section 8-8).

Photochemical oxidation in the presence of a sensitizing dye can be done under mild conditions and is frequently quite specific in its modification of imidazole groups. The dyes methylene blue or rose bengal appear to be best suited for the modification of imidazole groups. Typical procedures employ 0.01% dye (either methylene blue or rose bengal) at approximately neutral pH and at, or slightly below, room temperature. The solution, containing 0.5 to 10 mg/ml protein, is stirred rapidly to hasten exchange of oxygen with the atmosphere and is placed 3 to 15 cm in front of a relatively bright visible light source (i.e., a 150- to 300-watt spotlight, slide projector, etc.), and irradiated for a period of from a few seconds to a few minutes. The sample should be kept in the dark before and after this irradiation. The exact conditions to effect the desired loss of histidine will usually have to be determined by several trial experiments. Careful control of the reaction conditions is required to get reproducible results. A control containing protein and dye should be kept in the dark and run simultaneously. The dye can usually be removed by gel filtration, or more slowly by dialysis. Methylene blue can be absorbed on charcoal or Dowex 1-X10 (chloride form) and removed. For further information, the procedures of Ray (1967), Westhead (1965), or Ray and Koshland (1962) are recommended.

The reader is encouraged to investigate the possible usefulness of ethoxyformic anhydride (Section 5-1), haloacetates (Section 6-1), iodine (Section 9-1), and diazonium reagents (Section 9-3), as histidine reagents which are sometimes useful.

A-8   INDOLE GROUPS

The indole side chain of tryptophan residues absorbs strongly near 280 mμ. It is relatively reactive but is often protected from the solvent and reagents contained in the solvent by being buried in the interior of proteins along with other hydrophobic side chains. The lability of tryptophan to conditions normally used for acid hydrolysis of proteins has, in the past, made its quantitative determination relatively difficult. Treatment with N-bromosuccinimide (see Section 8-9) or 2-hydroxy-5-nitrobenzyl bromide (see Section 6-6) is recommended to effect its selective modification in proteins.

Treatment with N-bromosuccinimide is usually done in acetate or formate buffers (pH 3 to 4), but values closer to neutrality can also frequently be used. Higher pH has an advantage in that little or no peptide bond cleavage takes place. Also, higher pH is usually more selective, apparently differentiating between more and less reactive tryptophan residues (Spande et al., 1966). Reaction is usually relatively specific for tryptophan in the absence of sulfhydryl groups which are even more reactive. Oxidation of trypsin by N-bromosuccinimide was studied from pH 3 to 7 at 25°C in 0.1 M acetate or phosphate (pH 7) buffers. The reagent (0.01 M in 10-μl aliquots) was added with rapid stirring to a solution which had a total absorbance due to protein of 1.5 to 1.7 at 280 mμ. The extent of reaction was determined from the decrease in absorbance at least 30 seconds after addition of reagent. The moles of tryptophan oxidized were calculated by dividing the absorbance change by the molar extinction coefficient of tryptophan (5500) and multiplying the result by the empirical factor of 1.31 (Patchornik et al., 1958). At pH 7.0, 11 moles of reagent were required to effect a loss of 0.5 tryptophan residues; at pH 6.0, 5.0, 4.0, and 3.0, respectively, 8.7, 4.2, 3.1, and 2.6 moles of reagent gave losses of 1.0, 2.7, 3.6, and 3.8 tryptophan residues out of a total of 4.0.

Modification of tryptophan residues with 2-hydroxy-5-nitrobenzyl bromide ("Koshland's reagent") is best done by a variation of the procedures described by Koshland et al. (1964) or Horton and Koshland (1965, 1967) (see Section 6-6):

> The reagent is dissolved in acetone or dioxane at approximately 0.2 M and is rapidly mixed with an aqueous solution of protein. It has generally been found that a final acetone or dioxane concentration of 5% and a reagent concentration of 0.01 M gives a single-phase system, and this amount of organic solvent does not denature most proteins. The reagent is dissolved in the organic solvent immediately prior to its use and is protected from strong light. The pH of the solution will drop during the reaction, and a pH-stat or manual addition of alkali should be used to maintain a constant value. A low initial pH is preferred but is not essential for the reaction. At higher pH, cysteine and tyrosine residues become more reactive. The number of reactive tryptophan residues in α-chymotrypsin has been shown to vary with pH. At pH 2, for example, all four tryptophan residues react, whereas at pH 4 only one reacts (Oza and Martin, 1967). pH values between 3 and 5 are most commonly used.

The modified protein can be separated from low-molecular-weight reaction products (i.e., 2-hydroxy-5-nitrobenzyl alcohol) by standard techniques of gel filtration, dialysis, or various other procedures (see Horton and Koshland, 1967). The number of groups introduced into the protein can usually be determined by measurement of absorbance at 410 mμ after removal of hydrolyzed reagent and adjustment to pH > 10 ($\varepsilon = 18,000\ M^{-1}\ cm^{-1}$).

A-9 THIOETHER GROUPS

The thioether groups of methionine side chains are weakly nucleophilic, remain un-ionized from pH 1 to 14, and, like more typical hydrophobic side chains, usually have little access to the aqueous environment. Their resistance to protonization is in contrast to other nucleophilic groups in proteins and provides a basis for their selective substitution. The reactivity of methionine also increases in many proteins due to partial unfolding at low pH values, where most other groups become relatively unreactive. For proteins lacking sulfhydryl groups, reactions with a haloacetate or with hydrogen peroxide can be used to selectively modify methionine side chains.

Carboxymethylation. The carboxymethylation of methionine residues of pancreatic ribonuclease has been described by Neumann et al. (1962):

> To a solution of ribonuclease (6 mg/ml) was added an equal volume of a solution of iodoacetic acid (6 mg/ml). The pH was immediately adjusted to the desired value with 1 M HCl and the solution incubated at 40°C. At pH 2.7, most of the enzyme was converted to inactive di-, tri-, and tetraalkylated species after 3 hours. To terminate the reaction, the solution was poured onto a 0.9 × 5-cm bed of Amberlite IRC-50 pre-equilibrated with 5% acetic acid and eluted with 1-1 glacial acetic acid–water.

The reaction of iodoacetate with isocitrate dehydrogenase occurs under considerably milder conditions (Colman, 1968). When the enzyme (2 mg/ml) at pH 5.5 in 0.2 M sodium acetate buffer and 30°C was treated with $1.35 \times 10^{-2}\,M$ iodoacetate, 80% of the dehydrogenase and 60% of the reductase activity was lost in 90 minutes. Reaction was stopped by gel filtration (Sephadex G–25) at 5°C in 0.1 M Tris (pH 7.2). The loss of activity appears to result from the alkyation of one methionine residue.

Reaction with Hydrogen Peroxide. The following procedure has been used for the modification of chymotrypsin with hydrogen peroxide (Schachter and Dixon, 1964):

Chymotrypsin (4.4 mg/ml) at 30°C in 0.0005 M EDTA at pH 3.2 (maintained by adding 0.001 M perchloric acid as required) is adjusted to 0.38 M H_2O_2 with 30% H_2O_2 and allowed to react for 10 to 240 minutes. The reaction is stopped by adding the reaction solution to a solution containing a small amount of catalase. Only one of the two methionines reacted during this period and, although this one is only three residues removed from the active-site serine, its modification effected only a slight decrease in apparent catalytic activity.

Oxidation of the methionine adjacent to the active site of subtilisin (Carlsberg) can be accomplished at pH 8.8 (Stauffer and Etson, 1969). A 1% solution of the enzyme in 0.1 M borate buffer was treated with 10 μl of 30% H_2O_2 per milliliter of enzyme solution to give 0.1 M H_2O_2 and incubated at room temperature. The loss of apparent catalytic activity began to level off after 30 minutes, after which the reaction was stopped by the addition of 1 μg/ml catalase. The apparent loss in activity was due to a decrease in the rate of catalysis of all of the molecules and not to complete inactivation of part of the molecules. Only one of the five methionine residues reacted.

REFERENCES

Anfinsen, C. B., and E. Haber (1961): *J. Biol. Chem.*, **236,** 1361.
Azari, P. R., and R. E. Feeney (1961): *Arch. Biochem. Biophys.*, **92,** 44.
Balls, A. K., and H. N. Wood (1956): *J. Biol. Chem.*, **219,** 245.
Battell, M. L., C. G. Zarkadas, L. B. Smillie, and N. B. Madsen (1968): *J. Biol. Chem.*, **243,** 6202.
Bewley, T. A., and C. H. Li (1969): *Int. J. Prot. Res.*, **1,** 117.
Boyer, P. D. (1954): *J. Am. Chem. Soc.*, **76,** 4331.
Broomfield, C. A., J. P. Riehm, and H. A. Scheraga (1965): *Biochemistry*, **4,** 751.
Butler, P. J. G., J. I. Harris, B. S. Hartley, and R. Leberman (1969): *Biochem. J.*, **112,** 679.
Chen, C. C., A. L. Grossberg, and D. Pressman (1962): *Biochemistry*, **1,** 1025.
Chibnall, A. C., J. L. Mangan, and M. W. Rees (1958): *Biochem. J.*, **68,** 114.
Chu, F. S., E. Crary, and M. S. Bergdoll, (1969): *Biochemistry*, **8,** 2890.
Cole, R. D. (1967): *Methods Enzymol.*, **11,** 206.
Colman, R. F. (1968): *J. Biol. Chem.*, **243,** 2454.
Colman, R. F., and R. Chu (1970): *J. Biol. Chem.*, **245,** 601.

Crestfield, A. M., S. Moore, and W. H. Stein (1963): *J. Biol. Chem.*, **238,** 622.
Dixon, H. B. F., and R. N. Perham (1968): *Biochem. J.*, **109,** 312.
Ellman, G. L. (1959): *Arch. Biochem. Biophys.*, **82,** 70.
Fraenkel-Conrat, H. (1957): *Methods Enzymol.*, **4,** 247.
Fraenkel-Conrat, H., and H. S. Olcott (1945): *J. Biol. Chem.*, **161,** 259.
Gibbons, I., and R. N. Perham (1970): *Biochem. J.*, **116,** 843.
Grossberg, A. L., and D. Pressman (1968): *Biochemistry*, **7,** 272.
Habeeb, A. F. S. A. (1966): *Anal. Biochem.*, **14,** 328.
Habeeb, A. F. S. A., H. G. Cassidy, and S. J. Singer (1958): *Biochim. Biophys. Acta*, **29,** 587.
Hoare, D. G., and D. E. Koshland (1967): *J. Biol. Chem.*, **242,** 2447.
Horton, H. R., and D. E. Koshland (1967): *Methods Enzymol.*, **11,** 556.
Horton, H. R., and D. E. Koshland, (1965): *J. Am. Chem. Soc.*, **87,** 1126.
Kastenschmidt, L. L., J. Kastenschmidt, and E. Helmreich (1968): *Biochemistry*, **7,** 3590.
Klee, W. A., and F. M. Richards (1957): *J. Biol. Chem.*, **229,** 489.
Koshland, D. E., Y. D. Karkhanis, and H. G. Latham (1964): *J. Am. Chem. Soc.*, **86,** 1448.
Kress, L. F., and M. Laskowski (1967): *J. Biol. Chem.*, **242,** 4925.
Light, A., and N. K. Sinha (1967): *J. Biol. Chem.*, **242,** 1358.
Liu, W. H., G. Feinstein, D. T. Osuga, R. Haynes, and R. E. Feeney (1968): *Biochemistry*, **7,** 2886.
Liu, W. K., and J. Meienhofer (1968): *Biochem. Biophys. Res. Comm.*, **31,** 467.
Means, G. E., and R. E. Feeney (1968): *Biochemistry*, **7,** 2192.
Neumann, H., I. Z. Steinberg, J. R. Brown, R. F. Goldberger, and M. Sela (1967): *Eur. J. Biochem.*, **3,** 171.
Neumann, N. P., S. Moore, and W. H. Stein (1962): *Biochemistry*, **1,** 68.
Oza, N. B., and C. J. Martin (1967): *Biochem. Biophys. Res. Commun.*, **26,** 7.
Patchornik, A., W. B. Lawson, and B. Witkop (1958): *J. Am. Chem. Soc.*, **80,** 4748.
Ray, W. J. (1967): *Methods Enzymol.*, **11,** 490.
Ray, W. J., and D. E. Koshland (1962): *J. Biol. Chem.*, **237,** 2493.
Riordan, J. F., W. E. C. Wacker, and B. L. Vallee (1965): *Biochemistry*, **4,** 1758.
Schachter, H., and G. H. Dixon (1964): *J. Biol. Chem.*, **239,** 813.
Shapira, E., and R. Arnon (1969): *J. Biol. Chem.*, **244,** 1026.
Sokolovsky, M., J. F. Riordan, and B. L. Vallee (1966): *Biochemistry*, **5,** 3582.
Spande, T. F., N. M. Green, and B. Witkop (1966): *Biochemistry* **5,** 1926.
Sperling, R., Y. Burstein, and I. Z. Steinberg (1969): *Biochemistry*, **8,** 3810.
Stark, G. R., W. H. Stein, and S. Moore (1960): *J. Biol. Chem.*, **235,** 3177.
Stauffer, C. E., and D. Etson (1969): *J. Biol. Chem.*, **244,** 5333.
Swan, J. M. (1957): *Nature*, **180,** 643.
Swenson, A. D., and P. D. Boyer (1957): *J. Am. Chem. Soc.*, **79,** 2174.
Takahashi, K. (1968): *J. Biol. Chem.*, **243,** 6171.
Toi, K., E. Bynum, E. Norris, and H. A. Itano (1967): *J. Biol. Chem.*, **242,** 1036.
Vithayathil, P. J., and F. M. Richards (1961): *J. Biol. Chem.*, **236,** 1380.
Westhead, E. W. (1965): *Biochemistry*, **4,** 2139.
Wofsy, L., and S. J. Singer (1963): *Biochemistry*, **2,** 104.
Yankeelov, J. A., C. D. Mitchell, and T. H. Crawford (1968): *J. Am. Chem. Soc.*, **90,** 1664.

author index

Aanning, H. L. 128, *138*
Abadi, D. M. 83, *101*
Acher, R. 79, *101*
Adams, M. 42, *53*
Adams, P. 62, *65*
Ahlfors, C. E. 168, *171*
Albu-Weissenberg, M. 42, *54*
Alexander, E. R. 126, *134*
Alexander, J. 98, *101*
Alexander, N. M. 111, *134*
Alexander, P. 55, *64*
Allewell, N. 119, *137*
Allison, R. G. 63, *64*
Allison, W. S. 158, 159, 160, *173*, 180, 182, *192*
Anastasi, A. 153, *172*
Anderson, B. M. 19, *23*, 113, 114, *134*, *135*, *136*
Anderson, C. D. 19, *23*, 113, 114, *134*, *136*
Anderson, J. A. 132, 133, *134*
Anderson, P. J. 157, *171*
Ando, T. 59, *65*
Andreotti, R. E. 171, *172*
Andrews, P. 62, *64*
Anfinsen, C. B. 52, *53*, 78, 79, 80, 83, 84, *101*, *102*, 183, 185, 186, *191*, 221, *229*
Arens, A. 209, 210, *211*
Arnon, R. 38, 41, *53*, 79, *101*, *104*, 150, *174*, 200, 205, *210*, 221, *230*

Atassi, M. Z. 83, *101*, 125, *135*, 183, 185, *191*, 207, *210*
Azari, P. R. 38, *53*, 225, *229*
Azegami, M. 121, *135*

Bailey, J. L. 55, *64*, 152, *171*
Baker, B. R. 6, *22*, 25, *33*
Balls, A. K. 3, *22*, 71, 84, *101*, 225, *229*
Banks, T. E. 145, *148*
Bar-Eli, A. 78, *102*
Baret, R. 93, *103*
Barka, T. 157, *171*
Barman, T. E. 57, *64*, 124, 125, *135*
Barnard, E. A. 31, *34*, 109, *136*
Barnett, R. 127, *135*
Baronowsky, P. 28, *34*
Bates, C. 188, *192*
Battell, M. L. 219, *229*
Bauer, K. 209, 210, *211*
Baum, H. 155, *171*
Beaven, G. H. 184, 185, *191*
Becker, R. R. 78, 79, *103*
Beechey, R. B. 181, *192*
Belenkii, B. G. 208, *210*
Bellin, J. S. 166, 167, *171*, *173*
Bello, J. 183, *192*
Belman, S. 120, *135*

Benassi, C. A. 168, *171*, 206, *210*
Bender, M. L. 22, 63, *65*, 100, *103*
Bendet, I. J. 36, *53*
Benesch, R. 83, *101*, 111, 134, 135, 204, 210
Benesch, R. E. 83, 101, 111, 134, *135*, 204, *210*
Benisek, W. F. 52, *53*, 93, *101*
Bennett, H. S. 204, *210*
Bensusan, H. B. 181, *191*
Bergdoll, M. S. 215, *229*
Berger, A. 78, *102*
Bergmann, M. 59, *64*
Berliner, E. 176, *191*
Bernier, I. *171*
Bertoli, D. 177, *192*
Bertoli, D. A. 176, *192*
Bethune, J. L. 71, *101*
Bewley, T. A. 38, *53*, 150, *171*, 221, *229*
Beychok, S. 62, *64*, *65*
Beyreuther, K. 80, *101*
Bezkorovainy, A. 75, *101*
Bier, M. 70, 83, *104*
Bizzini, B. 42, *53*, 127, *135*
Blake, C. C. F. 4, *23*, 62, *64*
Blass, J. 42, *53*, 127, *135*
Block, R. J. 55, *64*
Blossey, B. K. 145, *148*
Boccú, E. 206, *210*

Bodlaender, P. *148*
Boesel, R. W. 183, *191*
Bond, R. P. M. 181, *192*
Borders, C. L. 143, *148*
Boroff, D. A. 167, *172*
Bosshard, H. R. 115, *135*
Boyer, P. D. 57, *64*, 203, 210, 218, *229*, 230
Bradbury, S. L. 107, *135*
Bradshaw, R. A. 19, *23*, 107, *135*, 199, 202, *211*
Brand, E. 59, *64*
Brand, L. 49, *53*
Brandon, B. A. 126, *135*
Braunitzer, V. G. 80, *101*
Braxton, H. 52, *54*
Brewer, C. F. 112, 114, *135*
Broomfield, C. A. 9, *23*, 139, 140, 141, *148*, 223, *229*
Brovetto-Cruz, J. 209, *210*
Brown, J. R. 152, *173*, 221, *230*
Brown, R. K. 83, *102*, 209, *211*
Bruening, G. E. 62, *65*
Bruice, T. C. 183, *191*
Brunings, K. J. 181, *191*
Buchert, A. R. 165, 167, *174*
Buck, F. F. 71, *103*
Bunnett, J. F. 119, 120, *135*
Burgess, R. R. 62, *64*
Burr, M. 49, 50, *53*, 110, *135*
Burstein, Y. 38, 41, *54*, 200, *211*, 221, *230*
Butler, P. J. G. 76, 77, *101*, 215, *229*
Butterworth, P. H. W. 155, *171*
Buttkus, H. 198, *210*
Bynum, E. 194, *211*, 224, *230*
Byrne, R. 92, *103*

Cahnmann, H. J. 205, *210*
Callan, T. P. 162, *174*
Calvin, M. 79, *103*
Camera, E. 150, *172*
Campbell, D. H. 43, *53*
Canfield, R. E. 4, *23*
Cann, J. R. 62, *64*
Carino, R. L. 127, *137*
Carney, A. L. 93, *102*

Carpenter, F. H. 46, *54*, 83, *103*, 117, *137*, 183, *191*
Carraway, K. L. 145, *148*
Carriuolo, J. 73, *102*
Carsten, M. E. 120, *135*
Caruso, D. R. 125, *135*
Cassidy, H. G. 75, *102*, 215, *230*
Cavins, J. F. 115, 116, 117, *135*
Cecil, R. 152, 153, 154, *171*
Cha, C. Y. 181, *191*
Chadwick, C. S. 48, *53*
Chambers, D. C. 57, *65*
Chan, T. L. 124, *135*
Chan, W. W. C. 153, *171*
Chang, H. F. W. 132, *134*
Changeux, J. P. 9, *23*
Chapman, A. 73, *101*
Chauvet, J. 79, *101*
Chen, C. C. 216, *229*
Chen, R. F. 97, *101*
Chervenka, C. H. 93, 94, 95, *101*
Chevallier, J. 71, *101*
Chibnall, A. C. 152, *171*, 223, *229*
Chignell, D. A. 155, *171*
Chinard, F. P. 157, *172*
Chio, K. S. 198, *210*
Chow, R. B. 93, 95, *102*
Christen, P. 183, *191*, *192*
Christen, Ph. 73, *103*
Christensen, H. N. 93, *101*, 134, *135*
Chu, F. S. 215, *229*
Chu, R. 219, *229*
Churchich, J. E. *134*
Clarke, H. T. 105, *136*
Clark-Waler, G. D. 112, *135*
Cleland, W. W. 149, 151, *172*
Coffey, J. 84, *104*
Cohen, J. A. 25, *33*
Cohen, L. A. 6, *23*, 49, *53*, 171, *173*
Cohen, W. 27, 28, *34*
Cole, R. D. 45, *54*, 85, *101*, 117, 118, *135*, *137*, 152, 154, *171*, *172*, 222, *229*
Colloms, M. 69, *102*

Colman, R. F. 109, *135*, 155, 156, *172*, 219, *229*
Colowick, S. P. 158, *172*
Connellan, J. M. 156, *172*
Cooke, J. P. 78, 79, 83, 84, *101*
Coombs, T. L. 204, *211*
Cooper, A. G. 28, *34*
Corey, E. J. 171, *172*
Cori, C. F. 155, *173*, 202, *211*
Corsi, A. 183, *192*
Coulter, C. L. 52, *54*
Covelli, I. 177, *179*, 181, *191*, *193*
Cowan, J. C. 115, *135*
Cox, A. C. 62, *65*
Crary, E. 215, *229*
Crawford, D. L. 197, *210*
Crawford, I. P. 132, *138*
Crawford, T. H. 196, 197, *211*, 224, *230*
Cremona, T. 120, *136*
Crestfield, A. M. 107, *136*, 150, *172*, 219, *230*
Cuatrecasas, P. 52, *53*, 183, 185, 186, *191*
Cunningham, L. W. 180, 182, 191, *193*

Dahlquist, F. W. 143, *148*
Damjanovich, S. 155, *172*
Danner, D. J. 129, *136*
DasGupta, B. R. 167, *172*
Davies, D. R. 52, *54*
Davies, G. E. 40, *53*
Davies, M. 182, *192*
Davies, R. C. 71, *101*
Davison, P. F. 62, *54*
Day, R. A. 42, *54*
Dedman, M. L. 162, *172*
Degani, Y. 6, 7, *23*, 39, *54*, 169, *173*, 205, *211*
Delaage, H. 183, 184, *193*
Delaney, R. 209, *211*
Delgado, C. J. 163, *173*
Delpierre, G. R. 142, *148*
Demsey, W. B. 133, 134, *135*
DeVincenzi, D. L. 62, *64*
Dixon, G. H. 71, *101*, 152, 154, 162, 163, *172*, *173*, 229, *230*

Author Index

Dixon, H. B. F. 77, *102*, 215, *230*
Dixon, J. S. 150, *171*
Doherty, D. G. 182, *193*
Dopheide, T. A. A. 124, 125, *135*
Doscher, M. S. 139, 141, 142, *148*
Doty, P. 62, *65*
Douraghi-Zadeh, K. 145, *148*
Doyle, R. J. 183, *192*
Dreyer, R. 59, *64*
Dreze, A. 57, *64*
Dube, S. K. 178, *192*
Ducay, E. D. 150, *172*
Dunker, A. K. 62, *64*
Dutton, A. 42, *53*
duVigneaud, V. 128, *135*, *138*

Edelhoch, H. 49, *54*, 98, *102*, 176, 178, 179, 182, *192*
Edelman, G. M. 49, *53*, 101, *102*, *103*
Edman, P. 89, *102*
Edsall, J. T. 41, *53*, 126, *135*, 200, *210*
Ehring, R. 158, *172*
Eisen, H. N. 120, *135*
Eldjarn, L. 150, *172*
Elfring, W. H. 127, *137*
Elliott, D. L. 100, *102*
Ellman, G. L. 57, *64*, 155, *172*, 220, *230*
Elödi, P. 202, *210*
Englander, S. W. 62, *64*
Englberger, F. M. 176, 181, 182, *192*
Englund, P. T. 41, *53*, 200, *210*
Enser, M. 134, *137*
Erlanger, B. F. 28, *34*
Erwin, M. J. 176, 181, 182, *192*
Etson, D. 162, *174*, 229, *230*
Ettinger, M. J. 119, *135*
Evans, R. L. 93, 96, *102*, *103*
Eyl, A. 146, *148*

Fahrney, D. 33, *34*, 82, 83, 98, 100, *102*, *103*

Fahrney, D. E. 27, *34*
Fanger, M. W. 80, *102*
Farmer, T. H. 162, *172*
Fasella, P. 167, 168, *173*
Fasman, G. D. 144, *148*
Fasold, H. 42, *53*, 94, *102*
Fattoum, A. 183, 184, *192*
Fava, A. 150, *172*
Fedorcsák, I. 82, *103*
Feeney, R. E. 19, *23*, 30, 31, *34*, 38, 39, *53*, 59, 62, 63, *64*, *65*, 72, 91, 92, *102*, *103*, 121, 122, 128, 130, 131, 132, 134, *136*, 156, *172*, 194, 195, 205, *210*, 217, 224, 225, *229*, *230*
Feinstein, G. 39, 43, 44, *53*, 63, *65*, *148*, 194, 195, 205, *210*, 224, *230*
Fernandez, J. E. 126, *135*
Fernandez-Diez, M. J. 19, *23*, 156, *172*
Ferrier, B. M. 128, *135*
Finn, F. 28, *34*
Fischer, E. H. 132, 133, *135*, *136*, *137*
Fish, W. W. 62, *64*, *65*
Flavin, M. 112, *137*
Fletcher, J. C. 115, *135*
Flodin, P. 62, *65*
Folk, J. E. 118, *137*, 156, *172*
Fonda, M. L. 113, *135*
Fontana, A. 206, *210*, 211
Foote, C. S. 169, *172*
Fowler, J. S. 126, *135*
Fraenkel, W. 141, *148*
Fraenkel-Conrat, H. 3, *23*, 42, *53*, 69, 70, 95, *102*, *126*, 127, *135*, 139, *148*, 150, *172*, 179, 180, 181, *192*, 214, 222, *230*
Franzblau, C. 144, *148*
Fraser, R. D. B. 62, *64*
Frattali, V. 97, *102*
Freedman, M. H. 76, 77, 78, *102*
Freedman, R. B. 16, *23*, 121, 123, *135*
Freisheim, J. H. 36, *53*, 75, *102*
French, D. 126, *135*

Frensdorff, A. 37, *53*, 147, *148*
Freude, K. A. 167, 168, *172*
Fridovich, I. 71, *102*
Frieden, E. H. 139, *148*
Friedman, E. 111, *135*
Friedman, M. 16, *23*, 115, 116, 117, *135*
Friedman, M. E. 176, *192*
Frist, R. H. 36, *53*
Fruton, J. S. 73, *102*, 111, 112, *137*, 142, *148*
Fry, K. T. 142, *148*
Fuchs, S. 52, *53*, 183, 185, 186, *191*
Fujiki, H. 80, *101*
Funatsu, M. 21, 22, *23*, 71, *104*, 171, *172*

Gaddone, S. M. 176, 181, 182, *192*
Galiazzo, G. 164, 167, 168, *171*, *172*, *173*
Gallop, P. M. 144, *148*
Galzigna, L. 80, 81, *103*
Gehrke, C. W. 6, *23*
Gerhart, J. C. 36, 37, *53*, 62, *64*, 202, *210*
Gertler, A. 209, *210*
Gervais, M. 71, *103*
Gerwin, B. I. 106, *135*
Geshwind, I. I. 94, 95, *102*
Gibbons, I. 215, *230*
Gibian, M. J. 100, *102*
Gill, T. J. 48, *53*, 147, *148*
Gladner, J. A. 118, *137*
Glazer, A. N. 6, *23*, 78, *102*
Gold, A. M. 27, 33, *34*, 97, 98, 100, *102*
Goldberger, R. F. 79, 80, *102*, 151, 152, *173*, 176, *192*, 221, *230*
Goldfarb, A. R. 121, 122, 123, *135*
Goodfriend, T. L. 144, *148*
Gorbonoff, M. J. 62, *64*
Gordon, W. G. 165, *174*
Gorecki, M. 132, *135*
Goren, H. J. 109, *136*
Gottlieb, A. J. 194, *210*

Gounaris, A. D. 37, *53*, 75, *102*
Granata, A. L. 183, *192*
Grassetti, D. R. 157, *172*
Gratzer, W. B. 155, *171*, 184, 185, *191*
Gray, W. R. 97, *102*
Green, N. M. 171, *172*, 227, *230*
Greenberg, D. M. 132, *136*
Greene, F. C. 62, *65*
Greenwood, F. C. 182, *192*
Gregory, J. D. 111, 114, *136*
Gregory, M. J. 183, *191*
Griffith, O. H. 51, *53*
Groff, T. 143, *148*
Grohlich, D. 75, *101*
Gross, E. 204, 205, *210*
Grossberg, A. L. 76, 77, *102*, 197, *210*, 216, 224, *229*, *230*
Grossweiner, L. I. 166, *172*
Grovenstein, E. 176, *192*
Gruen, L. 179, 182, *192*
Guldalian, J. 83, *102*
Gundlach, H. G. 106, 107, 108, 109, *136*
Gurd, F. R. N. 83, *103*, 105, *136*
Gutcho, M. 111, *135*
Gutte, B. 154, *172*
Gwynne, E. 182, *192*

Habeeb, A. F. S. A. 37, 42, *53*, 57, *65*, 75, 83, 94, 95, 96, *102*, 121, 122, 129, 132, *136*, 207, *210*, 215, 217, *230*
Haber, E. 221, *229*
Hachimori, Y. 163, *172*, 188, 190, *192*
Haefele, L. F. 171, *172*
Haimouich, J. 43, *54*
Halliday, K. A. 162, 163, *173*
Halmann, M. 119, *136*
Hamilton, C. L. 51, *55*
Hamilton, G. A. 142, *148*
Hammes, G. G. 11, *23*
Hammett, L. P. 16, *23*
Hand, E. S. 90, *102*
Harada, M. 121, *136*, 195, *210*

Harbury, H. A. 80, 93, 95, *102*
Hardman, K. 119, *137*
Hardy, W. R. 100, *102*
Harrington, K. J. 187, 191, *192*
Harrington, W. F. 48, *53*
Harris, J. I. 76, 77, 95, *101*, *102*, 215, *229*
Hartdegen, F. J. 182, *192*
Hartley, B. S. 17, *23*, 48, *53*, 76, 77, 97, *101*, *102*, 215, *229*
Hartman, F. C. 42, *53*, 92, *102*
Hass, G. M. 107, *135*
Hass, L. F. 36, *53*
Hatano, H. 107, *136*
Hauenstein, J. D. 15, *23*
Hayashi, K. 21, 22, *23*, 71, *104*, 171, *172*
Haynes, R. 30, 31, *34*, 43, *53*, 63, *65*, 91, 92, *102*, 122, *136*, 194, 195, *210*, 224, *230*
Hedrick, J. L. 62, *64*, *65*, 133, *137*
Hegyi, G. 82, *103*
Heiney, R. E. 83, *103*
Heinrikson, R. L. 107, 110, *136*
Heitz, J. R. 19, *23*, 113, 114, *136*
Hellerman, L. 157, *172*
Helmreich, E. 155, *172*, 220, *230*
Hermann, D. H. 119, *135*
Herriott, R. 182, *192*
Herriott, R. M. 3, *23*, 141, *148*
Hess, G. P. 70, 71, 83, *103*
Hettinger, T. P. 93, 95, *102*
Higgins, H. G. 187, 191, *192*
Hill, R. D. 98, *102*, 168, *172*
Hill, R. J. 41, *54*
Hill, R. L. 19, *23*, 107, *135*, 199, 202, 208, *211*
Hiramoto, R. 129, *136*
Hirs, C. H. W. 6, *23*, 55, *65*, 119, 120, *135*, *136*, 161, 162, *172*
Hlavka, J. J. 144, *148*

Hoare, D. G. 57, *65*, 145, 146, *148*, 223, *230*
Hoch, F. L. 204, *211*
Hockert, T. J. 176, *179*, *192*
Hodgson, C. F. 167, *172*
Hoffee, P. 26, *34*, 168, *172*
Hofmann, T. 209, *210*, *211*
Holeysovsky, V. 109, *136*, 162, *172*
Hollands, T. R. 73, *102*
Holloway, C. T. 181, *192*
Horanyi, M. 82, *103*
Horecker, B. L. 26, *34*, 73, 76, *103*, *104*, 120, 134, *136*, *137*, 168, *172*, 201, *211*
Horinishi, H. 57, *65*, 144, *148*, 163, *172*, 188, 190, *192*, 193, 195, *211*
Horowitz, M. G. 204, *210*
Horton, H. R. 49, *53*, 110, 124, 125, *136*, 228, *230*
Howard, A. N. 187, 191, *192*
Hubert, E. 134, *136*
Hughes, R. C. 132, *136*
Hughes, W. L. 41, *53*, 93, *102*, 178, 182, *192*, 199, *210*
Humbel, R. E. 115, *135*
Hunter, M. J. 90, 91, 92, *102*
Hunter, W. M. 182, *192*
Hvidt, A. 9, *23*, 62, *65*

Iliceto, A. 150, *172*
Imoto, T. 171, *172*
Inagami, T. 29, *34*, 107, *136*, 146, *148*
Irie, M. 166, *174*
Irion, E. 209, 210, *211*
Ishida, N. 208, 210, *211*
Ishimoto, K. 95, *103*
Itano, H. A. 194, 195, *210*, *211*, 224, *230*
Iwai, K. 59, *65*, 121, *135*

Jacobs, R. M. 128, *138*
Jaffé, H. H. 16, *23*
Jansen, E. F. 3, *22*
Jao, L. 143, *148*
Jardetzky, O. 61, 62, *65*
Jarvis, D. 128, *135*
Jencks, W. P. 73, 90, *102*, 126, 127, 135, *136*

Jenkins, W. T.   132, *136*
Jirgensons, B.   62, *65*
Johansen, J. T.   185, *192*
Johncock, P.   85, *102*
Johnson, P.   48, *53*, 98, *102*
Jolles, J.   63, *65*
Jolles, P.   63, *65*, *171*
Jones, G. M. T.   33, *34*, 80, *103*
Jones, W. M.   124, 125, *135*
Jorgensen, K. H.   115, *135*
Jori, G.   164, 167, 168, *171*, *172*, *173*
Jovin, T. M.   41, *53*, 200, *210*

Kabat, E. A.   84, *102*
Kagan, H. M.   187, *192*
Kallen, R. G.   126, 127, *136*
Kallos, J.   52, *54*
Kameyama, T.   49, *54*
Kanarek, L.   19, *23*, 199, 202, *211*
Kanaska, Y.   49, *54*
Kaplan, N. O.   180, 182, *193*
Karkhanis, Y. D.   49, *53*, 123, 125, *136*, 228, *230*
Kassab, R.   82, *103*, 183, 184, *192*
Kassell, B.   93, 95, *102*, *103*, 162, *172*
Kastenschmidt, J.   155, *172*, 220, *230*
Kastenschmidt, L. L.   155, *172*, 220, *230*
Katchalski, E.   42, 43, *54*, 78, 79, *102*, 151, *173*, 179, *192*
Kay, C. M.   41, *53*
Kelly, H.   49, *53*, 125, *136*
Kenkare, U. W.   164, *172*
Kenner, R. A.   185, *192*
Kent, A. B.   132, 133, *135*, *137*
Keresztes-Nagy, S.   36, *53*, 74, *103*
Khorana, H. G.   144, *148*
Kilby, D. C.   176, *192*
Kim, O. K.   142, *148*
Kimball, R. W.   166, 173
King, T. P.   198, *210*

Kirtley, M. E.   49, *53*, 110, *136*
Kitz, R.   98, *101*, *103*
Klapper, M. H.   204, *210*
Klee, W. A.   31, *34*, 94, *103*, 216, *230*
Klein, S. M.   133, 134, *136*
Kleppe, K.   155, *172*
Klotz, I. M.   36, *53*, 74, 83, *103*, 204, *210*
Knight, I. G.   181, *192*
Knopp, J.   48, *53*
Kochert, M.   197, *211*
Koenig, D. F.   4, *23*
Kohnstam, G.   85, *102*
Kolthoff, I. M.   152, 153, *172*
Koltun, W. L.   83, *103*
Komatsu, S. K.   72, *103*
Kominz, D. R.   202, *210*
Konigsberg, W. H.   41, *54*
Korman, S.   105, *136*
Kornberg, A.   41, *53*, 200, *210*
Kornberg, H. L.   132, *138*
Koshland, D. E.   9, *23*, 31, *34*, 49, 50, *53*, 57, *64*, *65*, 99, 100, *103*, *104*, 110, 123, 124, 125, *135*, *136*, 145, 146, *148*, 162, 164, 168, 169, *172*, *173*, 223, 227, 228, *230*
Koshland, M. E.   176, 181, 182, *192*
Kotaki, A.   121, 136, 195, *210*
Kowal, J.   120, *136*
Krebs, E. G.   132, 133, *135*
Kress, L. F.   150, *172*, 221, *230*
Kronman, M. J.   171, *172*
Kuczenski, R.   168, *173*
Kung, Y. T.   154, *172*
Kunz, H. W.   147, *148*
Kurihara, K.   163, *172*, 188, 190, *192*
Kurzer, F.   145, *148*
Kwon, T. W.   198, *210*
Kycia, J. H.   119, *136*

Labouesse, B.   70, 71, 83, *103*
Labouesse, J.   71, 101, *103*
Lai, C. Y.   26, *34*, 168, *172*

Laing, R. R.   98, *102*, 168, *172*
Lamkin, W. M.   84, *104*
Langdon, R. G.   94, *103*
Lange, R.   159, *173*
Lardy, H. A.   204, *210*
Laskowski, M.   150, *172*, 179, 182, *192*, 221, *230*
Latham, H. G.   49, *53*, 123, 125, *136*, 228, *330*
Laufer, L.   111, *135*
Lauffer, M. A.   36, *53*
Laursen, R.   28, *34*
Lawson, W. B.   28, *34*, 83, *102*, 169, *173*, *174*, 227, *230*
Lazdunski, M.   109, *136*, 162, *172*, 183, 184, *193*
Leach, S. J.   55, 65
Leberman, R.   76, 77, *101*, 215, *229*
Lee, M. J.   71, *103*
Lee, V.   207, *211*
Leikhim, E.   209, *211*
Lerman, L. S.   43, *53*
Leuthardt, F.   73, *103*
Levine, L.   144, *148*, 209, *211*
Levy, A. L.   95, *102*
Levy, D.   83, *103*
Levy, M.   57, *65*, 127, *136*
Lewin, S.   127, *136*
Li, C. H.   38, *53*, 94, 95, *102*, 150, *171*, 176, 182, *192*, 207, *210*, 221, *229*
Li, T. K.   20, *23*, 107, *136*
Liener, I. E.   95, *103*
Light, A.   150, *172*, 221, *230*
Lin, Y.   132, *136*
Lindblow, C.   185, *193*
Lindley, H.   45, *53*, 117, *136*
Lindsay, D. G.   81, *103*
Lindskog, S.   183, 184, 185, *192*
Link, T. P.   110, *136*
Lipscomb, W. N.   204, *210*
Lissitzky, S.   176, *193*
Lister, M. W.   89, *103*
Little, C.   163, *172*
Little, J.   41, *54*
Liu, A. K.   4, *23*
Liu, T. Y.   159, *173*

Liu, W. H.   63, *65*, 194, 195, *210*, 224, *230*
Liu, W. K.   150, *173*, 221, *230*
Loening, U. E.   153, *171*
Ludwig, M. L.   90, 91, 92, *102*, *103*
Ludwig, W.   178, *192*
Luescher, E.   43, *53*
Lundblad, R. L.   142, *148*
Lundgren, H. R.   55, *64*
Luppis, B.   120, *137*, 201, *211*

Machida, M.   49, *54*
Macleod, R. M.   41, *54*
MacRae, T. P.   62, *64*
Madsen, N. B.   155, *173*, 202, *211*, 219, *229*
Maeda, H.   208, 210, *211*
Maekawa, K.   95, *103*
Mahowold, T. A.   120, *136*
Mair, G. A.   4, *23*
Maizel, J. V.   62, *65*
Major, J. P.   156, *174*
Malmström, B. G.   202, *211*
Manecke, G.   40, *54*
Mangan, J. L.   223, *229*
Mann, K. G.   62, *64*, *65*
Mansour, T. E.   168, *171*
Marchalonis, J. J.   182, *192*
Marcus, F.   134, *136*
Mares-Guia, M.   27, *34*
Marfey, P. S.   41, *54*, 147, *148*
Margolis, S.   94, *103*
Marini, M. A.   127, *136*
Markus, G.   152, *173*
Marrian, D. H.   111, *135*, *136*
Martin, C. J.   125, 127, *136*, *137*, 228, *230*
Martinez-Carrion, M.   167, 168, *173*
Marzotto, A.   80, 81, *103*, 164, 167, 168, *172*
Mason, M.   132, *137*
Massey, V.   48, *53*, 97, 98, *102*
Mathew, E.   31, *34*, 83, *103*
Matsubara, H.   57, *65*
Matsuo, Y.   132, *136*
Maurer, P. H.   139, 140, *148*

Mayberry, W. E.   176, 177, 179, *192*
Maybury, R. H.   200, *210*
Mayer, H.   147, *148*
Mayer, M. M.   84, *102*
McClure, W. O.   49, *53*, 101, *102*, *103*
McConnell, H. M.   51, *53*
McCormick, D. B.   43, *54*
McCullagh, D. R.   182, *193*
McFarlane, A. S.   182, *192*
McKinney, L. L.   115, *136*
McPhee, J. R.   152, 153, *171*, *173*
McVey, E. B.   167, *172*
Means, G. E.   121, 128, 130, 131, 132, 134, *136*, 217, *230*
Mecham, D. K.   126, *135*, 150, *172*
Mechanic, G.   57, *65*
Meienhofer, J.   41, *54*, 150, *173*, 221, *230*
Meighen, E. A.   75, *103*
Melchior, W. B.   82, 83, *103*
Menger, F.   22
Meriwether, B. P.   31, *34*, 83, *103*
Merrifield, R. B.   154, *172*
Metzger, H.   29, *34*, 187, *192*
Michaeli, D.   33, *34*
Michel, O.   176, *193*
Michel, R.   176, *193*
Midgley, A. R.   41, *54*
Minato, S.   129, *137*
Mitchell, C. D.   196, 197, *211*, 224, *230*
Mohammad, A.   150, *172*
Monod, J.   9, *23*
Moore, G. L.   42, *54*
Moore, J. E.   41, *54*, 112, *136*
Moore, S.   31, *34*, 56, 57, 59, *65*, 83, 85, 86, 87, *89*, 95, *103*, *104*, 105, 106, 107, 108, 109, *136*, *137*, 142, *148*, 150, 161, 162, 164, 169, *172*, *173*, 215, 219, 228, *230*
Morihara, K.   164, *173*, 183, *192*
Morita, R. Y.   78, 79, *103*
Morris, C. J. O. R.   162, *172*
Morrison, M.   129, *136*

Mounter, L. A.   26, *34*
Mounter, M. E.   26, *34*
Mourgue, M.   93, *103*
Mühlrad, A.   82, *103*, 183, *192*
Munson, P. L.   162, 164, *174*
Murase, Y.   129, *137*
Murray, J. F.   157, *172*

Nagamatsu, A.   111, 112, *137*
Nagami, K.   183, *192*
Naidu, M. S. R.   181, *191*
Nakanishi, K.   129, *137*
Nakatani, M.   168, *173*
Nakaya, K.   144, *148*, 195, *211*
Nanci, A.   100, *103*
Neet, K. E.   100, *103*
Neuberger, A.   71, *101*
Neumann, H.   151, 152, *173*, 221, *230*
Neumann, N. P.   109, *137*, 162, *164*, 169, *173*, 228, *230*
Neurath, H.   36, *53*, 71, 75, *101*, *102*, 146, *148*, 185, *192*
Ng, L.   83, *103*
Nielsen, S. O.   9, *23*, 62, *65*
Nilsson, A.   183, 184, 135, *192*
Nishikawa, A. H.   78, 79, *103*
Nord, F. F.   70, 83, *104*
Norris, E.   194, *211*, 224, *230*
North, A. C. T.   4, *23*
Nuenke, B. J.   182, *191*

O'Brien, P. J.   163, *172*
Ohashi, M.   121, *137*
Okuyama, T.   121, 123, *137*
Olcott, H. S.   3, *23*, 42, *53*, 126, 127, *135*, 139, *148*, 198, *210*, 222, *230*
O'Leary, M. H.   31, *34*, 71, 83, *103*
Olofson, R. A.   147, *148*
Onodera, M.   157, *173*
Ontjes, D. A.   162, 1€4, *174*
Oppenheimer, H. L.   70, 71, 83, *103*
Orestova, V. A.   208, *210*
Osborn, M.   62, *65*

Oster, G. 166, *173*
Osuga, D. T. 19, *23*, 30, 31, *34*, 63, *65*, 92, *102*, 122, *136*, 156, *172*, 194, 195, *210*, 224, *230*
Ottesen, M. 62, *65*, 185, *192*
Ottewill, R. H. 48, *53*
Oza, N. B. 125, *137*, 228, *230*
Ozawa, H. 41, 42, *54*

Page, J. 197, *211*
Pajetta, P. 80, 81, *103*
Pardee, A. B. 202, *210*
Park, J. H. 31, *34*, 83, *103*
Parker, D. J. 158, 159, 160, *173*, 180, 182, *192*
Parsons, S. M. 143, *148*
Patchornik, A. 6, 7, *23*, 39, *54*, 79, *103*, 169, *173*, 205, *211*, 227, *230*
Patel, R. P. 42, *54*
Pauly, H. 187, *192*
Perez-Villasenor, J. 28, *34*, 100, *104*
Perham, R. N. 33, *34*, 77, 80, *102*, *103*, 215, *230*
Perlman, R. L. 176, *192*
Perlmann, G. E. 37, *53*, 75, 85, 87, *102*, *103*
Peterson, D. L. 168, *173*
Pflumm, M. N. 62, *65*
Phillips, A. T. 132, *137*
Phillips, D. C. 4, *23*
Phillips, J. H. 188, *192*
Pierce, G. B. 41, *54*
Pigiet, V. 75, *103*
Pihl, A. 73, *101*, 150, 159, *172*, *173*
Pitt-Rivers, R. 178, *192*
Plapp, B. V. 117, *137*
Polgar, L. 100, *103*
Pollara, B. 120, *137*
Polyanovsky, O. L. 36, *54*
Pontremoli, S. 120, 134, *137*, 168, *173*, 201, *211*
Porath, J. 62, *65*
Porter, J. W. 155, *171*
Potts, J. T. 48, *54*
Pradel, L. A. 82, *103*, 183, 184, *192*

Pressman, D. 17, *23*, 76, 77, 101, *102*, *104*, 176, 178, *192*, *193*, 197, *210*, 216, 224, *229*, *230*
Price, N. C. 184, 185, *192*
Price, S. 42, *54*
Pugh, E. 134, *137*
Pugh, E. L. 73, *103*, 168, *172*

Quiocho, F. A. 42, *54*, 129, *137*

Raab, O. 165, *173*
Racs, J. 143, *148*
Radda, G. K. 16, *23*, 121, 123, *135*, 184, 185, *192*
Radhakrishnan, T. M. 146, *148*
Rafter, G. W. 158, *173*
Raftery, M. A. 45, *54*, 117, *137*, 143, *148*
Rajagopalan, T. G. 142, *148*
Rall, J. E. 177, *192*
Ram, J. S. 41, *54*, 139, 140, *148*
Ramachandran, G. N. 8, *23*
Ramachandran, J. 207, *211*
Ramachandran, L. K. 169, 171, *173*
Ramsdell, P. A. 157, *172*
Randall, J. J. 119, *135*
Rao, G. J. S. 171, *173*
Ray, D. B. 84, *104*
Ray, W. J. 31, *34*, 57, *65*, 162, 164, 168, 169, *172*, *173*, 227, *230*
Raynaud, M. 42, *53*, 127, *135*
Rees, M. W. 152, *171*, 223, *229*
Reineke, E. P. 178, *192*
Reynolds, J. H. 91, *103*
Rice, D. W. 151, *174*
Richards, E. G. 48, *53*
Richards, E. G. 62, *65*
Richards, F. M. 31, *34*, 42, 52, *53*, *54*, 84, 93, 94, *101*, *103*, *104*, 108, 119, 129, *137*, *164*, *172*, *173*, 216, 223, *230*
Richmond, V. 152, *173*

Riehm, J. P. 9, *23*, 112, 114, 115, *135*, *137*, 139, 140, 141, 142, 144, *148*, 223, *229*
Rimon, S. 85, 87, *103*
Riordan, J. F. 6, 20, 22, *23*, 52, *54*, 69, 71, 72, 73, 74, 75, *103*, 183, 184, 185, 186, 191, *192*, *193*, 203, *211*, 225, 226, *230*
Rippa, M. 134, *137*, 168, *173*, 201, *211*
Riva, F. 167, 168, *173*
Robberson, B. 117, *137*
Robbins, F. M. 171, *172*
Robbins, J. B. 43, *54*
Roberts, E. 111, *137*
Roberts, G. C. K. 61, 62, *65*
Robinson, G. W. 19, *23*, 107, *135*, 199, 202, *211*
Robinson, H. C. 112, *135*
Robinson, J. D. 19, *23*
Robrish, S. A. 188, *192*
Rocchi, R. 206, *211*
Roche, J. 93, *103*, 176, *193*
Roe, A. 188, *193*
Roholt, O. A. 17, *23*, 101, *104*, 176, 178, 183, *192*, *193*
Rosch, D. 6, *23*
Rosen, C. G. 82, *103*
Rosnati, L. 154, *173*
Rothfus, J. A. 118, *137*
Rouser, G. 111, *137*
Rubenfield, S. 163, *173*
Rueckert, R. R. 62, *64*
Rupley, J. A. 182, *192*

Sabato, G. D. 62, *65*
Sagers, R. D. 133, 134, *136*
Saidel, L. J. 127, *137*
Sakaguchi, S. 57, *65*, 95, *103*
Sanger, F. 119, *137*, 161, *173*
Sanner, T. 73, *101*
Sarma, V. R. 4, *23*
Saroff, H. A. 93, 96, *102*, *103*
Sasaki, R. M. 57, *65*
Sasisekharan, V. 8, *23*
Satake, K. 121, 123, *137*
Satzman, J. S. 127, *137*
Saxena, V. P. 37, *54*
Schachman, H. K. 36, 37, *53*, 62, *65*, 75, *103*

Schachter, H.   162, 163, *173*, 229, *230*
Schallenberg, E. E.   79, *103*
Schellenberg, K. A.   124, *135*
Schelton, J. R.   117, *137*
Scheraga, H. A.   9, 11, *23*, 115, *137*, 139, 140, 141, 142, 144, *148*, 176, 179, 181, 182, *191*, *192*, 223, *229*
Schirich, L. G.   132, *137*
Schmid, A.   73, *103*
Schmir, G. L.   171, *173*
Schoellmann, G.   27, 28, *34*, 110, *137*
Schonbaum, G. R.   63, *65*
Schrader, M. E.   166, *173*
Schramm, H. J.   28, *34*
Schrank, B.   80, *101*
Schroeder, W. A.   55, *65*, 117, *137*
Schumaker, V.   62, *65*
Schutte, E.   93, *104*
Scoffone, E.   80, 81, *103*, 164, 167, 168, *171*, *172*, *173*, 206, *210*, *211*
Scrimger, S. T.   209, *211*
Segal, H. L.   57, *64*, 203, *210*
Siebles, T. S.   115, *137*, 150, 165, 167, 168, *174*
Seifen, S.   144, *148*
Sekine, T.   49, *54*
Sela, M.   37, 43, *53*, *54*, 78, 79, 83, 84, *101*, *102*, *104*, 147, *148*, 151, 152, *173*, 179, *192*, 205, *210*, 221, *230*
Seligman, A. M.   112, *137*
Seon, B. K.   150, *173*
Setzkorn, E. A.   115, *136*
Shafer, J.   28, *34*
Shafer, J. A.   145, *148*
Shalitin, Y.   79, *103*, 132, *135*
Shall, S.   81, *103*
Shaltiel, S.   120, *137*, 169, *173*
Shapira, E.   38, 41, *53*, 150, *173*, 200, *210*, 221, *230*
Shapiro, A. L.   62, *65*
Shapiro, S.   134, *137*
Sharpless, N. E.   112, *137*
Shaw, D. C.   59, *65*, 87, *104*

Shaw, E.   6, *23*, 27, 28, 29, *34*, 110, *137*, *148*
Sheehan, J. C.   144, *148*
Sherwood, M.   179, *192*
Shibata, K.   57, *65*, 144, *148*, 163, *172*, 188, 190, *192*, *193*, 195, *211*
Shields, G. S.   95, *104*, 208, *211*
Shinitzky, M.   152, *173*
Shino, H.   123, *137*
Shinoda, T.   121, *137*
Shipley, B. A.   26, *34*
Sia, C. L.   76, *104*
Sigler, P. B.   52, *54*
Silman, H. I.   42, 43, *54*, 78, *102*
Silverstein, E.   202, *211*
Simon, S. R.   41, *54*
Simon-Reuss, I.   111, *135*
Simpson, R. B.   200, *210*
Simpson, R. T.   22, *23*, 72, 73, *104*
Singer, S. J.   25, 29, 30, *34*, 42, 46, *53*, *54*, 75, 83, 90, 91, 92, *102*, *104*, 187, 188, *192*, *193*, 215, 216, *230*
Sinha, N. K.   150, *172*, 221, *230*
Sinnhuber, R. O.   198, *210*
Skidmore, I. F.   128, *138*
Skinner, H. C. W.   52, *54*
Skupin, J.   150, *172*
Slobodian, E.   57, *65*, 163, *173*
Sluyterman, L. A. E.   86, *104*, 106, *137*, 167, *173*
Slywka, O.   163, *173*
Smillie, L. B.   219, *229*
Smith, A. J.   62, *65*
Smith, E. L.   95, *104*, 208, *211*
Smith, K. M.   36, *53*
Smith, R. A.   151, 152, *173*
Smyth, D. G.   70, 71, 72, 86, 87, 88, 89, *104*, 111, 112, *137*
Smythe, L. E.   127, *137*
Snell, E. E.   133, *135*
Snyder, E. R.   132, 133, *135*
Sobotka, H.   189, *193*
Sokolovsky, M.   52, *54*, 57, *65*, 183, 184, 185, 186, 188,

189, 190, 191, *192*, *193*, 226, *230*
Song, P. S.   133, *134*
Soria, M.   120, *137*
Spande, T. F.   6, 7, 19, *23*, 39, *54*, 57, *65*, 124, *137*, 169, 171, *173*, 205, *211*, 227, *230*
Spanio, L.   134, *137*
Speight, D.   85, *102*
Sperling, R.   38, 41, *54*, 200, *211*, 221, *230*
Spies, J. R.   57, *65*
Spikes, J. D.   167, 169, *172*, *173*
Spona, J.   142, *148*
Sportorno, G.   63, *65*
Stalling, D. L.   6, *23*
Stark, G. R.   6, *23*, 40, *53*, 85, 86, 87, 88, 89, *104*, 106, 107, 110, *136*, *137*, 215, *230*
Stauffer, C. E.   162, *174*, 229, *230*
Steele, W.   129, *136*
Stein, W. D.   31, *34*
Stein, W. H.   31, *34*, 56, 57, 59, *65*, 83, 85, 86, 87, 89, 95, *103*, *104*, 105, 106, 107, 108, 109, *136*, *137*, 142, *148*, 150, 162, 164, 169, *172*, *173*, 215, 219, 228, *230*
Steinberg, I. Z.   38, 41, *54*, 151, 152, *173*, 200, *211*, 221, *230*
Steiner, R. F.   19, *23*, 49, *54*, 98, *102*
Stephenson, C. C.   151, *174*
Stewart, J. M.   6, *23*
Stevens, F. C.   59, *65*
Stimmel, B. F.   182, *193*
Stockmayer, W. H.   151, *174*
Straessle, R.   178, *182*, *192*, 200, *210*
Straight, R.   169, *173*
Strausbauch, P. H.   133, *137*
Stricks, W.   152, *172*
Strumeyer, D. H.   99, *104*
Strumeyer, D. N.   162, *172*
Stryer L.   62, *65*
Sulebele, G.   202, *211*
Suzuki, T.   57, *65*, 188, 190, *193*, 195, *211*

Author Index

Svendsen, I. 185, *192*
Swan, J. M. 152, *174*, 222, 230
Swanepoel, O. A. 153, *174*
Swenson, A. D. 217, *230*
Sykes, R. L. 41, *54*

Tabachnick, M. 189, *193*
Taft, R. 16, *23*
Takahashi, K. 31, *34*, 59, *65*, 105, *137*, 196, *211*, 224, 225, *230*
Takeda, A. 49, *54*
Takenaka, A. 57, *65*, 188, 190, *193*
Takenaka, O. 57, *65*, 188, 190, *193*, 195, *211*
Tamaoki, H. 129, *137*
Tan, B. H. 153, *172*
Tanaka, M. 123, *137*
Tanford, C. 10, 15, *23*, 62, *64*, *65*
Tani, A. 144, *148*
Tappel, A. L. 198, *210*
Tashjian, A. H. 162, 164, *174*
Tawde, S. S. 41, *54*
Teale, F. W. J. 49, *54*
Telka, M. 95, *104*
Terminiello, L. 70, 83, *104*
Thunberg, T. 105, *137*
Tiemeier, D. C. 168, *173*
Tietze, F. 118, *137*
Toennies, G. 162, *174*
Toi, K. 194, *211*, 224, *230*
Tomita, M. 166, *174*
Traniello, S. 120, *137*, 201, *211*
Traylor, P. S. 188, *193*
Triplett, R. B. 145, *148*
Trump, G. N. 46, *54*
Trundle, D. 180, *193*
Tsai, H. J. 109, *137*
Tsernoglou, D. 119, *137*
Tsunoda, J. N. 112, *137*
Tuppy, H. 112, *137*
Turano, C. 167, 168, *173*
Turba, F. 94, *102*
Turner, C. W. 178, *192*

Uhing, E. H. 115, *136*
Ukita, T. 166, *174*

Ulmer, D. D. 71, *101*
Underhill, E. J. 126, *134*
Uraki, Z. 70, *104*
Urnes, P. 62, *65*
Uyeda, K. 76, *104*
Uziel, M. 41, *54*

Valentine, R. C. 155, *171*
Vallee, B. L. 6, 20, 22, *23*, 52, *54*, 57, *65*, 69, 71, 72, 73, 74, 75, *101*, *103*, *104*, 107, *136*, 183, 184, 185, 186, 187, 188, 189, 190, 191, *192*, *193*, 203, 204, *211*, 225, 226, *230*
Van Holde, K. E. 62, *65*
Van Rensburg, N. J. J. 153, *174*
Van Slyke, D. D. 95, *104*, 207, 208, *211*
Van Vunakis, H. 209, *211*
Vaslow, F. 182, *193*
Veronese, F. M. 206, *210*
Verpoorte, J. A. 185, *193*
Vincent, J. P. 183, 184, *193*
Vinuela, E. 62, *65*
Viswanatha, T. 169, *174*
Vithayathil, A. J. 71, *103*
Vithayathil, P. J. 84, *104*, 108, *137*, 164, *174*, 223, *230*
Von Korff, R. W. 120, *137*
Von Mutzenbecker, P. 178, *192*

Wacker, W. E. C. 69, 72, 74, *103*, 225, *230*
Waggoner, A. S. 51, *53*
Wagner, O. 209, 210, *211*
Wake, R. G. 153, *171*
Wall, J. S. 16, *23*, 116, 117, *135*
Wall, L. L. 6, *23*
Walsh, K. A. 36, 43, *53*, 75, *102*, 146, *148*, 185, *192*
Walters, S. L. 183, *191*
Wang, J. H. 10, *23*
Wang, S. S. 46, *54*, 117, *137*
Ward, D. N. 84, *104*
Ward, W. H. 41, *54*, 112, *136*

Wardlaw, A. C. 152, 154, *172*
Warren, S. 71, *104*
Warringa, M. G. P. J. 25, *33*
Wassarman, P. M. 156, *174*, 180, 182, *193*
Webb, J. L. 199, *211*
Weber, G. 48, 49, *53*, *54*, 97, 98, *104*
Weber, K. 62, *65*
Weil, L. 95, *104*, 115, *137*, 150, 165, 166, 167, 168, *174*
Weitzman, R. D. J. 152, *171*
Werner, E. A. 89, *104*
Westhead, E. W. 166, 167, 168, *174*, 227, *230*
Westheimer, F. H. 28, 31, *34*, 71, 83, *102*, *103*, *104*
Westphal, A. 197, *211*
Wetlaufer, D. B. 37, *54*
Whitaker, J. R. 22, 28, *34*, 62, *65*, 100, *104*, 107, *137*
White, W. N. 99, *104*
Whitehouse, M. W. 128, *138*
Wilchek, M. 37, *53*, 147, *148*
Wilchek, N. 124, *137*
Wilcox, P. E. 83, 93, 94, 95, *101*, 139, 141, 142, 143, *148*
Wild, F. 187, 191, *192*
Williams, G. R. 109, *137*
Williams, J. N. 128, *138*
Wilson, D. A. 132, *138*
Wilson, E. M. 132, *138*
Wilson, I. B. 98, *101*, 103
Wirsching, W. 94, *102*
Witholt, B. 49, *53*
Witkop, B. 6, 7, 19, *23*, 39, *54*, 57, *65*, 124, *137*, 169, 171, *172*, *173*, *174*, 204, 205, *210*, *211*, 227, *230*
Wofsy, L. 29, 33, *34*, 83, 90, 91, 92, *104*, 187, *192*, 216, *230*
Wold, F. 40, 41, 42, *53*, *54*, 92, *102*, *104*
Wolff, J. 177, 179, 181, *191*, *193*
Wood, H. N. 71, 84, *101*, 225, 229
Wood, W. A. 120, 132, *137*
Woodward, R. B. 147, *148*

Wormall, A.   183, *193*
Wyckoff, H. W.   119, *137*
Wyman, J.   9, *23*

Yagi, K.   121, *136*, 151, *173*, 195, *210*
Yamada, S.   194, 195, *211*
Yamasaki, N.   21, 22, *23*, 71, *104*
Yamashiro, D.   128, *138*
Yang, J. T.   62, *65*

Yankeelov, J.   197, *211*
Yankeelov, J. A.   196, 197, *211*, 224, *230*
Yankus, C. A.   166, 167, *171*
Yasunobu, K. T.   112, *137*
Yon, J.   71, *101*
Yoo, T. J.   101, *104*
Young, D. M.   48, *54*
Young, J. D.   6, *23*
Yphantis, D. A.   62, *65*

Yy, C. I.   134, *135*
Yu, T. C.   198, *210*

Zahler, R. E.   120, *135*
Zahn, H.   41, *54*
Zarkadas, C. G.   219, *229*
Zerner, B.   63, *65*, 71, *104*
Zschocke, R.   75, *101*
Zumwalt, R. W.   6, *23*
Zwicker, E. F.   166, *172*

subject index

ANS (see 1-Anilinonaphthalene-4-sulfonate)
Acetaldehyde, used in reductive alkylation, 217
Acetamido-4-nitrophenol, as chymotrypsin label, 50
Acetic anhydride, 12, 32, 69, 70
  acetylation of amino groups with, 214
Acetoacetate decarboxylase
  reaction with acetic anhydride, 70
  substrate labeling, 25
Acetoacetylation
  of egg white lysozyme, 81
  of insulin, 81
  of pancreatic ribonuclease, 80
Acetyl chloride, 71
Acetylation, 69
  of amino groups with acetic anhydride, 214
  of phenolic hydroxyl groups, 225
  rates of, 70
  with N-acetylimidazole, 74
Acetylcholinesterase, reaction with sulfonyl halides, 98
N-Acetylcysteine, reaction with trinitrobenzenesulfonic acid, 121
N-Acetylhomocysteine thiolactone, 83
N-Acetylimidazole, 12, 18, 69, 72
  acetylation with, 74, 225
  rate of hydrolysis, 73
S-Acetylmercaptosuccinic anhydride, 83
α,N-Acetyl-3-nitrotyrosine, spectrum, 184
O-Acetylphenol, deacetylation of, 72
N-Acetyltryptophan, reaction with 2-hydroxy-5-nitrobenzyl bromide, 124

O-Acetyltyrosine
  deacylation of, 70
  in carboxypeptidase A, 73
  formation in carboxypeptidase A, 72
  TMV protein, 69
Acrylamide, 115
Acrylonitrile, 12, 114, 116
  alkylation of reduced lysozyme, 150
Active center
  special reactivity of, 31
  substrate protection of, 24
Active site, 20
  (see also Active center)
  definition, 20
  reagents for
    affinity labels, 27
    pseudosubstrate labeling, 26
    substrate labeling, 25
  selective reagents, 25
Acylation, 68
  (see also Acetylation, Succinylation, Maleylation)
  of amino groups, quantitation, 83
O-Acyltryosine, determination of, 84
Adrenocorticotropic hormone (ACTH)
  monosulfoxide derivative, 162
  reaction with 2-nitrophenylsulfonyl chloride, 207
Affinity labels (see Active site, reagents for)
Alanine, acetylation of, 70

241

Alcohol dehydrogenase, horse liver
  inactivation by maleimides, 113
  reaction with iodoacetic acid, 19, 107
Alcohol dehydrogenase, yeast, inactivation by N-alkylmaleimides, 19, 113, 114
Aldehyde/NaBH$_4$, 12
Aldolase
  and N-acetylimidazole, 73
  labeling with dihydroxyacetone phosphate, 26
  maleylation of, 76
Aldolase, rabbit muscle
  inactivation by dinitrofluorobenzene, 120
  photooxidation of, 168
  reaction with iodine, 180
  reduction of pyridoxal phosphate complex, 134
  S-sulfonation, 153
Alkylation, 105
Alkylation, reductive, of amino groups, 130, 216
N-Alkylmaleimides, inactivation of alcohol dehydrogenase, 113
N-Alkyl-5-phenylisooxazolium, for activation of carboxyl groups, 147
Amide-forming reagents, 144
Amidination, 44, 89
  of amino groups with ethyl acetimidate, 216
D-Amino acid oxidase
  inactivation by N-alkylmaleimides, 113
  modification with glyoxal, 195
Amino groups
  acetylation of, 68, 214
  acetylation rates of, 70
  acylation of, 68
    quantitation, 83
  amidination of, 216
  carbamylation of, 84, 215
  guanidination of, 93, 216
  ionic state of, 14
  maleylation of, 76, 215
    spectrum, 77
  nitroguanidination of, 96
  pK_a and nucleophilicity of, 16
  pK_a's of, 15
  quantitation, 57
    acylation, 83
  reaction with
    acrylonitrile, 114, 116
    aldehydes, 128
    aryl halides, 118
    N-carboxyanhydrides, 78
    citraconic anhydride, 215
    diacetyl trimer, 196, 197

Amino groups—cont.
  diazonium ions, 187
  diketene, 80
  ethoxyformic anhydride, 82
  ethyl thiotrifluoroacetate, 79
  formaldehyde, 125
  haloacetates, 109
  imidoesters, 90
  ketones, 129
  maleimides, 112
  malonaldehyde, 198
  S-methylglucosylisothiourea, 95
  O-methylisourea, 94
  nitrous acid, 207
  sulfonyl halides, 97
  TNBS, 30, 217
  reactivities of, 12, 13
  reductive alkylation of, 130, 216
  reductive isopropylation with, 217
  reversible reagents for, 32
  succinylation of, 74, 214
ε-Aminocaproic acid, reaction with methyl benzimidate, 91
S-(2-Aminoethyl)cysteine, 117
  bond cleavage by trypsin, 45
Amino-terminal determination, with cyanate, 89
3-Aminotyrosine, formation from 3-nitrotyrosine, 185
1-Anilinonaphthalene-4-sulfonate (ANS), 101
Anthracene isocyanate, 48
Antibody
  combining sites, 29, 46
  purification (see Solid supports)
Arginine (see Guanidino groups)
Arginine kinase, lobster muscle, reaction with tetranitromethane, 183
Aryl halides, 118
L-Asparaginase, E. coli, reaction with nitrous acid, 208, 210
Aspartate (see Carboxyl groups)
Aspartate amino transferase, pig heart, photooxidation with methylene blue, 167, 168
Aspartate transcarbamylase
  dissociation by sulfhydryl reagents, 36
  reaction with
    mercurials, 202
    succinic anhydride, 75
Avidin, purification on biotin-cellulose, 43
Azo derivatives, spectra, 190
Azotetrazole-N-acetylhistidine, spectrum, 190, 191

Subject Index

BDC (see 1-Benzyl-3-(3-dimethylaminopropyl)-carbodiimide)
BSA (see Serum albumin, bovine)
Benzil, 45, 194
Benzoquinone, 129
1-Benzyl-3-(3-dimethylaminopropyl)carbodiimide (BDC), 146, 223
Bifunctional reagents, 39, 41, 42
Bisazotetrazole-N-acetylhistidine, spectrum, 190, 191
Bisazotetrazole-N-acetyltyrosine, spectrum, 189, 190
Bisdiazobenzidine, 42
3,6-Bis(mercurimethyl)dioxane, reaction with mercaptalbumin, 41, 200
Bonds
  covalent, 7
  noncovalent, 8
    electrostatic interactions, 8
    hydrogen bonding, 8, 9
    hydrophobic forces, 8
  peptide
    chemical cleavage of, 38
    hydrolysis of, 45
Borohydride (see Sodium borohydride)
Bovine serum albumin (see Serum albumin, bovine)
N-Bromoacetamide, 169
2-Bromoacetamido-4-nitrophenol, 49, 50, 110
Bromoacetic acid, 105
α-Bromoacetophenone, 28
α-N-Bromoacetylarginine methyl ester, reaction with sulfenylthiosulfate derivative of streptococcal proteinase, 160
2-Bromoethylamine, 45, 118
N-Bromosuccinimide (NBS), 12, 169
  reaction with indole groups, 227
Buried and exposed groups, reagents for determination, 18
Buried groups, definition, 17
2,3-Butanedione, use in preparation of diacetyl trimer, 224

CMC (see 1-Cyclohexyl-3-(2-morpholinyl-4-ethyl)-carbodiimide metho-p-toluenesulfonate)
Carbamylation, 84
  equilibrium and rate constants for, 86
  of amino groups, 215
  of γ-globulin, rabbit, 216
S-Carbamylcysteine, 86

Carbamylimidazole, 88
O-Carbamyltyrosine, 87, 88
Carbodiimide, water soluble (WSC), 13, 144
  for quantitative determination of carboxyl groups, 223
Carbonic anhydrase
  reaction with tetranitromethane, 183
  reaction with haloacetates, 107
N-Carboxyanhydrides, 12, 78
  solubilization of reduced proteins, 37
Carboxyl groups
  and hydrogen bonding, 9
  conversion to amides, 222
  determination in proteins with water-soluble carbodiimides, 145
  esterification of, 139, 222
  ionic state of, 114
  pK_a's of, 15
  quantitation, 57, 223
  reaction with
    cyanate, 87
    diazoacetates, 141
    methanol/HCl, 139, 222
  reactivities of, 12
Carboxymethylation
  of reduced proteins, 219
  of sulfhydryl groups, 106
  of thioether groups of methionine, 108, 228
S-Carboxymethylcysteine, properties of, 107
Carboxymethylhistidine, properties of, 107, 108
S-Carboxymethylmethioninesulfonium
  determination of, 108
  properties of, 108
  stability in acid, 108, 109
Carboxypeptidase A
  acetylation by N-acetylimidazole, 22, 72, 73, 74
  nitration of, 184, 226
  photooxidation of, 168
  reaction with p-mercuribenzoate, 204
  reactivity of tyrosines, 18
Casein, iodination of, 178
Chloroacetic acid, 105
Chlorotyrosine, formation during peroxide treatment, 163
Cholinesterase, labeling with ^{32}P-diisopropylfluorophosphate, 25
α-Chymotrypsin
  acetamido-4-nitrophenol labeled, 50
  acetylation of, 70, 71
  active-site-specific reagents, 28

α-Chymotrypsin—*cont.*
  acylperoxide-promoted oxidation of tryptophan and cystine, 163
  column chromatography of, 44
  inactivation by TPCK, 27
  iodination of, 181
  monosulfoxide derivative of, 162
  reaction of active serine with water-soluble carbodiimides, 145
  reaction with
    ethoxyformic anhydride, 82
    formaldehyde, 127
    hydrogen peroxide, 229
    2-hydroxy-5-nitrobenzyl bromide, 123, 125
    sulfonyl halides, 27, 99
  treatment with sodium borohydride, 150
Chymotrypsinogen
  esterification by diazoacetates, 142
  esterification with methanol/HCl, 139
  maleylation of, 215
  reaction with trinitrobenzenesulfonic acid, 123
  reductive methylation of, 132
Circular dichroism, 61, 62
Citraconic anhydride, 32, 77
  modification of amino groups with, 215
Collagen, treatment with tetranitromethane, 183
Colorimetric determination of amino acid side chains, 57
Conalbumin, chicken (*see* Ovotransferrin, chicken)
Conformation, changes in, 35, 47, 60
Creatine kinase, rabbit muscle, reaction with iodine monochloride, 180
Cross-linking
  reaction, 39
  reagents, 39, 41, 42
Cyanate, 12, 32, 84
  carbamylation of amino groups with, 215
  removal from urea solutions, 88
Cyanogen bromide, 12, 39, 204
1,2-Cyclohexanedione, 12, 45, 194
  reaction with guanidino groups, 224
1-Cyclohexyl-3-(2-morpholinyl-4-ethyl)carbodiimide metho-*p*-toluenesulfonate (CMC)
  reaction with α-chymotrypsin, 145
  structure of, 146
Cystamine, sulfitolysis of, 153
Cysteine (*see also* Sulfhydryl groups)
  promotion of S-sulfonation, 154
  redox potential, 151
Cystine (*see* Disulfide groups)

Cytochrome c
  horse heart
    reaction of iodoacetate with methionine residues, 109
    reaction with salicylaldehyde, 128
  photooxidation of, 168
  reaction with benzoquinone, 129
  trifluoroacetylated, reactivation of, 80
  tuna heart, effects of guanidination, 93

DFP (*see* Diisopropylfluorophosphate)
DNA polymerase, *E. coli*, reaction with mercuric ion, 200
DTNB (*see* 5,5'-Dithiobis(2-nitrobenzoic acid))
Dansyl chloride, 97
Deacetylation, of O-acetylphenol, 72
Deamination, with nitrous acid, 207
Denaturation, 10
  detection of, 60
Diacetyl, 197
Diacetyl trimer, 12, 196
  preparation of, 224
  reaction with BSA, 224
N,S-Diacetylthioethanolamine, 83
Diagonal electrophoresis, and acylation, 23, 80
Diazoacetamide, preparation and properties of, 142
Diazoacetates, 12, 141
Diazoacetyl-D,L-norleucine methyl ester, reaction with pepsin, 142
Diazoacetylphenylalanine, inactivation of pepsin, 142
Diazoatization, 187
Diazomethane, 141
Diazonium ions, 186
Diazonium salts, 12
Diazonium-1-H-tetrazole, 188
1-Diazo-4-phenyl-2-butanone, inactivation of pepsin, 142
Dicarbonyls (*see* Cyclohexanedione, Glyoxal, Phenylglyoxal, Diacetyl trimer, Malonaldehyde)
Diethylpyrocarbonate (*see* Ethoxyformic anhydride)
1,5-Difluoro-2,4-dinitrobenzene, 41, 123
  reaction with aminotyrosyl staphylococcal nuclease, 186
p,p'-Difluoro-*m,m'*-dinitrodiphenyl sulfone, 41, 123
  reaction with aminotyrosyl staphylococcal nuclease, 186

Dihydroxydinaphthyl disulfide, 157
Diiodotyrosine, formation during iodination of proteins, 175, 176, 226
Diisopropylfluorophosphate (DFP)
  reaction with serine esterases, 26
  structure of, 26
Diketene, 12, 80
Dimethyl adipimidate, 42, 92
Dimethyl suberimidate, 40
Dimethyl sulfoxide
  acceleration of reaction between glycine and dinitrofluorobenzene, 119
  enhancement of reaction of acrylonitrile with amines, 115
4-(p-Dimethylaminobenzenazo)phenylmercuric acetate, 204
4-(4-Dimethylamino-3,5-dinitrophenyl)maleimide, 112, 113
Dimethylaminonaphthalenesulfonyl chloride, 48
  reaction with aminotyrosyl in staphylococcal nuclease, 186
ε-N,N-Dimethyllysine, formation by reductive alkylation of proteins, 130
2,3-Dimethylmaleic anhydride, 77
Dimethylmaleylarginine, deacylation of, 77
N-2,4-Dinitroanilinomaleimide, 112, 113
Dinitrobenzenesulfonic acid, 120
Dinitrochlorobenzene, 120
Dinitrofluorobenzene, 12, 32, 119
41-Dinitrophenyllysylribonuclease, photooxidation of, 169
Dioxane peroxide, oxidation of tryptophan residues, 163
Diphenylcarbamyl chloride, 28
Diphenyldiazomethane, inactivation of pepsin, 142
N,N'-(1,3-Diphenylene)-bis-maleimide, 112, 113
Disulfide bonds
  cleavage by sulfite, 152
  electrolytic reduction, 152
  reduction and reoxidation, 37
  reduction of, 149
Disulfide groups
  cleavage of, 221
  photooxidation of, 166
  reaction with
    performic acid, 161
    reducing agents, 149
    sulfite, 222
    thiols, 149
  reactivities of, 12, 13

Disulfide groups—cont.
  reduction of, 221
  reoxidation of following reduction, 221
  reversible reagents for, 32
  sulfitolysis of, 152
5,5'-Dithiobis(2-nitrobenzoic acid) (DTNB), 12, 32, 155
  reactivity with proteins, 19
  substitution of sulfhydryl groups, 220
  used for quantitative determination of sulfhydryl groups, 220
Dithioerythritol, for reduction of disulfide bonds, 149
Dithionite, for sulfitolysis of disulfide bonds, 152
4,4'-Dithiopyridine, 157
Dithiothreitol
  and o-iodosobenzoate inactivation of glyceraldehyde-3-phosphate dehydrogenase, 158
  for reduction of disulfide bonds, 149
  reactivation of 5,5'-dithiobis(2-nitrobenzoic acid)-inactivated phosphorylase b, 155
  redox potential, 151

EDC (see 1-Ethyl-3-(3-dimethylaminopropyl)-carbodiimide)
Effector sites, definition, 20
Elastase, porcine, reaction with nitrous acid, 209
Electron paramagnetic resonance, 49
Electrophilic reagents, 175
Electrophoresis, 62
Ellman's reagent (see 5,5'-Dithiobis(2-nitrobenzoic acid))
Enolase, muscle, reaction with mercurials, 202
Enolase, yeast, photooxidation of, 168
Environmentally sensitive labels, 47
  chromophoric groups (see Reporter groups)
  fluorescent labels (see Fluorescent labels)
  spin labels (see Spin labeling)
Essential groups, 20
Essential residues
  definition, 30
  determination of, 30
Ester-forming reagents, 139
Esterification, 139
  of carboxyl groups, 222
Ethoxyformic anhydride, 12, 32, 81
Ethyl acetimidate, amidination of amino groups with, 216
Ethyl thiotrifluoroacetate, 12, 32, 79

1-Ethyl-3-(3-dimethylaminopropyl)carbodiimide
    (EDC), 223
  structure, 146
N,N'-Ethylene-bis-(iodoacetamide), 41
Ethylenimine, 12, 117
N,-Ethylmaleimide (NEM), 13, 19, 111
  reaction with
    glutathione, 111
    isocitrate dehydrogenase, 219
  reactivity with proteins, 19
  substitution of sulfhydryl groups, 219
  treatment of ribonuclease A, 112
Exposed groups, definition, 17

Ferriheme, nitration with tetranitromethane, 183
Ficin
  activation by phosphorothioate, 152
  reaction with
    haloacetates, 107
    TLCK, 28
Fluoresceinisothiocyanate, 48
Fluorescence labeling, with dansyl chloride, 97
Fluorescence polarization, 47
Fluorescent labels, 47
Formaldehyde, 12, 42, 125, 131
  used in reductive alkylation, 217
Formalin (see Formaldehyde)
Formol titration, 126
Fructose-1,6-diphosphatase
  pig kidney, reduction of pyridoxal phosphate complex, 134
  rabbit liver, reaction with
    dinitrofluorobenzene, 120
    p-mercuribenzoate, 201
Fumarase, porcine heart
  inhibition by alkylmercury nitrates, 202
  reaction with haloacetates, 107
  sulfhydryl reactivity with alkylmercury nitrates, 19

Gel filtration, 62
γ-Globulin
  rabbit, carbamylation of, 216
  treatment with tetranitromethane, 183
Glutamate (see Carboxyl groups)
Glutamic dehydrogenase
  nitration by tetranitromethane, 185
  reaction with trinitrobenzenesulfonic acid, 123
Glutamic dehydrogenase, bovine liver, reduction of pyridoxal phosphate complex, 134

Glutaraldehyde, 42, 129
Glutathione, reaction with N-ethylmaleimide, 111
Glyceraldehyde-3-phosphate dehydrogenase
  inactivation by o-iodosobenzoate, 158
  oxidation by hydrogen peroxide and cytochrome c, 163
  pig muscle, inactivation by tetrathionate, 159, 160
  rabbit muscle
    inactivation by
      hydrogen peroxide, 163
      iodine, 180
    reaction with dinitrofluorobenzene, 120
  reaction with
    acetylating agents, 31
    5,5'-dithiobis(2-nitrobenzoic acid), 156
Glycine, reaction with
  dinitrofluorobenzene in 80% dimethyl sulfoxide, 119
  α,β-unsaturated compounds, 116
Glycogen phosphorylase (see Phosphorylase)
Glycylglycine, reaction with methyl benzimidate, 91
Glyoxal, 12, 195
  inactivation of ribonuclease A, 196
Growth hormone, human
  reaction with 2-nitrophenylsulfenyl chloride, 207
  reduction by dithiothreitol, 150
Guanidination, 93
  of amino groups with O-methylisourea, 216
  of pancreatic ribonuclease, 216
Guanidino groups
  ionic state of, 14
  pK_a's of, 15
  quantitation of, 57
  reaction with
    1,2-cyclohexanedione, 194, 224
    diacetyl trimer, 196, 197, 224
    dicarbonyls, 194, 224
    glyoxal, 195
    malonaldehyde, 198
    phenylglyoxal, 195, 224
  reactivities of, 12, 13
  reversible reagents for, 32
1-Guanyl-3,5-dimethyl pyrazole, 94, 95

HNBB (see 2-Hydroxy-5-nitrobenzyl bromide)
Haloacetates, 12, 105
Haptene, 43, 46

Hemerythrin
  reaction with p-mercuribenzoate, 204
  succinic anhydride, 36, 74
Hemoglobin, interaction with pyridoxal phosphate, 134
Hexamethylenediisocyanate, 42
Histidine (see also Imidazole groups)
  reaction with NBS, 169
Histidine reagents, 227
Homoarginine, properties of, 95
Homocitrulline, formation from lysine and cyanate, 85
Hydantoins, formation from α-carbamylamino acids, 89
Hydrogen exchange, 62
Hydrogen peroxide, 12, 32, 162
  reaction with thioether groups, 229
Hydrolysis of proteins, 56
Hydroxylamine
  for deacetylation of O-acetylphenol, 72
  for deacylation of O-acyltyrosine, 71
N-(4-Hydroxy-1-naphthyl)maleimide, 112, 113
2-Hydroxy-5-nitrobenzyl bromide (HNBB), 12, 110, 123
  reaction with tryptophan residues, 228
3-Hydroxypyrene-5,8,10-trisulfonic acid, 48

Imidazole groups
  iodination of, 177, 181
  ionic state of, 14
  photooxidation of, 165, 226
  pK_a's of, 15
  quantitation, 57
  reaction with
    diacetyl trimer, 196
    diazonium ions, 187
    ethoxyformic anhydride, 82
    haloacetates, 107
    maleimides, 112
  reactivities of, 12, 13
  reversible reagents for, 32
Imidoesters, 89
Immunogenicity, augmenting, 46
Immunoglobulin G
  cleavage by cyanogen bromide, 205
  solubilization during reduction by polypeptidylation, 78
Indole groups
  photooxidation of, 165
  quantitation 57

Indole groups—cont.
  reaction with
    N-bromosuccinimide, 227
    hydrogen peroxide, 163
    2-hydroxyl-5-nitrobenzyl bromide, 123, 228
    iodine, 181
    performic acid, 161
    sulfenyl halides, 206
    TNM, 183
  reactivities of, 12, 13
Insulin
  and N-acetylimidazole, 74
  bovine, reactivity of tyrosines, 18
  cleavage of disulfide bonds with performic acid, 161
  iodination of, 179
  porcine, reaction with cyanate, 85
  reaction with
    O-methylisourea, 93
    tetranitromethane, 183
    trinitrobenzenesulfonic acid, 123
  reduction by dithiothreitol, 150
Iodination, 175
  determination of relative rates, 176
  of phenolic groups of ovotransferrin, 225
  of tyrosine derivatives, 177, 179
Iodine, 12, 175
  incorporation into
    ribonuclease, 177
    serum albumin, 178
  reactivity with proteins, 19
  use in iodination of phenolic groups, 225
Iodoacetamide, 107
  inactivation of trypsin, 29
Iodoacetate
  reaction with
    methionine residues, 107, 228
    sulfhydryl groups, 218
  use in reduction and carboxymethylation of proteins, 219
Iodoacetic acid, 105
Iodohistidine, 181
o-Iodosobenzoate, 12, 157
  for sulfitolysis of disulfide bonds, 152
  inactivation of glyceraldehyde-3-phosphate, 158
Iodotyrosine
  pK values, 179
  spectral properties, 178, 179
Ionic state, modification of, 35

Isocitrate dehydrogenase
  pig heart, inactivation by 5,5'-dithiobis(2-nitrobenzoic acid), 155
  reaction of iodoacetate with methionine residues, 109, 229
  reaction with N-ethylmaleimide, 219
Isomorphic replacement, 51
Isopropylation, amino groups, 217
ε-N-Isopropyllysine, formation during reductive alkylation of proteins, 130

Koshland's reagent (see 2-Hydroxy-5-nitrobenzyl bromide)
Kynurenine, formation from tryptophan by photooxidation, 166

Labeling
  affinity, 27
  pseudosubstrate, 26
  radioactive, 16
  substrate, 25
Labels
  chromophoric groups, environmentally sensitive, 48
  environmentally sensitive, 47
  fluorescent, 47
  heavy atom, 51
  spin, 49
    nitroxide radicals, 50
β-Lactoglobulin
  reaction with acrylonitrile, 115
  reactivity of sulfhydryls, 18, 19
β-Lipoprotein, reaction with O-methylisourea, 94
Lysozyme
  chicken
    acetylated, pH activity curve, 22
    reduction and reoxidation, 37
  disulfide cleavage by phosphorothioate, 151
  egg white
    acetylation of, 21, 71
    esterification with methanol/HCl, 139
    modification with N-bromosuccinimide, 171
    reduction of sulfoxide derivative, 164
    structural formula, 4
  esterification with triethyloxonium fluoroborate, 143
  iodination of, 179
  proflavin sensitized photooxidation, 168
  reaction with 2-nitrophenylsulfenyl chloride, 207
  reduced reoxidation of, 37

Lysozyme—cont.
  reduction and alkylation, 150
  reduction by dithiothreitol, 150
  reduction of 3-nitrotyrosine, 186

Malate dehydrogenase, pig heart, reaction with p-mercuribenzoate, 202
Maleic anhydride, 12, 32, 76
  maleylation of
    amino groups with, 215
    chymotrypsinogen with, 215
Maleylation, 76, 215
Malonaldehyde, 45, 197
Mercaptalbumin, reaction with mercuric ion, 199
β-Mercaptoethanol
  for reduction of disulfide bonds, 149, 219, 221
  reaction with α,β-unsaturated carbonyl compounds, 115
  reversal of S-sulfonation, 154
β-Mercaptoethylamine, promotion of S-sulfonation, 154
Mercurials, 32, 41, 198
p-Mercuribenzoate (PCMB), 13
  effect of fructose-1,6-diphosphatase, 201
  formula, 201
  qualitative determination of substitution, 218
  quantitative determination of substitution, 218
  reaction with sulfhydryl groups, 218
  reactivity with proteins, 19
  spectrum of, 203
  titration of 3-phosphoglyceraldehyde dehydrogenase, 203
Mercuric chloride, 199
Mercuric ion
  cross-linking of sulfhydryls, 41
  promotion of sulfitolysis, 153
p-Mercuriphenylazo-β-naphthal, 204
Meromyosin, heavy, rabbit muscle, reaction with dihydroxydinaphthyl disulfide, 157
Methanol/HCl, 13
  for esterification of carboxyl groups, 139, 222
Methionine
  (see also Thioether groups)
  reaction with cyanogen bromide, 204
Methionine sulfoxide
  decomposition in hot acid, 164
  determination of, 164, 169
  reduction of, 164
2-Methoxy-5-nitrobenzyl bromide, 49, 110, 125
2-Methoxy-5-nitrotropone, 13, 32, 129

Subject Index

Methyl acetimidate, 13, 32
   reaction with ε-aminocaproic acid and glycylglycine, 90
Methyl acrylate, 115
Methyl benzimidate hydrochloride, 90, 91
Methyl diazoacetate, preparation of, 142
Methyl iodide, 110
Methyl picolinimidate, 93
Methylation, reductive, of pancreatic ribonuclease, 131
Methylene blue, photosensitizing dye, 166, 226
S-Methylglycosylisothiouruea, 95
Methylglyoxal, inactivation of ribonuclease A, 196
S-Methylisothiourea, 93
O-Methylisourea, 13, 93
   guanidination of
      amino groups with, 216
      pancreatic ribonuclease with, 216
   reaction with
      BSA, 95
      ribonuclease, 94
ε-N-Methyllysine, formation during reductive methylation of proteins, 130
Methylmercuric ion, formula, 201
S-Methyl-1-nitro-2-thiopseudourea, 96
Mono(p-azobenzenearsonic acid)-chloroacetyltyrosine, spectrum, 189
Monoazotetrazole-N-acetylhistidine, spectrum, 191
Monoazotetrazole-N-acetyltyrosine, spectrum, 189
Monoiodotyrosine, formation during iodination of proteins, 176
Myoglobin
   apomyoglobin, reaction with 2-hydroxy-5-nitrobenzyl bromide, 125
   sperm whale
      nitration by tetranitromethane, 185
      reactivity of tyrosines, 18
Myosin adenosine triphosphatase, reaction with mercurials, 202

NBS (see N-Bromosuccinimide)
NEM (see N-Ethylmaleimide)
N → O acyl shift, during esterification in methanol/HCl, 59, 139
Neocarzinostatin, deamination by nitrous acid, 210
Nitration, of tyrosine residues, 183, 226
Nitroarginine, 96

Nitroguanidination, 96
1-Nitroguanyl-3,5-dimethylpyrazole, 96
p-Nitrophenyldiazoacetate, 28
3-Nitrotyrosine, 183
Nitrous acid, 13, 207
Nitroxide radicals, use for spin labeling, 50
Nuclear magnetic resonance, 61, 62

Optical rotatory dispersion, 62
Orosomucoid, succinylation of, 75
Ovalbumin, chicken
   and N-acetylimidazole, 74
   reactivity of
      sulfhydryls, 18, 19
      tyrosines, 18
   reduction and reoxidation, 37
Ovomucoid, turkey
   reaction with
      cyanogen bromide, 39
      trinitrobenzenesulfonic acid, 30, 122
   reductive methylation of, 131
Ovomucoid-Sepharose, for chromatography of trypsin, α-chymotrypsin, 44
Ovotransferrin
   chicken
      iodination of, 226
      reduction of disulfides, 38
   succinylation of, 75
Oxidizing agents, 149
Oxytocin, reaction with acetone, 128

PCMB (see p-Mercuribenzoate)
pH, effect on modification, 14
pK values, 14, 15
PLP (see Pyridoxal-5-phosphate)
Papain
   activation by phosphorothioate, 152
   reaction with
      cyanate, 86
      haloacetates, 106
      TLCK, 28
   reduction and formation of mercury bridge, 200
   reduction by thiols, 150
Paraformaldehyde, 126
Parathyroid hormone
   monosulfoxide derivative, 162
   reduction of sulfoxide derivative, 164
Pepsin
   and N-acetylimidazole, 73
   esterification with diazomethane, 141
   inactivation by

Pepsin—*cont.*
  diazoacetylphenylalanine, 142
  1-diazo-4-phenyl-2-butanone, 142
  diphenyldiazomethane, 142
  porcine, reaction with 2-hydroxy-5-nitrobenzyl bromide, 125
  reaction with
    cyanate, 85
    diazoacetyl-D,L-norleucine methyl ester, 142
    ethoxyformic anhydride, 82
Pepsinogen
  effects of carbamylation, 85
  porcine, reactivity of tyrosines, 18
  reaction with cyanate, 87
  succinylation of, 37
Performic acid, 13, 160
Phenanthroquinone, determination of arginine residues, 195
Phenol-2,4-disulfonyl chloride, 42
Phenolic hydroxyl groups, acetylation of using N-acetylimidazole, 225
Phenolic groups
  acetylation of, 71, 225
  acylation of, 68
  and hydrogen bonding, 9
  carbamylation of, 86
  iodination of, 175, 177, 179, 225
  ionic state of, 14
  nitration of, 183, 226
  photooxidation of, 165
  pK_a's of, 15
  reaction with
    carbodiimides, 145
    diazonium ions, 187
  reactivities of, 12, 13, 18
  reversible reagents for, 32
N,N'-(1,3-Phenylene)bismaleimide, 41
Phenylglyoxal, 14, 32, 195
  reaction with guanidino groups of ribonuclease A, 196, 224
3-Phenyl-7-isocyanatocoumarin, 48
Phenylmethanesulfonyl fluoride, 27
  reaction with
    papain, 100
    serine proteases, 98
Phenylmethanesulfonyl-chymotrypsin
  formation of anhydrochymotrypsin, 99
  reaction with
    hydrogen peroxide, 100
    2-mercaptoethylamine, 99
  reformation of chymotrypsin from, 100

Phenylmethanesulfonyl-subtilisin, conversion to thiol-subtilisin, 100
Phosphofructokinase
  and N-acetylimidazole, 73
  maleylation of, 76
  sheep heart, photooxidation of, 168
Phosphoglucomutase, photooxidation of, 31, 168
6-Phosphogluconate dehydrogenase, photooxidation of, 168
Phosphogluconate dehydrogenase, *C. utilis*, reduction of pyridoxal phosphate complex, 134
3-Phosphoglyceraldehyde dehydrogenase
  reaction with mercurials, 201
  titration with *p*-mercuribenzoate, 203
Phosphorothioate, 151
  redox potential, 151
Phosphorylase, glycogen
  labeling with pyridoxal phosphate, 26, 133
  reaction with
    5,5'-dithiobis(2-nitrobenzoic acid), 153
    mercurials, 202
Phosphorylase b, muscle
  carboxymethylation of, 219
  substitution of sulfhydryl groups using DTNB, 220
Photooxidation, 13, 32, 165
  of aspartate amino transferase, 167
  of imidazole groups, 226
Physical structure, modification to change, 35
Physical techniques, table, 62
Polyalanylation, 37
Polypeptidylation, quantitation of, 83
Polypeptidyl enzymes, 78
Procarboxypeptidase, succinylation of, 36, 74
Proflavin, photooxidation of lysozyme, 168
Prolactin, reduction by dithiothreitol, 150
Protein function, as affected by structure variation, 63
Protein stability, 9
Protein structure, 7
  effect of environment, 10
  mobility of, 9
  variations into study function, 63
Proteolysis, changing susceptibility to, 45
Pseudosubstrate labeling (*see* Active site, reagents for)
Pyrenebutyric acid anhydride, 48
Pyridoxal-5-phosphate (PLP)
  as a photooxidation sensitizer, 168
  interaction of proteins with, 132

Quantitation of modifications, 55

Reactivity, effects of neighboring groups on, 106
Reagents
  acylating, 68
  alkylating, 105
  cross-linking, 39
  electrophilic, 175
  for amidination, 89
  for esterification, 139
  reducing and oxidizing, 149
Reducing agents, 149
  redox potentials, 141
Reduction
  and carboxymethylation of proteins, 219
  of disulfide bonds of ribonuclease, 221
Reduction and reoxidation of disulfide bonds, schematic, 37, 38, 149
Reductive alkylation (*see* Alkylation, reductive)
Reductive methylation (*see* Methylation, reductive)
Relative reactivity of amino acid side chains, 11
  buried groups, 17, 18, 19
  determination of, 11
  exposed groups, 17, 18, 19
  pK values (*see also* pK values), 14, 15
  side-chain specificity, 11, 12, 13
  similar groups, 15
Rennin
  photooxidation of, 168
  reaction with dansyl chloride, 98
Reoxidation of reduced disulfide groups of ribonuclease, 221
Reporter groups, 49, 50
Residues, essential, quantitation of, 30
Reversible reagents, 32, 33
Ribonuclease A, bovine pancreatic
  acetamidination of, 91
  disulfide cleavage by phosphorothioate, 151
  esterification with
    diazoacetylglycinamide, 141
    methanol/HCl, 139, 140, 222
  guanidination of, 216
  inactivation by *o*-iodosobenzoate, 158
  iodination of, 177
  monosulfoxide derivative of, 162
  oxidation by hydrogen peroxide, 162
  photooxidation with methylene blue, 167, 168
  pK values of side chains, 15
  reaction of
    haloacetates with amino groups, 110

Ribonuclease A—*cont.*
    iodoacetate with methionine residues, 31, 109, 228
  reaction with
    bromoacetate, 107, 109
    cyanate, 85
    dinitrofluorobenzene, 119
    ethoxyformic anhydride, 82
    glyoxals, 196
    O-methylisourea, 93, 94
    nitrous acid, 209
    phenylglyoxal, 197, 225
    trinitrobenzenesulfonic acid, 123
  reactivity of
    lysine-41, 17
    tyrosines, 18
  reduction
    and formation of mercury bridge, 200
    and reoxidation, 37
  reduction of
    disulfide bonds of, 150, 221
    pyridoxal phosphate complex, 134
    sulfoxide derivative, 164
  reductive
    isopropylation, 131
    methylation, 131
  saponification of esterified derivatives, 140, 142, 223
  WSC-promoted reaction with glycine-N-phthalimidomethyl ester, 147
Ribonuclease T1, reaction with iodoacetate, 31, 105
Ribonuclease-S-peptide, reduction of sulfoxide derivative, 164
Rose bengal, photosensitizing dye, 166, 226

Salicylaldehyde, 32, 128
Saponification of methoxyl groups, 223
Serine proteases, reaction with cyanate, 87
Serum albumin
  bovine (BSA)
    and N-acetylimidazole, 74
    complexation with pyridoxal phosphate, 134
    effect of
      succinylation on, 75
      sulfite on, 153
    esterification with methanol/HCl, 139
    modification of guanidino groups of, 224
    reaction with
      diacetyl trimer, 196

Serum albumin—*cont.*
    ethoxyformic anhydride, 82
    1-guanyl-3,5-dimethylpyrazole nitrate, 95
    O-methylisourea, 95
    reactivity of sulfhydryls, 19
    reduction by dithiothreitol, 150
    succinylation of, 37
  human
    iodination of, 178
    reaction with trinitrobenzenesulfonic acid, 121, 123
Serum transferrin, succinylation of, 75
Side reactions, 58
Silver salts, 204
Site-specific reagents, definition, 25
Sodium borohydride
  determination with trinitrobenzenesulfonic acid, 132
  for reductive alkylations, 130, 217
  redox potential, 158
  reduction of
    PLP-protein complexes, 133
    protein disulfide bonds, 13, 32, 150
Solid supports
  attachment of proteins to, 40
  purification of
    antibody on, 43
    avidin on, 43
    bovine trypsin on, 43, 44
Special-purpose groups, 43
Specificity of side chains, 11
Spin labeling, 49
Staphylococcal enterotoxin B, succinylation of, 215
Staphylococcal nuclease, nitration and reduction, 186
Streptococcal proteinase
  inactivation by tetrathionate, 159
  reaction with haloacetates, 106
Substrate
  labeling (*see* Active site, reagents for)
  protection of active-center groups, 24
Substrate-binding sites, definition, 20
Subtilisin
  nitration by tetranitromethane, 185
  oxidation by hydrogen peroxide, 162, 229
Succinic anhydride, 13, 32, 36, 74
  succinylation of amino groups with, 214
Succinylation, 74, 214
  to alter ionic state, 35
Succinylcarboxypeptidase A, 75
  from succinylprocarboxypeptidase, 74

O-Succinyltyrosine, hydrolysis of, 75
Sulfenyl halides, 13, 206
Sulfhydryl groups
  buried, reactivities of, 18
  carbamylation of, 86
  carboxymethylation of, 105
  ionic state of, 14
  oxidation by TNM, 183
  pK_a's of, 15
  qualitative determination of, 218
  quantitation, 57, 218, 220
  reaction with
    acrylonitrile, 114, 116
    aryl halides, 119
    carbodiimides, 145
    DTNB, 155
    ethylenimine, 117
    N-ethylmaleimide, 219
    haloacetates, 105
    hydrogen peroxide, 162
    iodine, 179
    iodoacetate, 218
    o-iodosobenzoate, 157
    maleimides, 110
    p-mercuribenzoate, 199
    performic acid, 160
    silver salts, 204
    sulfenyl halides, 206
    tetrathionate, 159
  reactivities of, 12, 13, 19
  reversible reagents for, 32
  substitution with
    DTNB, 220
    p-mercuribenzoate, 218
Sulfite
  cleavage of disulfide bonds, 13, 32, 152, 222
  interaction with trinitrophenylamines, 121
Sulfitolysis, 152, 222
Sulfonate esters, 98
Sulfonates, 96
S-Sulfonation, of aldolase, rabbit muscle, 153
Sulfonyl halides, 13, 96
β-Sulfopropionyl chloride, 83
Super-reactivity of protein side chains, 59

TLCK (*see* Tosyllysinechloromethyl ketone)
TNBS (*see* Trinitrobenzenesulfonic acid)
TNM (*see* Tetranitromethane)
TNS (*see* 2-Toluidinylnaphthalene-6-sulfonate)
TPCK (*see* Tosylphenylalaninechloromethyl ketone)

Taka-amylase A, reaction with 2-methoxy-5-nitrotropone, 129
Tetrofluorosuccinic anhydride, 80
3,3′-Tetramethyleneglutaric anhydride, 83
Tetranitromethane (TNM), 13, 18, 183
  use in nitration of phenolic groups, 226
Tetrathionate, 13, 159
  blocking of sulfhydryl groups, 159, 183
  inactivation of glyceraldehyde-3-phosphate dehydrogenase, 160
Thioether groups
  carboxymethylation of, 108, 228
  photooxidation of, 165
  reaction with
    haloacetates, 105, 108
    hydrogen peroxide, 162, 229
    iodine, 182
    performic acid, 161
  reactivities of, 12, 13
  reversible reagents for, 32
Thioglycolate, reduction of disulfide bonds, 149
Thiols, 13, 32
Thyroglobulin, iodination of, 179
Thyroxine
  formation during iodination, 178
  spectral properties, 179
Tobacco mosaic virus protein
  acetylation of, 69
  effect of iodine, 179
  reaction with water-soluble carbodimides, 146
Toluenesulfonyllysinechloromethyl ketone (see Tosyllysinechloromethyl ketone)
Toluenesulfonylphenylalaninechloromethyl ketone (see Tosylphenylalaninechloromethyl ketone)
2-Toluidinylnaphthalene-6-sulfonate (TNS), binding to α-chymotrypsin, 101
Tosyllysinechloromethyl ketone (TLCK), 27, 29, 110
Tosylphenylalaninechloromethyl ketone (TPCK), 27, 28
Transaldolase, C. utilis, reaction with dinitrochlorobenzene, 120
Transglutaminase, guinea pig liver, inactivation by 5,5′-dithiobis(2-nitrobenzoic acid), 156
Triethyloxonium fluoroborate, 143
Trifluoroacetic anhydride, 45, 79
Triglycine, acetylation of, 70
2,4,6-Trinitrobenzaldehyde, reaction with bovine serum albumin, 128

Trinitrobenzenesulfonic acid (TNBS), 13, 30, 121
  reaction with amines, 16, 30, 217
Trinitrochlorobenzene, 120
Trypsin
  acetylation of, 71
  activation from amidated trypsinogen, 146
  column chromatography of, 44
  disulfoxide derivative, 162
  inactivation by
    iodoacetamide and methylguanidine, 29
    TLCK, 27
  iodination of, 181
  nitration by tetranitromethane, 185
  reaction of iodoacetate with methionine residues, 109.
  reaction with
    N-bromosuccinimide, 227
    iodoacetate, 107
    S-methylglucosylisothiourea, 95
    nitrous acid, 209
  reduction by sodium borohydride, 150
Trypsin inhibitors
  egg white (see Ovomucoid)
  formulas, 29
  Kunitz bovine pancreatic
    monosulfoxide derivative of, 162
    oxidation by hydrogen peroxide, 162
    reaction with N-bromosuccinimide, 19
    reduction by sodium borohydride, 150
  lima bean, acetamidination of, 91
  pancreatic, reduction by dithiothreitol, 150
  treatment with 1,2-cyclohexanedione, 195
Trypsinogen
  nitration by tetranitromethane, 185
  reaction of iodoacetate with methionine residues, 109
  reduction by sodium borohydride, 150
  WSC-promoted amidation, 146
Tryptophan
  (see also Indole groups)
  determination with 2-hydroxy-5-nitrobenzyl bromide, 125
  reaction with N-bromosuccinimide, 169
Tyrosine
  (see also Phenolic groups)
  acylation of, 69, 70
  derivatives, iodination of, 177, 179
  residues, 18
Tyrosyl groups (see Phenolic groups)

Ultracentrifugation, 62

α,β-Unsaturated compounds, reaction with amino acids, 16
Urea, formation of ammonium cyanate, 88

Viscosimetry, 62

WSC (*see* Carbodiimide, water soluble)

Water-insoluble proteins, 40
Water-soluble carbodiimide (*see* **Carbodiimide**, water soluble)
Woodward's reagent K, 42

X-ray diffraction, 62